Diamond Based Composites

and Related Materials

NATO ASI Series

Advanced Science Institutes Series

A Series presenting the results of activities sponsored by the NATO Science Committee, which aims at the dissemination of advanced scientific and technological knowledge, with a view to strengthening links between scientific communities.

The Series is published by an international board of publishers in conjunction with the NATO Scientific Affairs Division

A Life Sciences **B Physics**	Plenum Publishing Corporation London and New York
C Mathematical and Physical Sciences **D Behavioural and Social Sciences** **E Applied Sciences**	Kluwer Academic Publishers Dordrecht, Boston and London
F Computer and Systems Sciences **G Ecological Sciences** **H Cell Biology** **I Global Environmental Change**	Springer-Verlag Berlin, Heidelberg, New York, London, Paris and Tokyo

PARTNERSHIP SUB-SERIES

1. Disarmament Technologies	Kluwer Academic Publishers
2. Environment	Springer-Verlag / Kluwer Academic Publishers
3. High Technology	Kluwer Academic Publishers
4. Science and Technology Policy	Kluwer Academic Publishers
5. Computer Networking	Kluwer Academic Publishers

The Partnership Sub-Series incorporates activities undertaken in collaboration with NATO's Cooperation Partners, the countries of the CIS and Central and Eastern Europe, in Priority Areas of concern to those countries.

NATO-PCO-DATA BASE

The electronic index to the NATO ASI Series provides full bibliographical references (with keywords and/or abstracts) to more than 50000 contributions from international scientists published in all sections of the NATO ASI Series.
Access to the NATO-PCO-DATA BASE is possible in two ways:

- via online FILE 128 (NATO-PCO-DATA BASE) hosted by ESRIN, Via Galileo Galilei, I-00044 Frascati, Italy.

- via CD-ROM "NATO-PCO-DATA BASE" with user-friendly retrieval software in English, French and German (© WTV GmbH and DATAWARE Technologies Inc. 1989).

The CD-ROM can be ordered through any member of the Board of Publishers or through NATO-PCO, Overijse, Belgium.

3. High Technology – Vol. 38

Diamond Based Composites

and Related Materials

edited by

Mark A. Prelas

University of Missouri,
Columbia, MO, U.S.A.

Andrew Benedictus

University of Missouri,
Columbia, MO, U.S.A.

Li-Te Steven Lin

The Institute of Physics, Academia Sinica,
Taipei, Taiwan, R.O.C.

Galina Popovici

University of Illinois,
Urbana, IL, U.S.A.

and

Peter Gielisse

Florida State University,
Florida A&M University,
Tallahassee, FL, U.S.A.

Kluwer Academic Publishers

Dordrecht / Boston / London

Published in cooperation with NATO Scientific Affairs Division

Proceedings of the NATO Advanced Research Workshop on
Diamond Based Composites
St. Petersburg, Russia
June 21–22, 1997

A C.I.P. Catalogue record for this book is available from the Library of Congress

ISBN 0-7923-4667-X

Published by Kluwer Academic Publishers,
P.O. Box 17, 3300 AA Dordrecht, The Netherlands.

Sold and distributed in the U.S.A. and Canada
by Kluwer Academic Publishers,
101 Philip Drive, Norwell, MA 02061, U.S.A.

In all other countries, sold and distributed
by Kluwer Academic Publishers,
P.O. Box 322, 3300 AH Dordrecht, The Netherlands.

Printed on acid-free paper

ACKNOWLEDGMENT

NATO ADVANCED RESEARCH WORKSHOP

DIAMOND BASED COMPOSITES

Sponsored by:

North Atlantic Treaty Organization

Office of Naval Research (London)

University of Missouri-Columbia

Laboratory, A.F.Ioffe Phys.-Technical Institute

TABLE OF CONTENTS

FOREWORD

The objective of this meeting was to discuss the state of the art in the development of advanced composite materials using wide band gap materials (diamond and cubic boron nitride) and potential applications for these materials. The goal of this workshop was to bring together the scientific and industrial communities. It is well known that the level of materials science in the Former Soviet Union was very high. Specifically, there has been a great deal of work done on diamond and other wide band gap based composites. The Diamond (and other wide band gap) Based Composites (DBC) workshop was designed to bring together leaders in the scientific community and industry of the western and former eastern block countries to engage in discussions in this important technical area. Only with this free exchange can both the western and eastern communities fully understand the state of the discipline, to understand what research avenues to pursue, and to advance the development of the DBC technology for applications.

Diamond-based composites offer the advantages of diamond and related materials like hardness, high Young's modulus as well as other properties. These material have demonstrated new and unexpected features like unusual stability against high temperature and pressure shocks, a large internal surface that can be technologically controlled, a possibility to control electric, magnetic and optical properties etc... A feasibility of controlling internal surface variation over wide limits offers potential for DBC applications as efficient filters, absorbent, sensors and other tools of environmental control and monitoring.

The DBC NATO Advance Research Workshop covered topics including problems of synthesis of the materials, their characterization and properties. Significant attention was paid to trends in high pressure and high temperature technology. Additionally, low pressure technologies for synthesis of DBC and related materials was addressed. One of the highlights of the conference is a description of achievements in nanocomposite technology. Basic principles of DBC material science was considered with an accent on future developments in the fields of electronics, optics, industrial tools and components, biotechnology and the medicine.

The workshop focused on the issues of development of wide band-gap materials in composites. The topics ranged from molecular clusters, nanophase materials, growth, processing, and synthesis. Commercial composite materials have been on the market for a number of years specifically as abrasives.

The processing of composite based materials can be categorized by six basic methods: 1) In situ growth - high pressure high temperature "catalytic" conversion; 2) mix and sinter such as c-BN plus metal (ceramic polymer) mix; 3) direct sintering; 4) direct polymorphic conversion; 5) shock detonation; 6) SHS sintering.

One of the highlights of the meeting was the work by Guseva and Babaev in which linear chain N-doped carbon films were created with s-p hybridization. These films underwent shock wave compression to transform the C(sp):N to C(sp^3):N. This work detailed the characterization of the material which indicated that carbon nitride was formed.

Mark A. Prelas
University of Missouri
Columbia, MO 65211 USA

Andrew Benedictus
University of Missouri
Columbia, MO 65211 USA

Li-Te Steven Lin
The Institute of Physics, Academia Sinica
Taipei 11529, Taiwan, R.O.C.

Galina Popovici
University of Illinois
Urbana, Illinois 61820 USA

Peter Gielisse
Florida State University
Florida A&M University
Tallahassee, Florida 32310

ADVANCE COMPOSITE MATERIALS ON THE DIAMOND BASE

S.K. GORDEEV
Laboratory of heavy-melted non-metal composite materials.
Central Research Institute of Materials, *St. - Petersburg,*
Paradnaya str. 8, *St.Petersburg, 191014, Russia.*

Abstract. This report considers the non-traditional areas of application of diamond composites. New directions of diamond applications have been proposed. There are engineering materials, sorption-active materials and biocompatibl e materials.
Key-words: diamond, nanodiamond, composite materials.

1. Introduction.

The main reason of this Workshop is the excellent properties combined in diamond. But only part of these properties are used in the traditional fields of diamond applications (Figure 1) [1,2]. It is well-known that the diamond is noted for its hardness, peculiar optic and electronic properties. Today human actives use these features of diamond in traditional applications such as grinding and electronics.

But diamond possesses a very high Young's modules, low thermoexpansive coefficient and special chemistry of its surface. These properties are prized for producing the new types of diamond materials for the new areas of application (Figure 1). In this case Young's modules, strength and thermoexpansiveness are important for engineering materials. The large specific surface and adsorptional activity are important for sorption-active and bio-materials.

We propose to focus on non-traditional areas which are exhibited by our classification (Figure 1).

2. Engineering composite materials on the diamond base.

Recently carbides have been seen as a perspective matrix for diamond composite materials [3]. The carbide of specific interest is silicon carbide. It shows the best correlation with the diamond properties. Silicon carbide possesses some excellent properties. Some of them are the high mechanical and thermomechanical characteristics, thermostability, hardness, etc.

The compatibility of diamond and silicon carbide in the single material would give it very important characteristics. Thus, these materials will have a very high

M. A. Prelas et al. (eds.), Diamond Based Composites, 1–11.

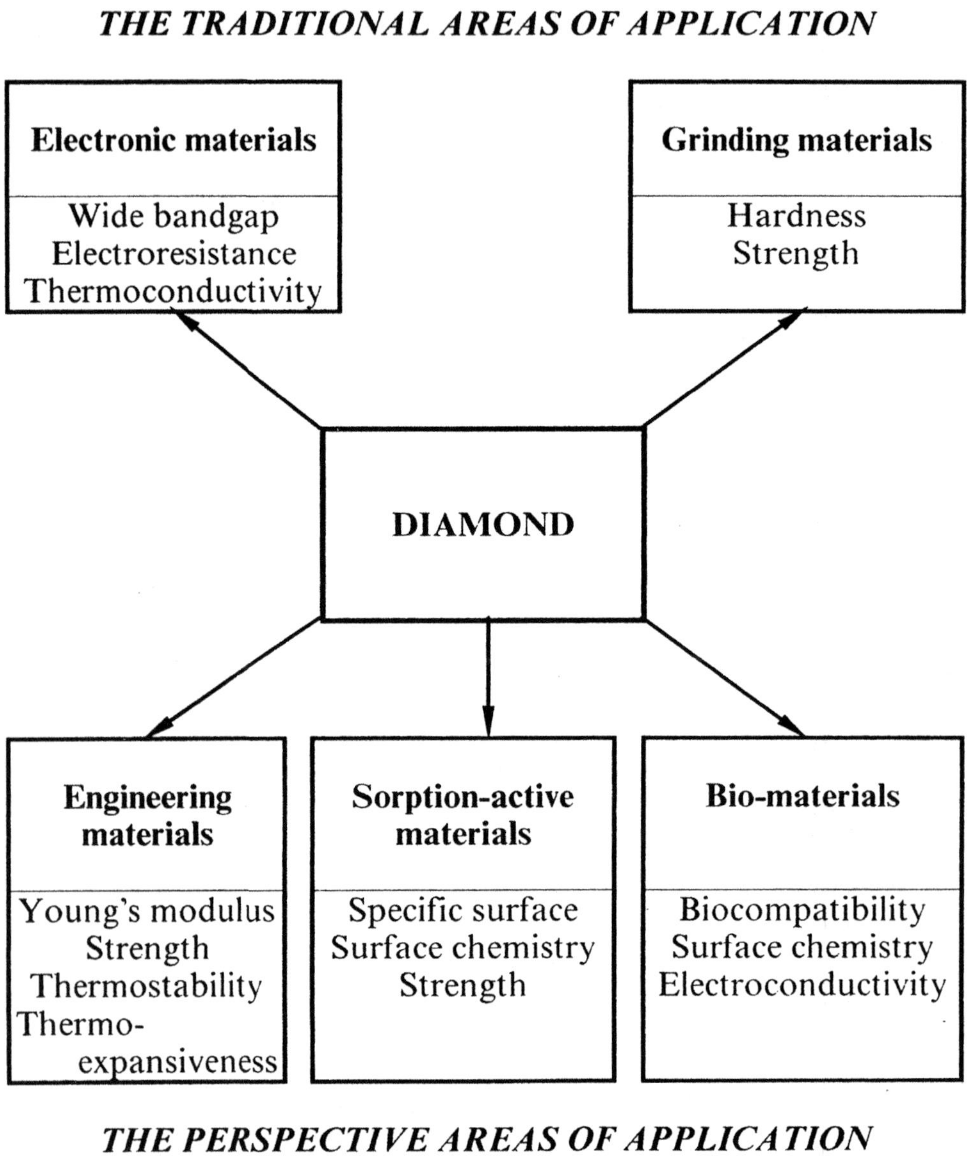

Figure 1. The diamond areas of application.

1. Structure:

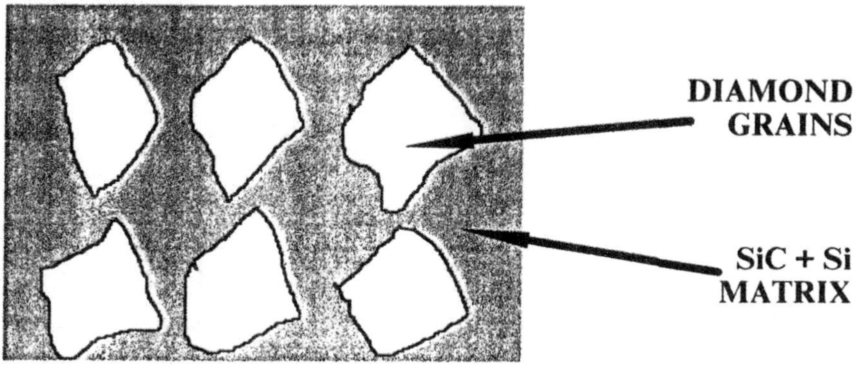

2. Calculated properties:

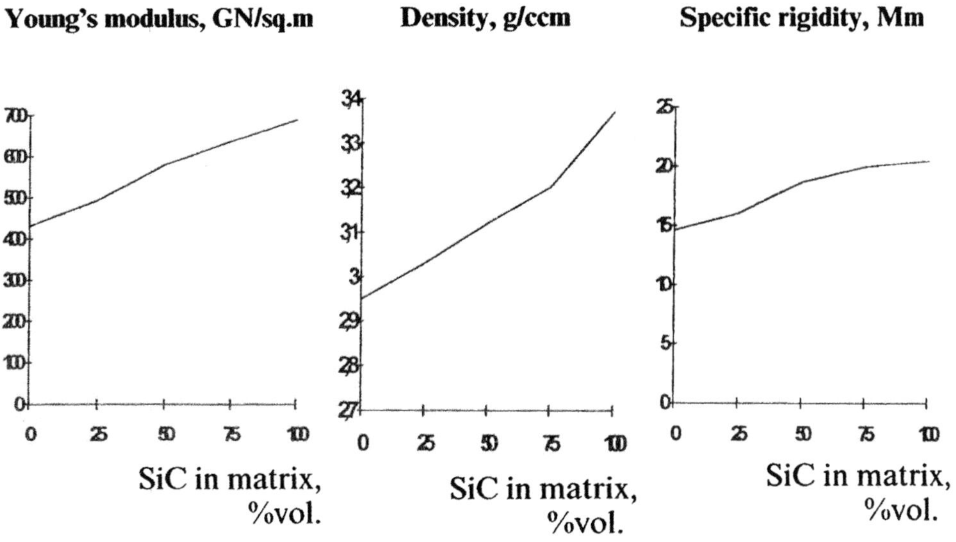

Figure 2. Engineering composite materials "Diamond-SiC-Si."

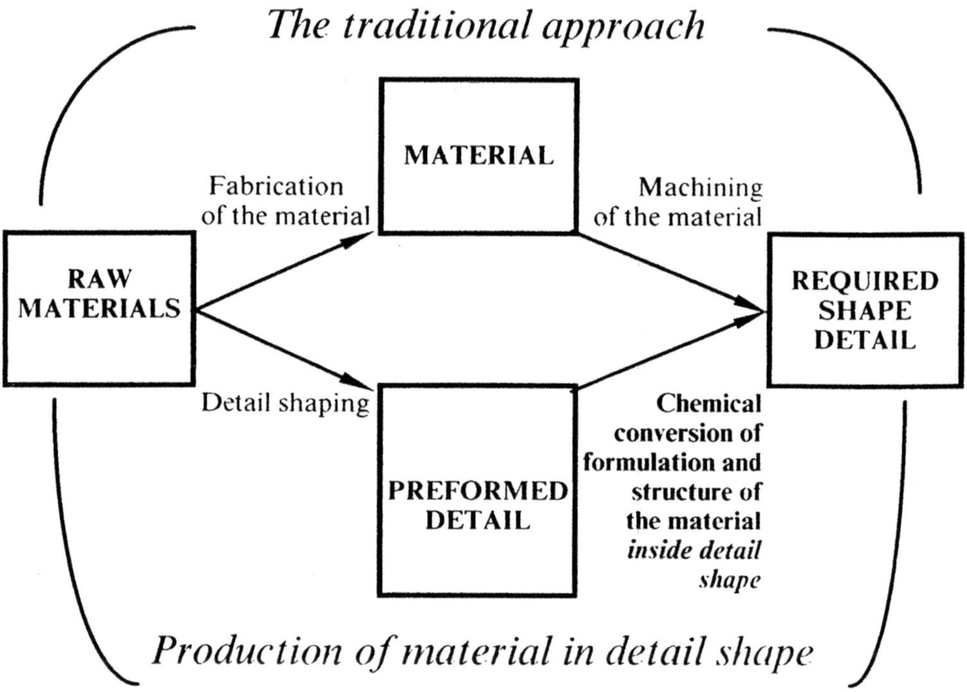

Figure 3. The approaches for required shape detail production.

Table 1. Properties of"Diamond-SiC-Si"-materials.

Formulation (volume percentage) Diamond	SiC	Si	Density, g/ccm	Young's modulus GN/sq.m	Thermoexpansive coefficient 1/K
57	37	6	3.29	650 660*	2.1 10-6
56	22	22	3.11	550 580*	

* - calculated dates

Table 2. Thermostability of the "Diamond-SiC-Si"- material.

	Heat treatment (vacuum)				
	without	1000°C, 40 min	1600°C 40 min	1600°C 80 min	1800°C 40 min
Young's modulus	516	516	492	492	19
Stability	100%	100%	95%	95%	4%

Young's modules and specific rigidity (Figure 2). Calculated modules values are more than 600 GN/sq.m. It is a very high level and realization of such values of Young's modules that is important for engineering applications, for example gravity stable details.

In technology a very high temperature of formation of the silicon carbide matrix complicates the production of these materials. This temperature is higher than that of diamond stability. The preparation of such materials maybe realized in high temperature and pressure conditions. However in this case we can only prepare very small samples having a very simple shape (the cylinder shape).

For production of complex shaped details and bulky details we have proposed non-traditional approach (Figure 3). We prepare the formulation and the structure of materials by means of a chemical process. This process takes place inside a detail's shape. We use the reactions which doesn't change the shape and the size of the detail.

By means of such an approach we have obtained non-porous diamond - silicon carbide - silicon materials. Their properties are shown on Tables I and 2.

Table 3. Young's modulus (F) and specific rigidity (Q) of materials.

Material	E, GN/sq.m	Density (ρ), g/ccm	Q=E/(ρ*g), km	Temperature of stability
Berillium	300	1.95	16000	700 °C
Diamond-SiC-Si	650	3.3	20000	1600°C
Diamond	1000	3.51	29000	1200°C

g - acceleration due to gravity

The features of the materials are low density, very high Young's modules, low thermo-expansive coefficient and high thermostability. The details don't change their shape and properties after heat treatment at 1600 degrees.

You can see in Table 3 that the specific rigidity of our material is higher then that of beryllium. I would note that the berylliun has the highest specific rigidity for engineering materials. Thus, we have produced the record rigidity engineering materials having, in addition, the high thermostability.

3. Composite materials on the base of nanodiamond.

The next part of the report is dedicated for composite materials on the nanodiamond base. Using the nanodiamond we have prepared the new type of materials [4]. They are carbon materials. The structure of nanocomposites is showed on the scheme (Figure 4). One can see that these materials include three types of nanofragments. There are nanodiamond particles with the size about 6 nm, carbon matrix with the size about 1 nm and nanopores with the size about 5 nm. The volume percentage of each phase can be changed on a large scale.

The features of such materials are large specific surface, thin porous structure and electroconductivity. Unusual porous structure gives adsorption activity to the nanocomposites. This activity develops at adsorption of organic compounds from gaseous mixtures and at extraction of metals from the dilute solutions (Figure 5).

4. Conclusion.

In conclusion, the perspectives of elaborating and applying of diamond composites materials in non-traditional areas of applications are being discussed.

1.Formulation: CARBON.

2.Structure:

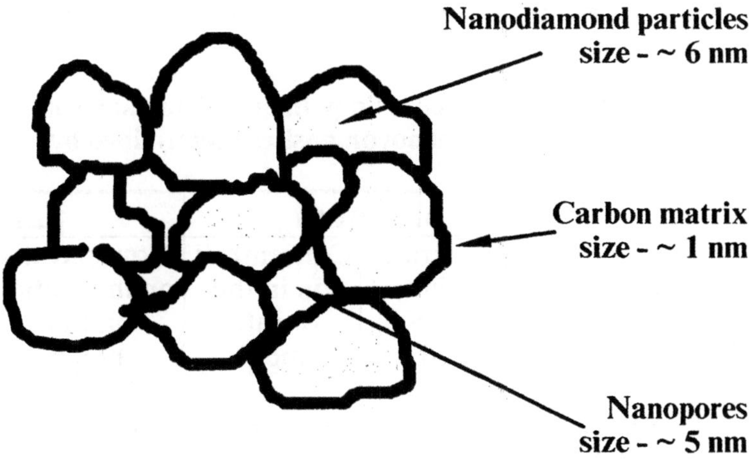

3.Features:

- **High specific surface - up to 300 sq.m/ccm.**
- **Thin porous structure.**
- **Electroconductivity.**

Figure 4. Nanocomposites on the diamond base.

1. Sorption of benzene from gaseous mixtures.

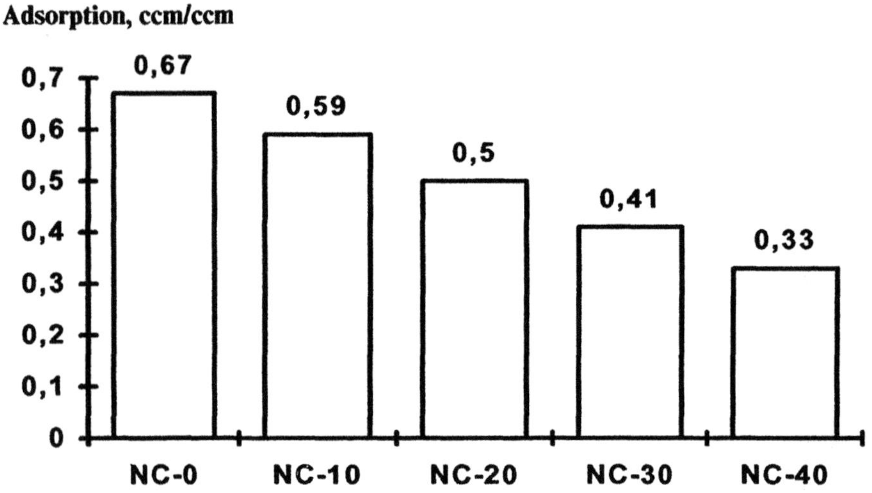

2. Extraction of platinum and palladium from aqueous solutions by means of nanocomposites (determined by V.Distler).

Material	platinum		palladium	
	Concentration in a initial solution, mol /l	Concentration in the filtrate, mol/l	Concentration in a initial solution, mol /l	Concentration in the filtrate, mol/l
1. Porous substrate .	$2.5*10^{-5}$	$1.3*10^{-5}$	$2.5*10^{-5}$	$1.5*10^{-5}$
2. Substrate + **nanocomposite layer.**	$2.5*10^{-5}$	$\mathbf{1.5*10^{-7}}$	$2.5*10^{-5}$	$\mathbf{2.0*10^{-7}}$

Figure 5. Sorption properties of the diamond nanocomposites.

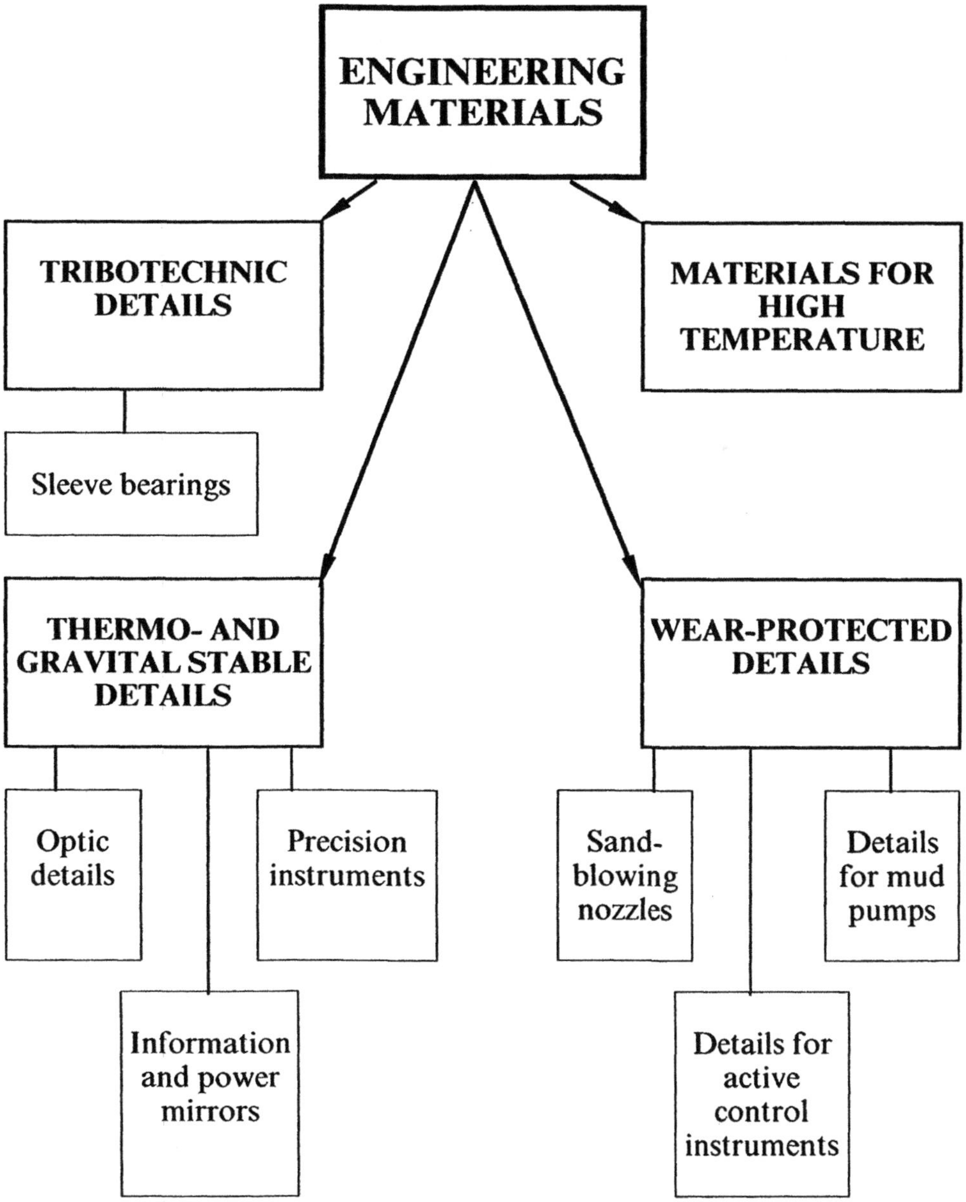

Figure 6. The perspectives of elaborating and applying of diamond composite materials.

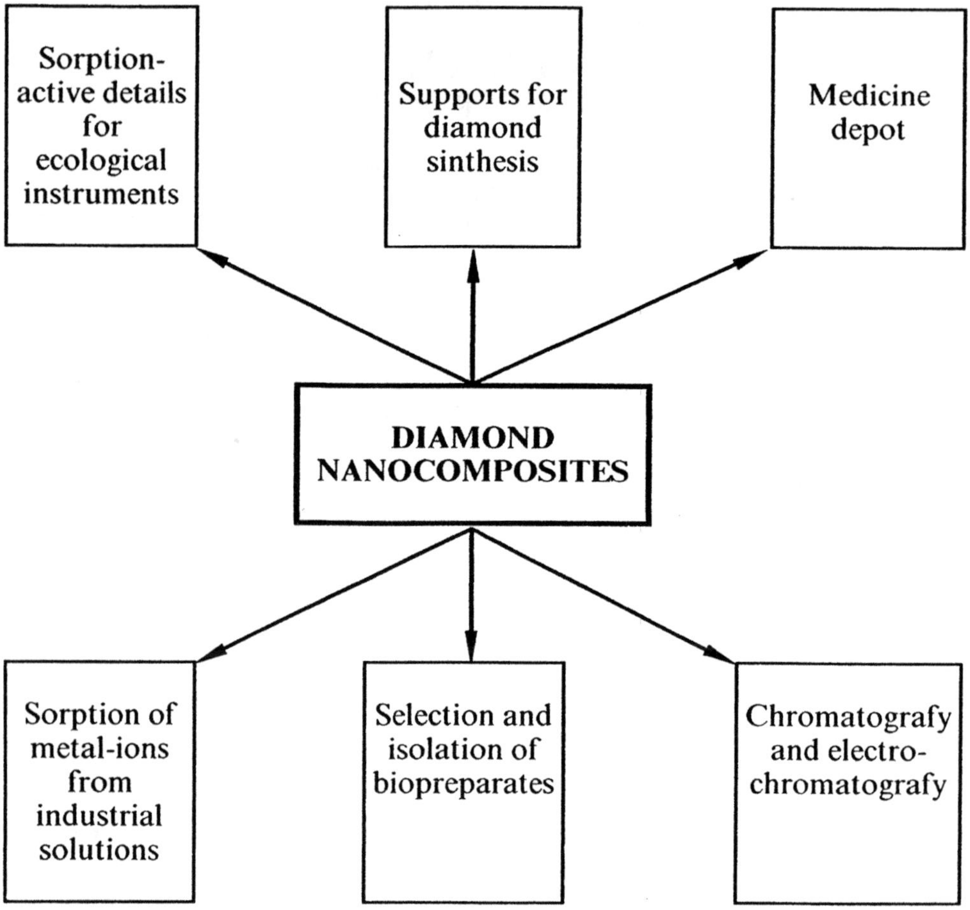

Figure 7. The perspectives of elaborating and applying of nanodiamond composites.

So we have proposed new directions of diamond applications . One of them is diamond - silicon carbide - silicon material. We hope that such new engineering materials will be elaborated and applied in tribotechnic details, thermo - and gravity stable details, wear-protected details, high temperature designs (Figure 6).

The possible applications of nanodiamond composites are in medical and ecological applications, extraction of noble metals, selection and chromatography of biopreparates and so on (Figure 7).

I am sure that the investigations directed on diamond's applications in non-traditional areas represented in my report will be developed very energetically. New results will give an additional perspectives for diamond excellent life.

Acknowledgments

The author thanks S.Zhukov, L.Katilenkova A.Smoljaninov, A.Puzyr and P.Belobrov for productive discussions.
This report have been supported by Central Research Institute of Materials , Company "Scandinavian memoiy metals" and Grant #96152 on program 'Fullerenes and others carbon clusters".

References.

1. ShulzhenkoA.A., Gargin V.G., Shishkin V.A., Bochenko A.A. (1988) *Polycrystaline materials on the diamond base,* Naukova dumka(USSR).
2. *Synthetic superhard materials,* (1986) v.2 Naukova dumka(USSR).
3. Voronin GA., Osipov A.S. (1990). Mechanism of formation and fisico-chemical properties of diamond-silicon carbide composites. In *Polycrystaline materials on the huse of synthet ic diamond and cubic boron nitride* Naukova dumka (USSR).
4. Gordeev S.K., Zhukov S.G., Nikitin Y.I., Poltoratskii V.G. (1995) Composite materials ultradispersed diamond - pyrocarbon. *Inorganic materials (Russia)* **31,** 470-474.

Diamond Composites for Grinding Applications

S.Ramanath, S.T. Buljan and R. D. Grieger
Norton Company
1 New Bond Street
Worcester, MA 01606

Abstract

A significant portion on industrial activity is centered around component forming, starting from a blank or near net shape and then using appropriate material removal processes to finish them. Some of the basic material removal processes include cutting, turning, milling, boring, as well as the use of loose and bonded abrasives. Bonded abrasive tools are made in a variety of shapes, but most commonly appear in the form of integral or segmented wheels or cylinders or blocks. The abrasive sections of such tools are resin, glass or metal matrix composites that contain abrasive as a dispersed phase. The choice of abrasive is governed primarily by its wear resistance, toughness and relative hardness with respect to work material as well as economic considerations. The outstanding mechanical properties of diamond make it an excellent abrasive and the best choice for very demanding applications. In general, resin, glass and particularly metal bonded diamond abrasive composites represent the largest family of diamond composites . Using specific examples, the present paper reviews factors governing design of diamond composites for grinding applications.

1. Introduction

The constitution of bonded abrasive composites used in grinding and cutting tools is typical of any particulate composite (Figure 1). In addition to abrasive grains such as diamond, the composite consists of a matrix (bond) that represents a continuous phase which may or may not contain additional modifiers in the form of distinct dispersed phases or intimately mixed or reacted components and porosity. The matrix, dispersoids and any porosity are collectively referred to as the bond. In addition to primary abrasive grains (diamond) such composites may also contain a secondary abrasive. The distinction between the secondary abrasive and a particulate matrix modifier is its relative hardness with respect to the work material. A sharp particulate phase substantively harder than the work material and capable of abrading it but less hard than the primary abrasive would be considered a secondary abrasive. Matrix modifiers are of lower hardness and do not contribute significantly to work material removal.

In difference to a number of other composites where porosity is an unwanted remnant of incomplete densification, diamond composites for grinding may contain deliberately introduced porosity as a means of composite wear resistance tailoring.

M. A. Prelas et al. (eds.), Diamond Based Composites, 13–29.

1.1 Abrasive selection

In bonded abrasives the choice of abrasives is governed by their wear resistance and their relative hardness with respect to work material. The hardness of selected work materials and abrasives is compared in Figure 2.(4-6)

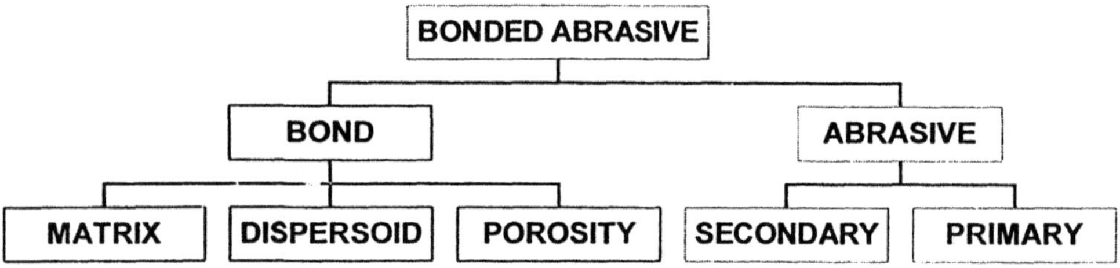

Figure 1. Constitution of bonded abrasives

Since diamond is, by far, the hardest it would be expected that its application would be only limited by the economic considerations. However, diamond is not commonly used for grinding of ferrous alloys due to relatively poor chemical wear resistance.

Abrasive wear resistance of diamond varies depending on presence of flaws, impurities and shape. For example, the amount of nitrogen strongly influences the abrasion wear resistance, as shown in Figure 3. Figure 4 shows two different diamond shapes; a strong-blocky shape and a weak irregular shape with significant differences in mechanical wear behavior.

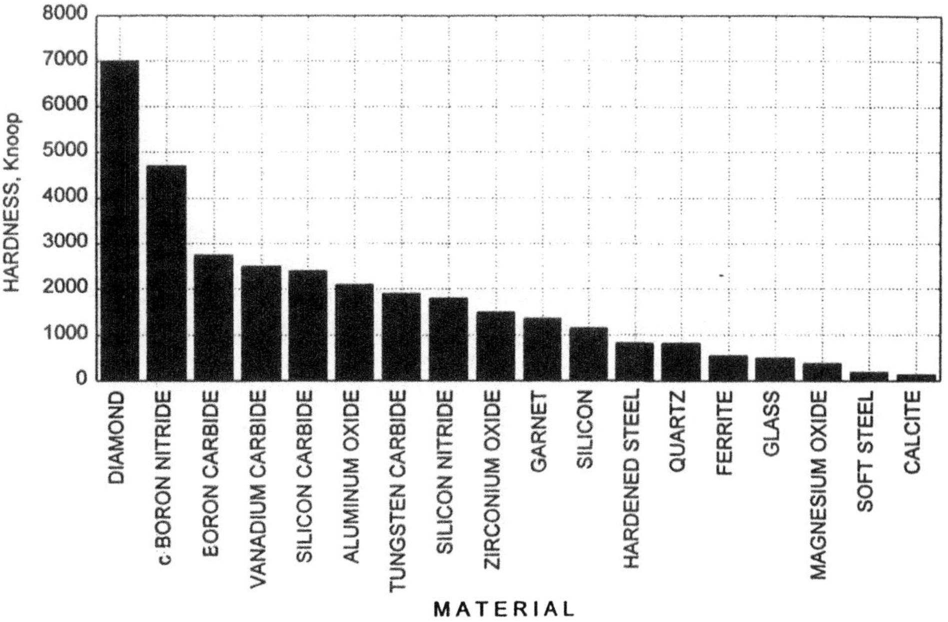

Figure 2. Hardness (Knoop) of selected work materials and abrasives

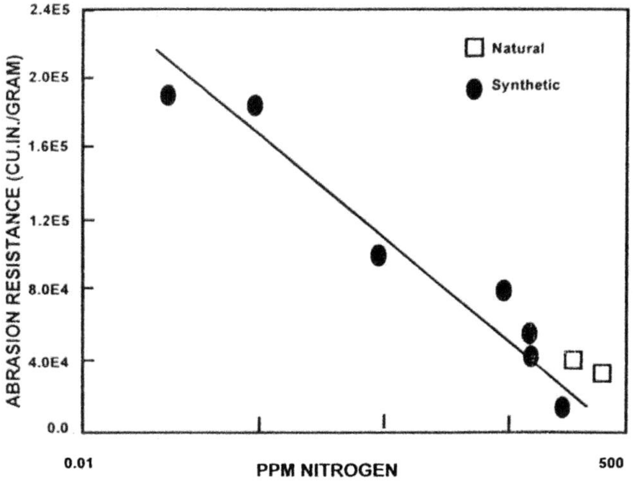

Figure 3. Abrasion resistance of diamond as a function of nitrogen content. Expressed as amount of Alumina removed per unit wear.

STRONG- BLOCKY

FRIABLE

Figure 4. Diamonds of different shapes and wear properties.

1.2 Wear

Wear occurring between two surfaces in dynamic contact, such as a grinding wheel and the work piece being ground, is a consequence of simultaneous action of mechanical (abrasion, delamination, fatigue) and thermally activated chemical processes such as diffusion, or oxidation. The dominant process among them determines the wear mode . It is therefore apparent that in the presence of a dominant chemical wear component any prediction of the material's wear resistance based solely on mechanical properties is meaningless. It is precisely, due to decreased chemical wear resistance, that diamond effectiveness as an abrasive for ferrous alloy grinding is severely diminished. Abrasives such as cubic Boron Nitride and Aluminum Oxide, despite their considerably lower hardness values, are substantially more effective.

Material removal by cutting and grinding using bonded abrasives is an enhanced example of a typical abrasion wear process. Since, by design, bonded abrasives operate in the regime of mechanical wear, it is to be expected that wear resistance of both , work material and the abrasive, will be reflected in the material's mechanical properties. Studies of brittle materials wear suggest that wear can be related to the materials resistance to penetration (hardness) and its fracture (Fracture toughness). Wear resistance of brittle materials is directly proportional to the product of hardness (**H**) and fracture toughness (**K_{ic}**):[7,8]

$$\text{Wear Resistance} \approx K_{ic}^{3/4} \times H^{1/2}$$

Fracture toughness is considerably more important than hardness. The other major component of the composite is the matrix or bond. Bonds in diamond composites for grinding are commonly Resin, Glass or Metal. Wear resistance of these bonds and their modifications (by use of porosity, dispersoids, alloying elements) span a broad range and sometimes overlap one another. Metal bonds are typically, the most and resin based bonds the least wear resistant

2. Abrasive Composite Design

Optimization of a diamond composite for best grinding performance depends on the ability to match bond and diamond wear resistance. Figure 6 shows wheel life as a function of bond wear resistance for two systems I and II. Wheel life in the context of this discussion is defined as useful life of the grinding wheel in the absence of intervention such as dressing which is an action designed to resharpen the wheel. For system I, the optimum bond wear resistance corresponds to that of point "A". If bond has lower wear resistance diamond is undercut and expelled too early . On the other hand if bond has higher wear resistance diamonds are held in place too long and allowed to dull. Dulling of the abrasive results in a reduced load per abrasive grain, sliding of the wheel and burning due to increased friction. Wheel ceases to cut and would require forced bond wear using an abrasive stick ("dressing process"). For system II consisting of different and more resistant diamond type, amount, size, etc., the optimum is at "B". If this diamond is used with optimum bond for system I, the operating point would shift to " B_1". Bonding the diamond from system I with the more wear resistant bond of system II would also reduce the life of the wheel. wheel life would shift from point A to point A_1. The wear resistance of either diamond or a bond is affected by the aggressiveness of the work medium. Since the effect on the diamond wear resistance is not equal to that of the bond, the development of wheels for substantially different work materials requires separate tailoring of bond wear properties.

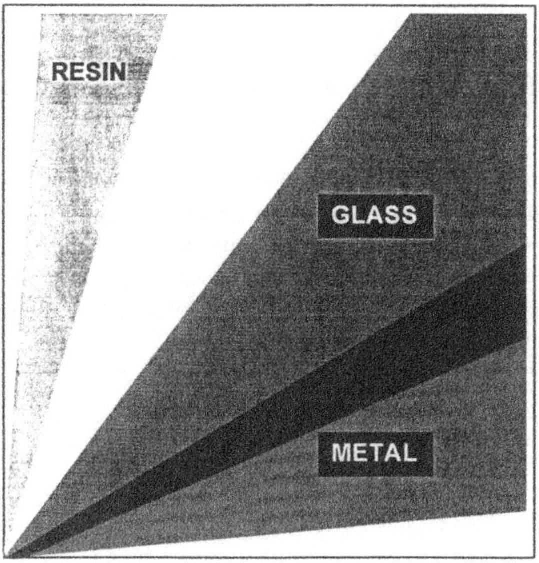

Figure 5. Relative wear resistances of various bonds. (Material loss with time in a Pin-on-Disc wear test).

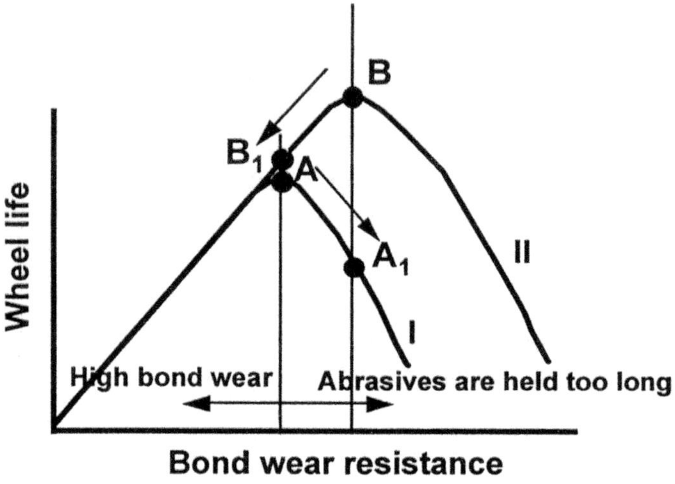

Figure 6. Optimum bond wear resistance for given system

3. Composite Property Tailoring

The means of composite property modifications and their relationship to the grinding process is discussed in continuation using an example of development of diamond composite for a specific application, namely grinding of advanced ceramics. The development of advanced, more efficient automobile engines has brought about the ever increasing need for high temperature wear resistant materials for structural applications such as turbocharger rotors, parts for turbine engines and valves for automotives. Silicon Nitride, SiAlON, and Zirconia ceramics of considerably improved strength and fracture toughness have been produced in the past two decades to meet these needs. These materials, in majority of cases, have to be finished to final shape, by grinding. Grinding of such ceramic materials requires wheels with durability and form retention and low and steady spindle power and grinding forces to achieve specified tolerances as well as acceptable surface quality on ceramic work pieces. Surface quality requirements include good surface finish and low sub-surface damage

The requirement for low sub-surface damage is particularly important in view of the fact that toughness of ceramics is 5 to 10 times lower than that of structural metals. As flaw size i.e. sub - surface damage increases , strength, σ_f, decreases: The Griffith's equation defines this relationship;

$$\sigma_f = \left(\frac{2E\gamma}{\pi c}\right)^{0.5} = \frac{K_{1c}}{(\pi c)^{0.5}}$$

Where E is the Young's Modulus, γ the surface energy, c the flaw size and K_{1c} the Fracture Toughness of the work material.

wheel life and form retention depends on optimum bond selection, type of abrasives used, which in present case is diamond. Low power and forces are obtained using sharp abrasives, low wheel-work interface friction, and adequate clearance for grinding chips. Steady power and forces are realized by maintaining a constant supply of sharp cutting points, matching abrasive grain-bond wear resistance, and minimizing ductile smearing of the work material and the bond.

Resin and metal bonds are commonly used in wheels for grinding ceramics. Resin bonded wheels with friable diamond abrasives grind with low and steady power producing good and acceptable surface quality. However, they are prone to excessive wear and frequent loss of form. Due to the very low bond wear resistance, they operate at sub-optimum region to the left of "A" in Figure 6. Not only does this increase the cost of manufacture but also it necessitates frequent operator intervention. On the other hand conventional metal bonded wheels are too durable and operate to the right of "A". They also cannot grind with steady power due to generation of wear flats on abrasives. Unsteady power from part to part also leads to inconsistent sub-surface damage in ceramic work piece. Hence a bond that is durable enough to hold the abrasives for their useful life but provides low and steady spindle power is desired. Current metal bonded wheels use tough and blocky abrasives. Such abrasives also develop wear flats that could increase the sub-surface damage to the ceramic. Hence, it was decided to try two types of weak and friable abrasives, typical example of which is shown in Figure 7.

Figure 8 shows the difference in the toughness (friability) of a blocky MBG II diamond versus a weak RVG which was chosen for this application. (9) Toughness is measured as the number of cycles required to reduce the average size of the abrasive by half on a known weight, while they are shaken and impacted by carbide balls in a steel container.

The Copper-Tin alloy was selected for the matrix material because of low temperatures needed for its processing as shown in phase diagram (Figure 9), green strength sufficient for handling of powder compacts, ability to wet diamond and tailorability to a range of application requirements.

Two compositions were explored: one in the low range of Sn content (Cl) and another at high range (C2), as shown in Figure 9. Due to the need for modifying properties based on grinding test results, high tin composition was explored. In order to match the bond wear rate to that of abrasives, hollow Mullite,

Figure 7. Typical abrasive type used in new metal bond.

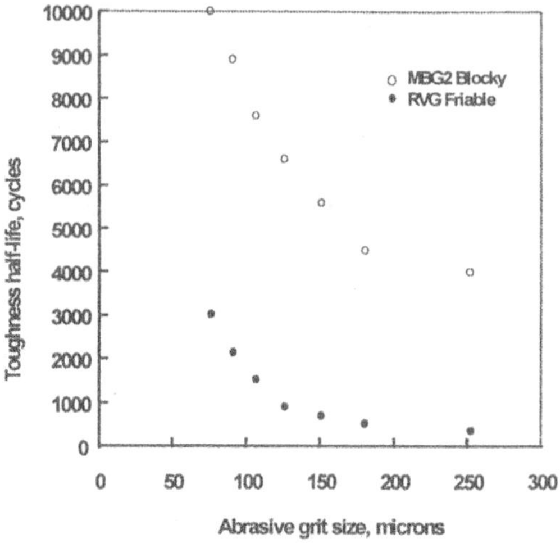

Figure 8. Toughness of blocky and friable diamond abrasive types.

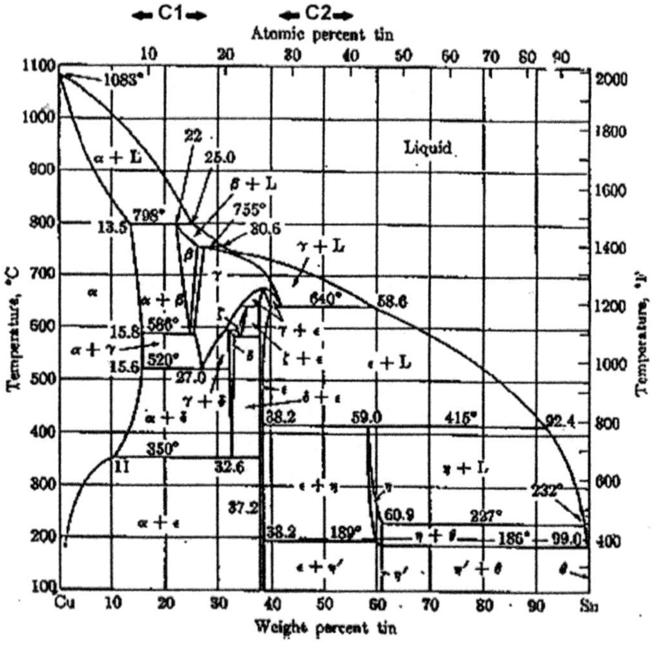

Figure 9. Copper - Tin Phase diagram (10)

Aluminum- Silicate (3A1203 25i02), as a dispersoid was added to low tin content bond (Cl). Hollow mullite also increases the chip clearance which is essential for efficient grinding action. Up to 85% by volume of each sphere was hollow. The average size of the filler was around 100 microns.

Using 3" (76 mm) diameter wheels and 4.5" (114 mm) diameter SiAlON ceramic work pieces, grinding tests were conducted in cylindrical plunge mode with a water based coolant. Figure 10 shows the schematic of the grinding test, the set-up of which is shown in Figure 11. Spindle power, grinding forces, wheel wear, and work piece surface finish were measured.

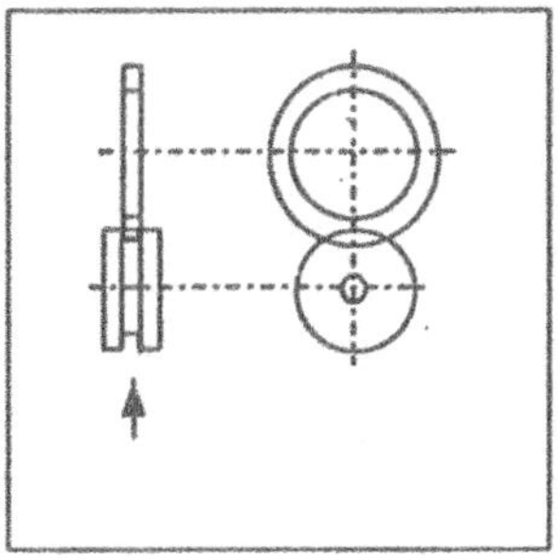

Figure 10. Schematic of grinding tests

Figure 11. Set-up of grinding test.

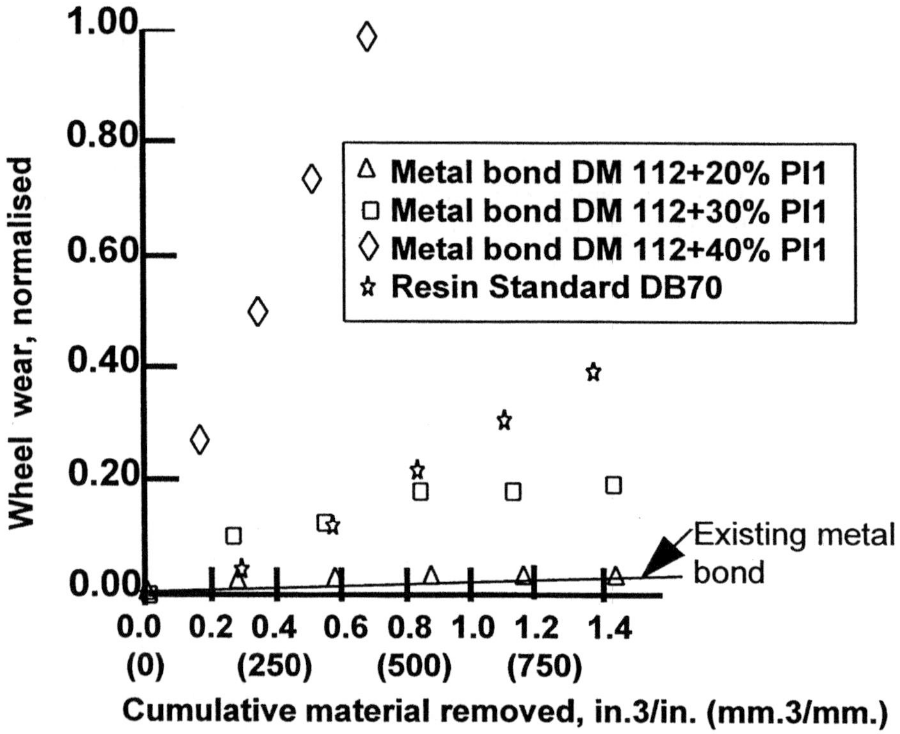

Figure 12. Wheel wear of metal bond with low tin (Cl) and different dispersoid levels.

4. Results and Discussion

As can be seen in Figure 12, the wear increased with mullite or porosity content and even wheels with wear rates higher than the reference resin bond were achieved. In grinding tests using a SiAlON ceramics as a work piece material it was found that induced porosity increased the overall power consumption, mainly due to abrasive pull out and bond-workpiece friction. (Figure 13)

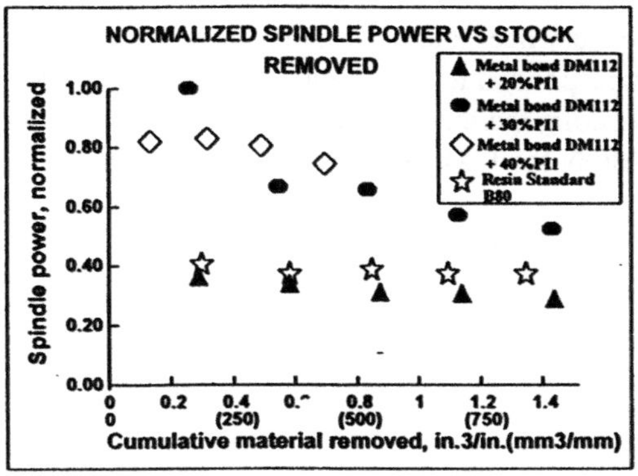

Figure 13. Spindle power vs. material removed.

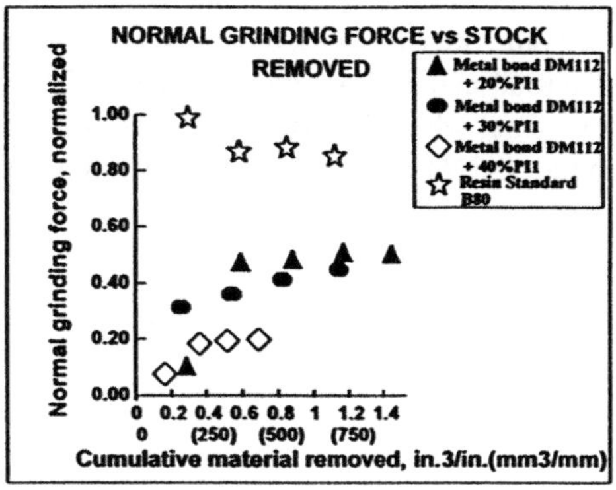

Figure 14. Normal grinding force vs. material removed.

As shown in Figure 14, normal grinding forces were rising with amount of material removed, in wheels containing all levels of dispersoids tested, which is not desirable since this would lead to excessive subsurface damage on the ceramic work piece. The force build up is due to the ductility of the bond that leads to its smearing on the wheel face preventing exposure of abrasive grains for grinding. A bond with same wear rate but of lower ductility was needed. Lower ductility can be achieved through decrease in fracture toughness, a measure of resistance to impact. By varying the tin content, the hardness and fracture toughness in Cu - Sn system can be tailored over a considerable range. Above 38% by weight of tin in a copper-tin system results in brittle inter-metallic phases as shown in the phase diagram earlier. Higher tin content also results to higher hardness. Bond wear, being a function of its hardness and fracture toughness raised to certain powers, was maintained at the same level by selecting tin values associated with regime 'C2' instead of 'Cl' in Figure 15. In short, wear resistance of the bond was held nearly the same while its toughness was reduced by changing the tin content in the composition.

Grinding wheels made with this new composition were tested as before in cylindrical plunge grinding. Such wheels consumed low and steady power versus fluctuating power with the reference resin bond as shown in Figure 16. The resin bonded wheel required periodic dressing with an abrasive stick to clear wheel loading. Under the test conditions, the total material removed with SiAlON corresponded to five hours of grinding without dressing, when the test was discontinued. Among the three ceramics with the new metal bond, Si3N4 drew the most power and SiAlON the least, as shown in Figure 17.

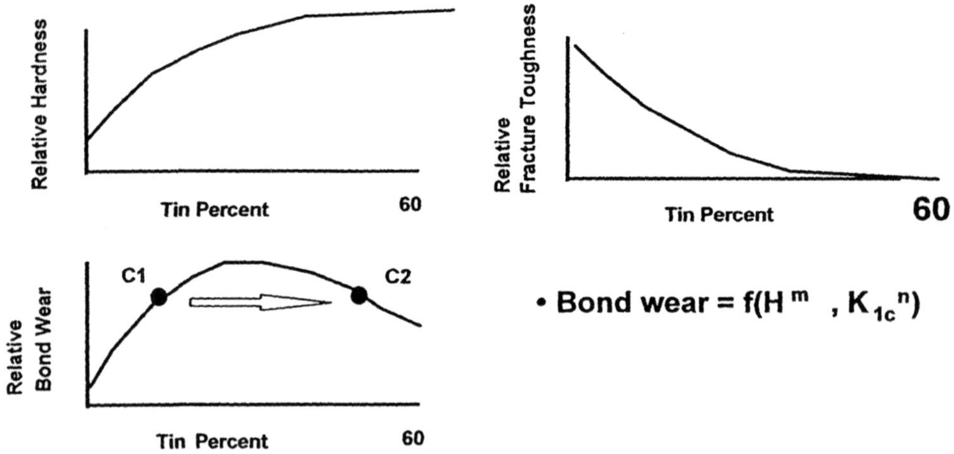

Figure 15. In order to reduce its ductility while maintaining a bond wear of level A, tin content was increased that resulted in higher hardness and lower fracture toughness.

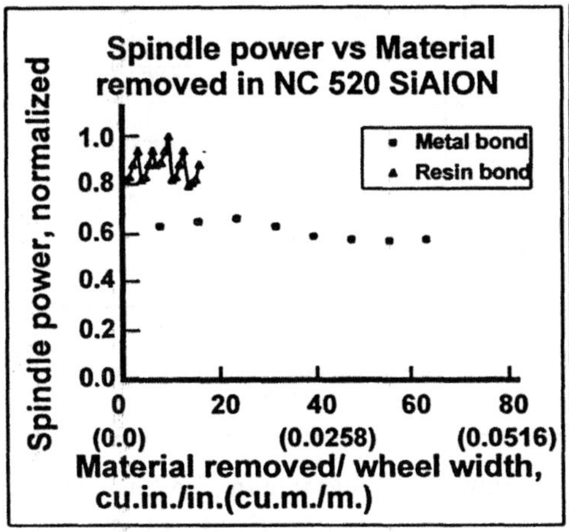

Figure 16. A nearly constant spindle power with the new metal bond vs. a fluctuating value that required frequent dressing with current resin bond in cylindrical plunge grinding of SiAlON ceramic.

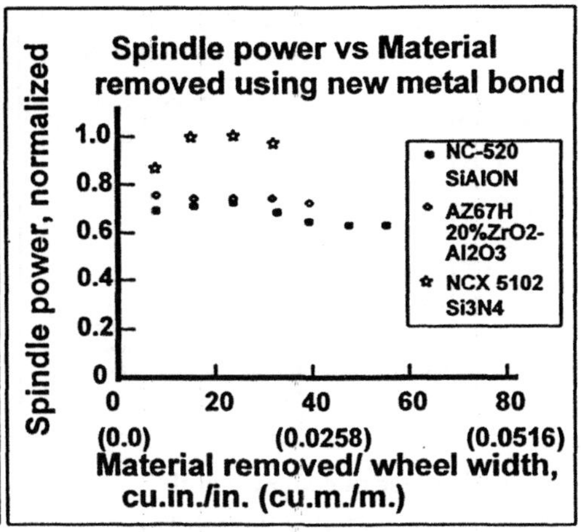

Figure 17. Nearly constant was grinding power drawn in grinding three typical advanced ceramics using the new metal bond.

Figure 18 shows the cumulative wear in grinding using the new metal bond and current resin bond. The wear rate of the new metal bond is significantly less than the current resin bond and also nearly constant. Among the three ceramics tested, NC520-SiAlON caused minimum wheel wear as shown in Figure 19. The reason for differences in wheel wear rates while grinding the three ceramics are due to the degree of difficulty associated in grinding them which in turn depend on their hardness and fracture toughness.

Wear resistance of SiAlON is the lowest of the three and hence it is the easiest to grind. Sub-surface grinding damage to SiAlON ceramic was determined indirectly by measuring the strength of C-shaped work pieces as shown in Figure 20.

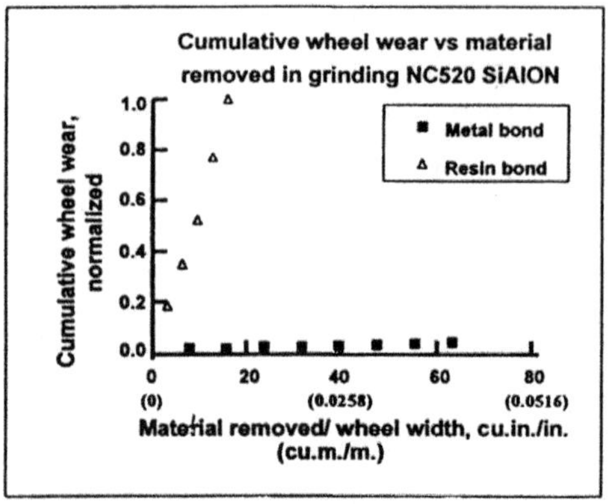

Figure 18. Wear rate significantly lower with the new metal bond vs. current resin bond in cylindrical plunge grinding of SiAlON ceramic.

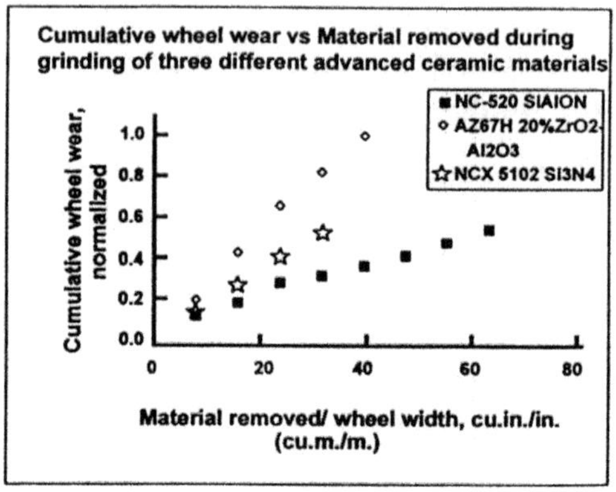

Figure 19. Wear rate in grinding Si3N4 is highest and SiAlON lowest among the three ceramics ground with the new metal bond.

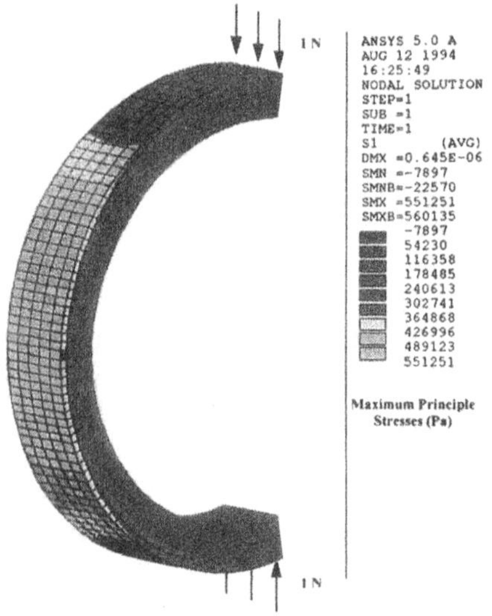

Figure 20. Finite element analysis of stresses break on ground ceramic work pieces.

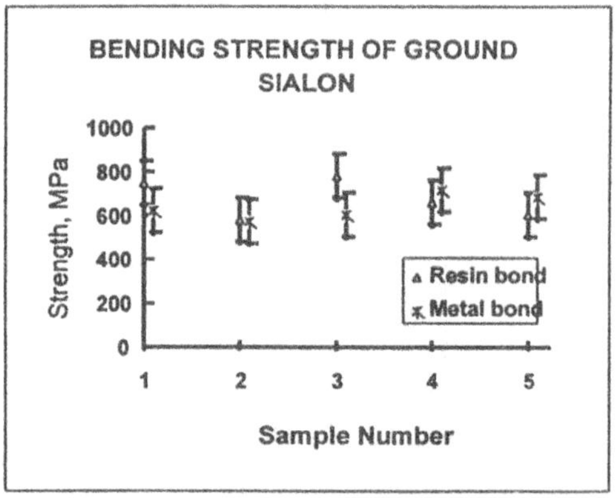

Figure 21. Bending strength of ceramic work pieces ground using current resin or new metal bonded diamond wheel showing statistically equal values.

The results indicate that the strength of samples ground by reference resin bonded wheel and new improved metal bonded wheel are statistically equal. Hence, we conclude that the damage level is the same. The surface finish of SiAlON work piece ground using resin and the new metal bonded wheel are shown in Figure 22. Similar levels of Ra were achieved with both wheels.

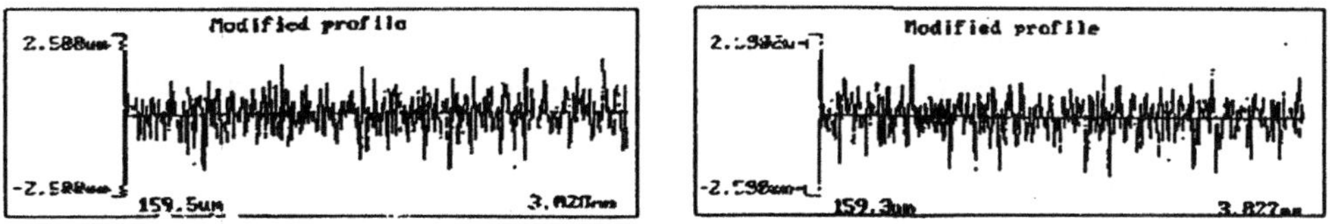

Resin

16.4 ± 1.2 microinch, Ra
(0.41 ± 0.03 micron)

Metal

18.4 ± 2.8 microinch, Ra
(0.46 ± 0.07 micron)

Figure 22. Surface finish of ground SiAlON ceramic with resin and new metal bonded wheels showing similar values.

Thus, using the principles of diamond abrasive composite design, a new metal bond has been developed, that is far superior to existing resin bonded wheels in durability and improved bond-wheel wear match relative to current metal bonds for grinding advanced ceramics. Grinding wheels made using this new metal bond has been shown to grind a range of such materials with acceptable quality, including subsurface damage. The use of dispersoids to alter bond properties that changes grinding action significantly has been demonstrated.

Conclusions

The usage of diamond composites for grinding, is due to the increasing demand for low cost finishing of high tolerance components, constantly growing Diamond grinding tools are designed to cause high abrasion rates on work materials while wearing in a slow controlled fashion. The matrices of such tools are complex multi-component systems where with appropriate selection of dispersoids and alloying elements it is possible to not only control the rate of wear but also change and manipulate the mode of wear of the bonded abrasive tool. The successful development of wheels for grinding of advanced ceramics discussed in the present paper attempts to point out the intricacy and importance of understanding material's microstructure/ property relationships as well as clear definition of design factors. Since the selection of diamond and particularly the bond needs to correspond to the aggressiveness of the work medium, diamond composites for grinding applications are encompassing a continually growing family of materials and properties.

Acknowledgments

Authors are grateful for the support received from the Northboro Research Center in characterization of sub-surface damage and World Grinding Center for their assistance in conducting grinding tests. This research was, in part, funded by the U.S. Department of Energy, Assistant Secretary for Energy Efficiency and Renewable Energy, Office of Transportation Technologies, as part of the Ceramic Technology Project of the Propulsion System Materials Program, under DE-ACO5-840R21400 with Martin Marietta Energy Systems Inc. Authors would also like to thank Mr. Harry Williston without whose able assistance in the preparation of bond and wheel samples this work would not be possible.

References

1. Subramanian, K., Ramanath, S., and Matsuda, Y. (1990) "Precision Production Grinding of Fine Ceramics", Proceedings of the First International Conference on New Manufacturing Technology, Chiba, Japan, pp 309-316.

2. Subramanian, K., and Ramanath, S.(1989)"Superabrasive Applications: An Expanding Horizon", Twenty-Seventh Abrasive Engineering Society Conference, pp 194-214.

3. Subramanian, K., Redington, P.D., and Ramanath, S.(1993) "A Systems Approach for Grinding of Ceramics", NIST Special Publication 847, Machining of Advanced Materials, Proceedings of the International Conference on Machining of Advanced Materials, Gaithersburg, MD, pp 43-53.

4. Gardinier, C. F.(1988) "Physical Properties of Superabrasives", Ceramic Bulletin, vol. 67(6), pp 1006-1009.

5. Ratterman, E., and Cassidy, R.(1990) "Abrasive Machining Process", Abrasives Section 5. ASM.

6. Subramanian, K.(1987) "Superabrasives for Precision Production Grinding- A case for interdisciplinary effort", Proceedings of the Symposium on Interdisciplinary Issues in Materials Processing and Manufacturing - ASME, New York, vol.2, pp.665-676.

7. Buljan, S.T. and Wayne, S.F. (1989) "Wear and Design of Ceramic Cutting Tool Material", Wear, vol.133(2), pp.1-13.

8. Buljan, S.T. and Wayna, S.F.(1994) "Microstructure and Wear Resistance of Silicon Nitride composites", Friction and Wear of Ceramics, Edited by Said Jahanmir, Marcer Dekker Inc., pp.261- 285.

9. Saint Gobain-Norton Internal Work.

10. Brooks, C.(1986) "Heat Treatment, Structure and Properties of Non-Ferrous Alloys", ASM Metals Park, Ohio, p 306.

HOT PRESSING OF NANODIAMOND POWDER

Galina Popovici[1], **M. A. Prelas and F. Golshani**
University of Missouri - Columbia, Nuclear Engineering Department, Columbia, MO 65211, USA

Pengdi Han, K. E. Huggins
University of Illinois at Urbana-Champaign, Materials Science Laboratory, Urbana, Illinois 61801, USA

Nanophase diamond is a material obtained in an explosion of carbon containing charge in a closed vessel. It consist of two phases: diamond and metastable carbon phase which can be transformed into diamond upon annealing. Nanodiamond powder was hot pressed at 7.7 GPa in the temperature range 750-1400 ^{0}C. The metastable carbon phase was transformed into diamond by this method. Hot pressing binds the nanocrystals and transforms the powder into a bulk material approaching diamond quality. There will be a variety of applications for such a diamond material in various industries from aerospace to manufacturing.

1. Introduction

Nanophase diamond is a material obtained in a detonation of carbon containing explosives in a closed vessel. It consist of two phases: diamond and metastable carbon phase. The metastable carbon phase can be obtained in highly non equilibrium conditions, like explosion, chemical vapor deposited diamond under electric bias, or neutron irradiation in a reactor. [1-3] It can be transformed into diamond by annealing. This carbon phase may serve as a binder for pressing diamond powders of different sizes in order to obtain material approaching diamond quality with high electrical resistivity, resistance to chemical attack and resistance to nuclear radiations In this paper the feasibility of obtaining such a material by hot pressing of the nanodiamond powder is studied.

---

[1] Present address: University of Illinois at Urbana-Champaign, Coordinated Science Laboratory, 1101 West Springfield Avenue, Urbana, IL 61801, e-mail: popovici@uiuc.edu

M. A. Prelas et al. (eds.), Diamond Based Composites, 31–38.

2. Experimental

Nanodiamond powder consists of spherical particles with a diameter between 2 and 50 nm with a peak of distribution at 6 nm. The nano particles were ultrasonically mixed with alcohol in order to obtain the most uniform distribution and prevent clustering. Then alcohol was evaporated and the powder was pressed in Pt container under 7.7 GPa for 20 mm.

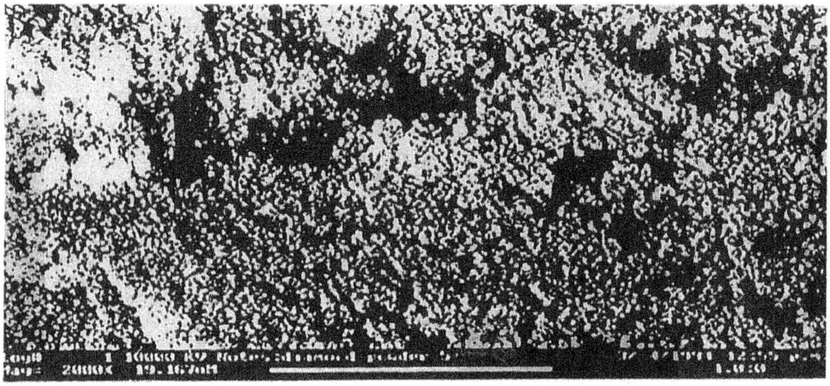

a)

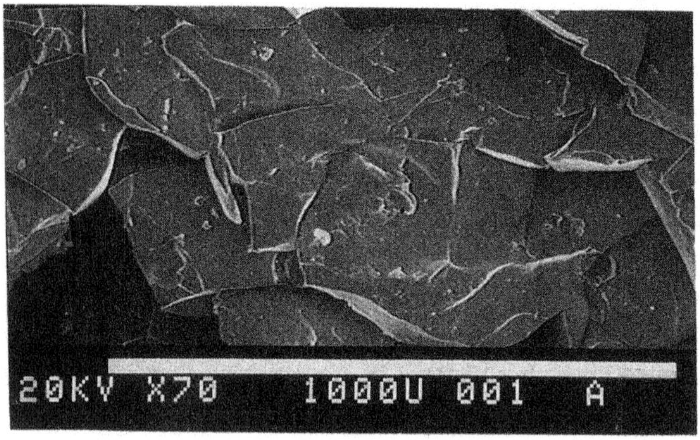

b)

Fig. 1. SEM photographs of nanodiamond powder. a) before pressing and after pressing at different temperatures: b) 750 ^{0}C; c) 1050 ^{0}C; d) 1260 ^{0}C; e) 1400 ^{0}C.

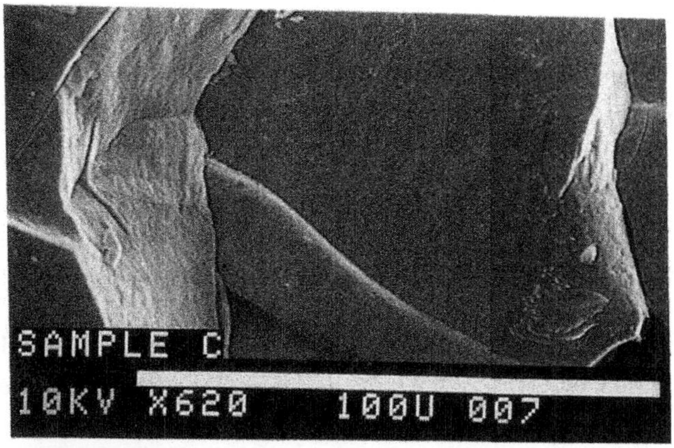

c)

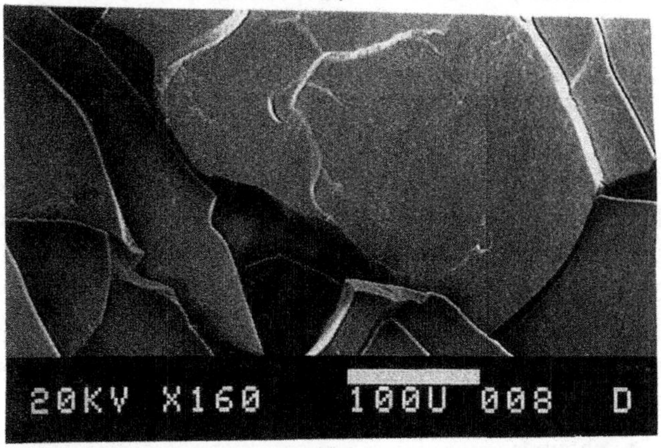

d)

e)

Fig. 1. SEM photographs of nanodiamond powder. a) before pressing and after pressing at different temperatures: b) 750 ^{0}C; c) 1050 ^{0}C; d) 1260 ^{0}C; d) 1400 ^{0}C.

Graphite tubing was used as a heater.

The pressure temperatures for different samples were following:

Sample	Temperature ('C)
A	750
B	900
C	1050
D	1260
E	1400

X-ray diffraction (XRD), scanning electron microscopy (SEM) and Raman measurements were performed before and after pressing. The Raman spectra were excited by an argon ion laser operated at 514.5 nm and 457.9 nm. The laser was focused to a spot size of approximately 0.3 mm^2. In the spectrograph, a 600 groove/mm grating dispersed the signal onto a CCD detector or a photomultiplier cooled by liquid nitrogen.

3. Results and Discussions

Fig. 1 presents photographs of the nanodiamond powder before (Fig. la) and after hot pressing (Fig. lb to Fig. le). The hot pressed samples have a bulk appearance. SEM pictures were taken after the surface of bulk material was broken by a diamond intender.

The XRD pattern is given only for sample D, because there was no qualitative difference among the XRD patterns for all five samples. The XRD patterns of nanodiamond before (Fig. 2a) and after pressing (Fig. 2b) have diamond and Pt lines only. The diamond lines are labeled by the letter "D'. The narrow lines in Fig. 2b belong to the Pt container.

The Raman spectrum of the nanophase diamond powder shown in Fig. 3 has two peaks. A symmetrically broadened diamond line is displaced from 1332 cm^{-1} characteristic for the bulk diamond [4] to 1326 cm-^{1} due to small size of particles.[5]. The second peak at 1629 cm^{-1} belongs to a carbon phase. This peak was also observed in the neutron irradiated diamond samples [2-3]. The same peak position (1629 cm^{-1}) was obtained at two excitation wavelengths, 514.5 nm and 457.9 nm, indicating that both peaks have Raman origin. The phase appears to be formed in highly non equilibrium conditions like the explosion method or neutron irradiation. This carbon phase converts to diamond during annealing at normal pressure[2-3] showing that it is a metastable carbon phase. This line is not a graphite line displaced from 1590 cm^{-1}. As graphite is a stable carbon phase, its line would not disappear on annealing, hence the line does not belong to graphite [2-3]. Raman spectra for the hot pressed nanodiamond bulk material are shown in Fig. 4. There is no peak at 1629 cm^{-1} which belongs to the

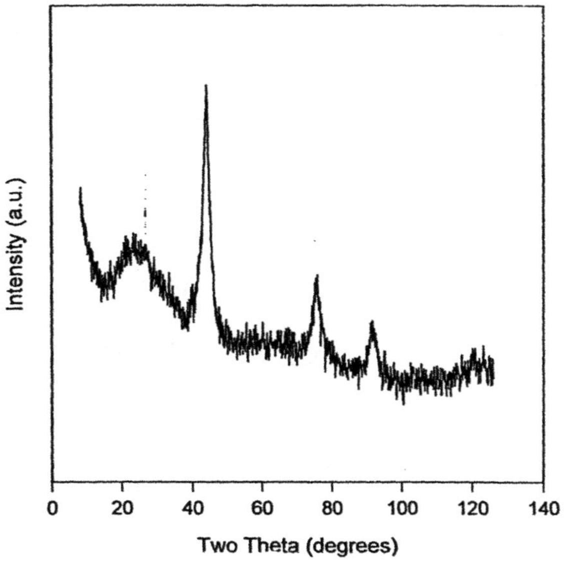

a)

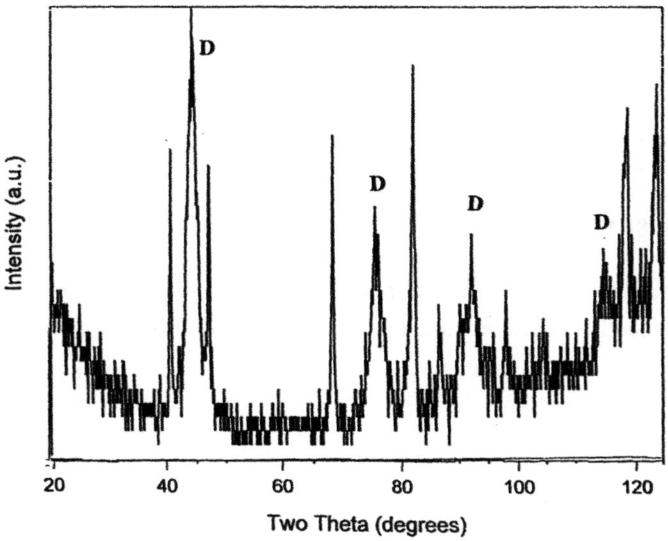

b)

F ig. 2 XRD patterns: a) before pressing and b) after pressing.

metastable diamond phase. The diamond peak at 1331 cm^{-1} is broadened. There are generally two possible causes for the broadening and shifting of Raman lines: small crystallite size and macro- and micro-stresses (random in-plane compressive and tensile stresses). [5-6] The stress is most probably responsible for the broadening of the Raman lines.

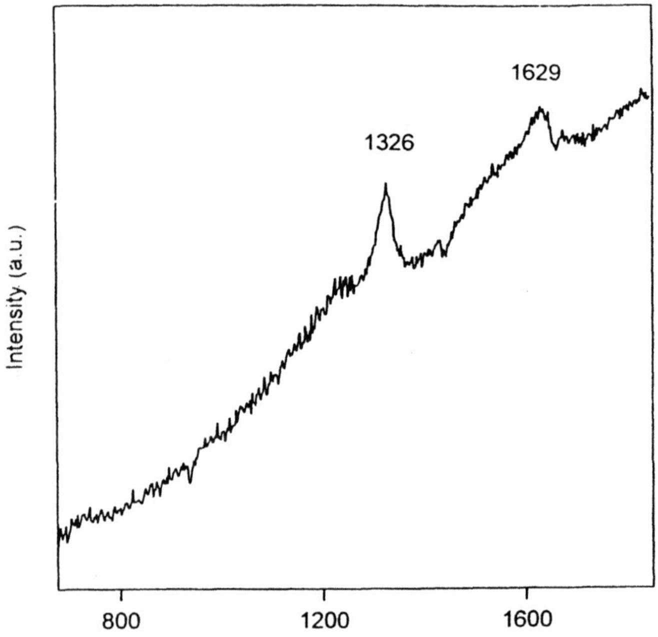

Fig. 3. Raman spectrum of the nanodiamond powder

Although pressing was performed at different temperatures, the results do not depend on temperature dramatically. Therefore, a temperature as low as 750 C, or even lower, can be used.

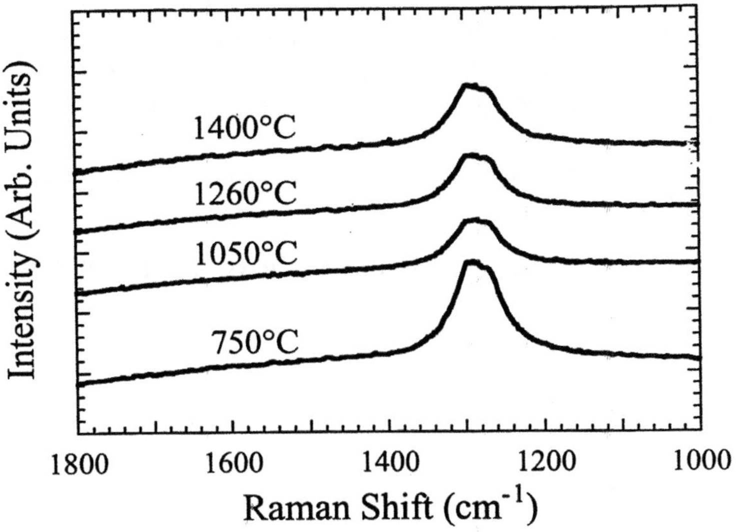

Fig. 4. Raman spectra of the materials obtained by hot pressing.

The above experiments show the feasibility of obtaining diamond bulk material by the method of hot pressing using nanodiamond powder. A larger grit together with nanodiamond can also be used for pressing to optimize the density of the bulk material. Neutron irradiation of nanodiamond also can be used to increase the ratio of the metastable carbon phase and give more energy to the lattice.

4. Conclusions

Nanodiamond powder was hot pressed at 7.7 GPa in the temperature range 750-1400 ^{0}C. The metastable carbon phase usually present in nanodiamond powder was transformed into diamond by this method. Hot pressing binds the nanocrystals and transforms the powder into a bulk material.

References

1. G. Popovici, C. Chao, M. A. Prelas, E. J. Charlson, J. Meese, J. *Mater. Research,* **19,2011** (1995)
2. S.A. Khasawinah, Galina Popovici, J. Farmer, T. Sung, M. A. Prelas, J. Chamberlain, and H. White, J. *Mater. Research,* **10,** 2523 (1995).

3.G. Popovici , A. A. Melnikov , V.5. Varichenko, S. Khasawinah, T. Sung, M. A. Prelas, A. B. Denisenko, N.M. Penina, V.A. Martinovich, E. N. Drozdova' A .M. Zaitsev, and W. R. Fahrner, to be published *Diamond Relat. Mater.*
4. D. S. Knight and W. B. White, J. Mater. Res. *4, 385* (1989)

5. E. Oheeraert, A. Deneuville, A. M. Bonnot, and L. Abello, Diamond and Related Materials, **1,** *525* (1992)

6. M. H. Grimsch, E. Anastakis, and M. Cardona, Phys. Rev. B18, 901(1991)

CHEMICAL VAPOR DEPOSITION OF DIAMOND FILMS ON DIAMOND COMPACTS

V.G. RALCHENKO, S.M. PIMENOV,
V.G. PEREVERZEV, I.I. VLASOV,
S.V. LAVRISCHEV, E.D. OBRAZTSOVA and
V.I. KONOV
General Physics Institute, ul. Vavilova 38, Moscow 117942, Russia

E.V. SHOROKHOV, D.M. LEBEDEV,
M.A. LEBEDEV and B.K. VODOLAGA
Russian Federal Nuclear Center - Research Institute of Technical Physics, Snejinsk 454070, Chelyabinsk Region, Russia

E.N. LOUBNIN
Institute of Physical Chemistry, Leninsky prosp.31, Moscow 117915, Russia

V.A. SPIVAK
Moscow plant "Crystal-Brilliants", 12 Smolnaya str., Moscow 125493, Russia

Abstract

Polycrystalline diamond films were grown on different types of diamond composites in order to modify their morphology and surface properties. A DC arc-jet deposition system operated at $Ar/CH_4/H_2$ gas mixture was used to obtain diamond coatings on diamond-copper composites produced by explosive compression technique, while a DC plasma reactor operated at CH_4/H_2 feed gas was used to grow diamond films on B_4C-carbon composites and sintered diamond compacts. Morphology, quality and stress in diamond crystallites before and after film deposition were studied with scanning electron microscopy and microRaman spectroscopy. Homoepitaxially grown films showed less defected structure with smaller content of nondiamond carbon impurities than the diamond deposited on copper grains. In all cases well-crystallized CVD diamond was produced with a good adhesion to the substrates, including those containing copper islands. It is concluded that the mechanical and thermal properties of diamond composites can be improved with CVD diamond layers.

M. A. Prelas et al. (eds.), Diamond Based Composites, 39–52.

1. Introduction

Diamond compacts find various applications in cutting of hard materials, wire drawing, heat sinks, etc. Diamond compacts and composites can be prepared by high pressure-high temperature sintering [1] or by shock consolidation of diamond powder [2]. A surface modification of such polycrystalline material may be required for specific usages in order to improve surface morphology, diamond quality, increase thermal conductivity or to form a top insulating layer on electroconductive composites. In many cases a chemical vapor deposition (CVD) of diamond film could provide a solution of the problem. For example, Yoshikawa et al. [3,4] used epitaxial diamond growth in microwave plasma to cover a sintered diamond insert which served for cutting of ceramic materials. It was found that the wear resistance of the film-coated cutting tool was greatly improved, probably because of better quality (lower internal stress, smaller defect concentration...) of CVD diamond. In the present paper we report on diamond film growth on diamond-metal composites and sintered diamond ceramics using a DC plasma and DC arc-jet deposition systems.

2. Experimental details

2.1. SAMPLES

Three types of diamond compacts used as the substrates for diamond film deposition are listed in Table 1. Diamond-copper and diamond-boron carbide samples were formed by a shock explosion technique. The schematic view of shock wave apparatus for compact production is shown in Fig. 1. The feedstock material (a mixture of diamond and metal powder) was compressed during ≈50 μs by a piston, which provides a plane shock wave load from explosive. A buffer plate was used to control the shock pressure and temperature.

TABLE 1. Diamond composites used as substrate for CVD diamond deposition

Type #	Compact composition	Preparation method
1	diamond-copper	shock explosion
2	diamond-graphite-B_4C	shock explosion
3	diamond ceramics	HPHT static compression

The receipts for preparation of diamond-Cu pellets is shown in Table 2. A commercial diamond powder of 30 μm average grain size produced by high pressure-high temperature (HPHT) synthesis was mixed with smaller copper powder (<6 μm) in volumetric proportion 70/30. The final pellet size was 20 mm in diameter and 7 mm height, the sample density was increased from 3.30 g/cm^3 to 3.62 g/cm^3 after compression. The resistivity of the compact was 0.1 Ohm • cm, so it can be considered as a conductive material. A rough estimate of shock compression parameters for diamond-Cu porous system made on the basis of one-dimensional model showed that the maximum pressure did not exceeded 5 GPa and maximum temperature was about 1000°C.

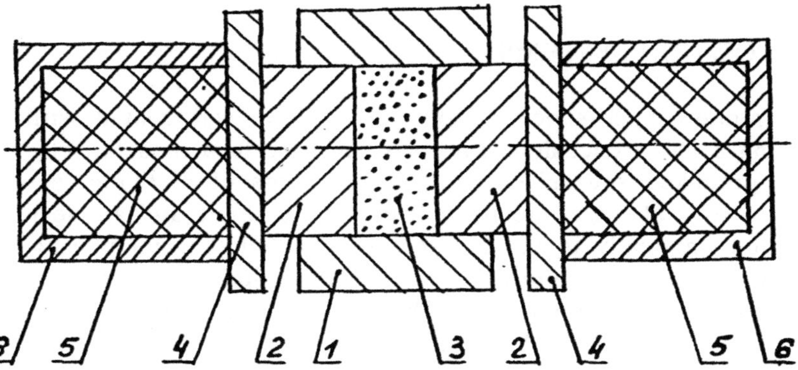

Fig. 1. Schematic view of shock press apparatus for diamond compact preparation: (1) matrix, (2) piston, (3) feedstock material, (4) buffer plate, (5) explosive with ignition system, (6) case.

The carbon-B_4C sample prepared according to Table 3 was composed of boron carbide powder with grain size less than 10 μm, amorphous carbon (a-C) and ultrafine diamond powder. The starting carbon material synthesized by shock explosion technique [5] was composed of 90% a-C, 6-8% diamond and a few percents of metal and organic impurities.

TABLE 2. Composition of diamond-copper compact

	diamond powder	copper powder
Percentage, % vol.	70	30
Grain size, μm	≈30	2-6

TABLE 3. Composition of carbon-B_4C compact

Component	B_4C	a-C	diamond
grain size	1-10 μm	---	5 nm

Polycrystalline diamond compact in shape of a triangle prism (5 mm side, 3.5 mm height) was sintered at high pressure-high temperature conditions for commercial application as the die for wire drawing. A DC plasma CVD reactor was used to grow diamond film onto that sample.

2.2. DEPOSITION SYSTEMS

The diamond-copper shock-consolidated sample was coated with diamond film using a DC arc-jet deposition system, described in detail elsewhere [6]. Briefly, premixed source gases, a few percents of methane in argon and hydrogen, flow at high velocity between the cathode and anode where they are heated ohmically in the arc region. The hot thermal plasma emanates from the orifice, and collides with a water-cooled substrate inserted in the stream to form diamond. The DC arc-jets are attractive for diamond synthesis, as they deliver the highest growth rate reported. The deposition parameters are shown in Table 4. The methane content was 1% with respect to hydrogen, Ar/H_2 flow rate ratio was ≈2, and substrate-to-nozzle distance was 50 mm. Maximum film diameter of 3 cm were obtained with this arc-jet apparatus. No seeding of diamond particles on substrate surface to enhance the diamond nucleation rate was used.

TABLE 4. Diamond deposition parameters for DC arc-jet and DC plasma systems

Parameter	DC arc-jet	DC plasma
Current (A)	90	1.4-2.2
Voltage (V)	60	200
Reaction gas	CH_4-Ar-H_2	CH_4-H_2
CH_4/CH_4+H_2 (%)	1	4
Gas flow rate (slm)	27	0.1
Gas pressure (Torr)	38	100
Substrate temperature (°C)	900	850
Growth time (min)	75	60 and 120

Carbon-B_4C compact and sintered diamond sample were used as substrates for diamond film growth in a DC plasma reactor with low flow rate [7]. The system operated at CH_4-H_2 mixture with 4% methane concentration, providing growth rates of ≈5 μm/h at the substrate of ≈1 cm^2 area. Other deposition parameters are shown in Table 4. Either well faceted polycrystalline films or nanocrystalline deposits could be formed depending on substrate temperature regulated by the current of the discharge.

2.3. CHARACTERIZATION

The substrate before and after diamond deposition were examined with microRaman spectrometer (model S3000, Instruments S.A.) to identify different carbon phases. The spectra were taken in backscattering geometry at 514.5 nm wavelength line of an Ar^+ ion laser source. The probe beam spot was approximately 2 μm in diameter. The surface morphology of compacts and films was observed with a scanning electron microscope (SEM).

3. Results

3.1. DIAMOND-COPPER COMPOSITE

Fig. 2a illustrates a surface morphology of diamond-Cu sample before diamond film deposition. Diamond grains of ca. 30 μm are surrounded with agglomerates of Cu particles of a few microns sized which look like bright domains in the picture. Note that typically the diamonds have irregular shapes and rounded edges. After diamond growth with DC arc-jet plasmatron the surface relief dramatically changed as a result of homoepitaxial growth of well faceted crystals mostly of octahedral shapes (Fig. 2b). On the basis of increase in particle sizes we roughly estimate the deposition rate to be of the order of 10 μm/h in the present experiments. Important that the copper fraction was also completely covered with diamond film with grain size of 3-10 μm. Those nice crystal shapes are typical for a very central region of deposition area near the axis of the arc-jet. Due to flow of reactants in radial direction along the substrate surface the local deposition parameters (temperature, radical concentration...) are also changed radially, leading to a gradual decrease in diamond quality with the distance from the substrate center. Figs. 2c,d show how worsens the film morphology as observed at 5 mm and 10 mm away from the center, respectively. A ball-like shape of diamond in Fig. 2c evidences a nanocrystalline structure of the deposit at the sample's periphery.

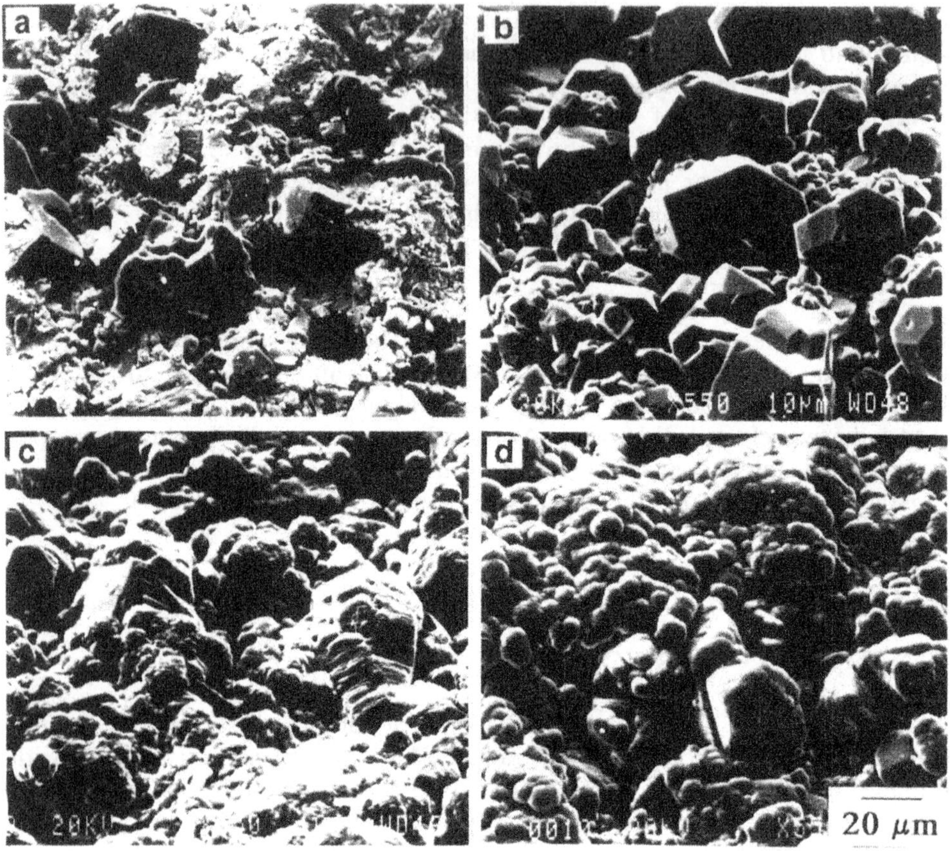

Fig. 2. SEM pictures of copper-diamond compact surface before (a) and after diamond film deposition by DC arc-jet at the center of deposition region (b), 5 mm away (c), and 10 mm away (d) from the film center.

The diamond quality is definitely higher in case of homoepitaxial growth as compared to the film on copper as was deduced from Raman spectra taken at different locations (Fig. 3). The film grown on a diamond particle in composite near the sample center (Fig. 3a, site 1) shows a sharp Raman peak at 1332 cm^{-1} characteristic of diamond structure. The linewidth (FWHM) of 4.6 cm^{-1} and very small signal from amorphous carbon phase at around 1500 cm^{-1} are very similar to the spectral features of HPHT crystals in the composite (linewidth of 4.0 cm^{-1}). In contrast the film deposited on copper fraction shows much wider diamond peak (8.1 cm^{-1}) and more amorphous carbon (see Fig. 3b, site 1). We note that the micro Raman facility allows to record spectra in neighbor spots of a few microns apart only, that's why we consider the two spectra discussed as taken from the single site.

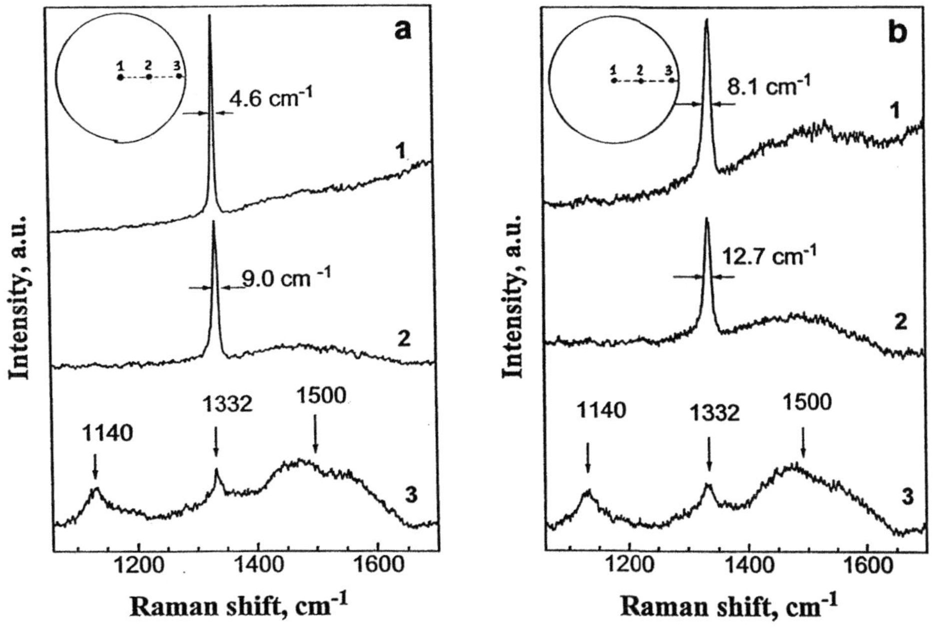

Fig. 3. Raman spectra of diamond film on diamond-Cu composite taken from the film parts grown homoepitaxially (a) and on Cu particles (b) at different locations on the sample surface (see the map at insert): (1) substrate center, (2) 5 mm away, and (3) 10 mm away from the substrate center. The sites 1-3 correspond to surface morphologics shown in Fig. 2b-d, respectively.

With increase in distance from the substrate center the diamond peak becomes wider both for film on diamond and Cu reflecting a more defected structure of the deposit (Fig. 3a,b, site 2). This tendency correlates with modification of surface morphology as shown in Fig. 2. Still the homoepitaxial film seems to be less degraded than that on copper as one can deduce from comparison of the peak widths (9.0 vs 12.7 cm^{-1}). The spectra taken from the very edge of the pellet (10 mm away from the center) display an additional peak at 1140 cm^{-1} which is typical for nanocrystalline diamond [8,9]. The reduced diamond peak at 1332 cm^{-1} and enhanced a-C peak at 1500 cm^{-1} confirm the further worsening of the film quality as was expected from the poor morphology observed in that region.

The position of diamond peak at Cu islands was shifted to about 1335 cm^{-1} from its normal position at 1332.5 cm^{-1} characteristic of stress-free diamond. This indicates a presence of compressive stress at the level of 2 GPa [10]. The stress obviously has a thermal nature, being generated during sample cooling after deposition process. The observed stress is much lower than that predicted from the difference in thermal expansion coefficients of diamond and thick Cu substrate (≈18 GPa [10]). The discrepancy can be explained by a partial relaxation of the stress due to plasticity of copper, and/or by the fact that the finite thickness of the substrate (i.e. copper grain) should be taken into account in the model when considering film deposition on small metal particles. The presence of the noticeable stress indicates that there is some adhesion of diamond on Cu. Generally the adhesion of diamond film is difficult to provide since copper does not form a carbide when contacts with carbon, but in our case it is promoted probably by anchor effect of diamond grains initially incorporated into compact.

The CVD diamond crystals can be easily differentiated from diamond grains in compact by their morphological features, first, by regular shapes with clear {100} and {111} facets, and second, by the presence of numerous penetration twins on those planes, as shown in more detail in Fig. 4. The shape of diamond crystal and appearance of twins are determined by the relative growth rates of {100} and {111} facets, V_{100} and V_{111}, respectively, as was analyzed by Wild et al. [11]. If the growth parameter is defined as $\alpha=\sqrt{3}\cdot V_{100}/V_{111}$, it was found that for $\alpha<1.5$ twins grow only on {100} facets, for $\alpha>2$ twins grow only on {111} facets, and in the range of $1.5<\alpha<2$ twins appear both on {100} and {111} facets. It is the latter case which corresponds to morphology observed in Fig.4.

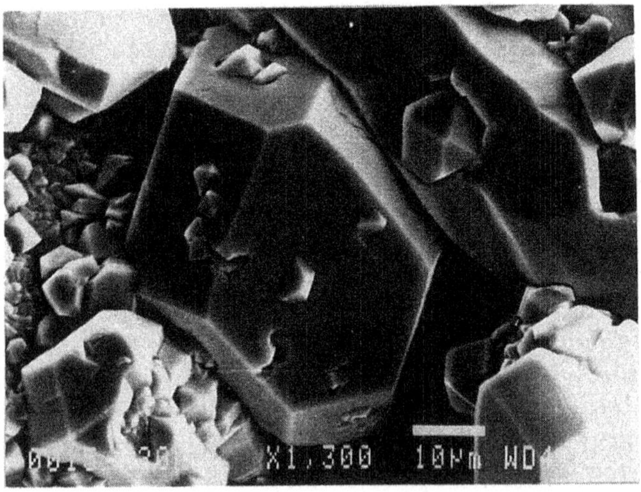

Fig. 4. Homoepitaxially grown crystal on diamond-Cu composite. Note the presence of penetration twins both on {100} and {111} facets.

As the diamond particles in compact were subjected to shock load during explosion one could attempt to search the presence of hexagonal diamond particles or domains with hexagonal structure within cubic diamond, as was discussed in relation to shock wave synthesis of diamond [12]. It is known that Raman peak of *h*-diamond is shifted to lower wavenumbers, laying between 1316-1326 cm^{-1} [13]. We did observe sometimes the peaks shifted down to 1314 cm^{-1} when the laser beam was focussed on selected diamond crystals sticking from the compact surface. We found that the shift increased with incident power of probe beam as shown in Fig. 5, indicating that actually we observed the temperature-depended peak shift [14] rather than the presence of hexagonal phase. At laser power of 200 mW (peak position at 1314 cm^{-1}) the local temperature of ≈850°C was achieved according to estimate based on tabulation of Herchen and Capelli [14]. A further increase in laser power resulted in fast oxidation of crystals. Interestingly, that after diamond film deposition the Raman peak remained always at 1332 cm^{-1} independently on laser power. This means that the film provided an effective spread of laser-generated heat, while there was a high thermal resistance in uncovered compact because of poor contact of diamond and copper particles. Thus a CVD diamond film can be used to improve thermal conductivity of a top layer of composites.

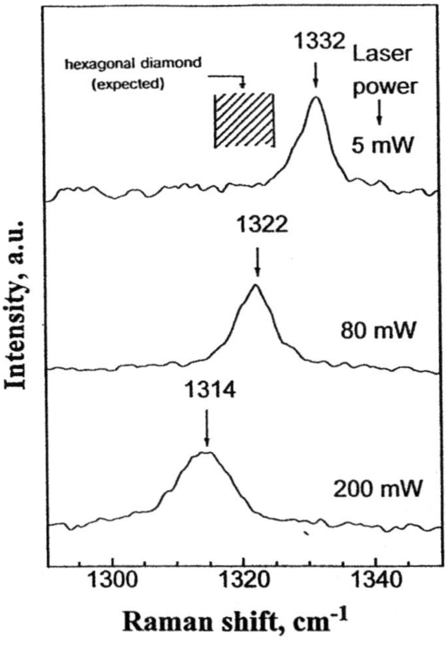

Fig. 5. Dependence of Raman diamond peak position on power of Ar^+ laser beam. The beam absorbed in diamond crystal on the surface of diamond-Cu composite causes local heating and the peak shift to lower frequencies. Dashed region shows position of peak of hexagonal diamond.

3.2. B_4C-CARBON COMPOSITE

Diamond film was deposited on ≈1 mm thick plate of B_4C-carbon sample using the DC plasma CVD reactor. Raman spectrum of the composite recorded before deposition (see Fig. 6) revealed the presence of boron carbide phase (peaks at 975 and 1076 cm^{-1}) and amorphous carbon. A signal from diamond phase was not detected because of two reasons: (i) the fraction of diamond powder in the feedstock was by order of magnitude lower compared to graphitic component, and (ii) cross-section of Raman scattering for diamond is by two orders of magnitude lower than that for graphite. The presence of B_4C has been confirmed also by X-ray diffraction analysis.

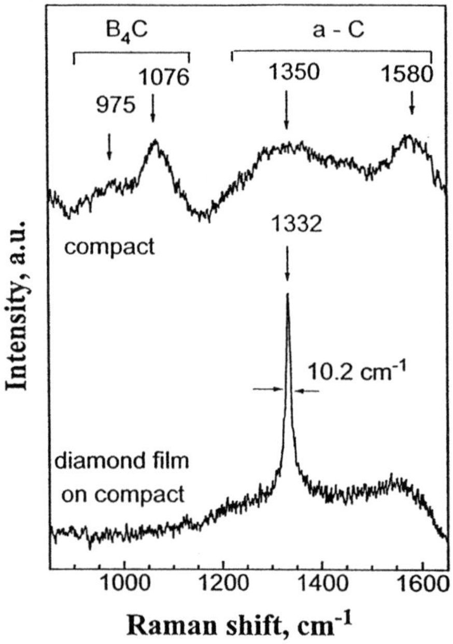

Fig. 6. Raman spectra of B_4C-carbon composite before (top spectrum) and after (bottom) deposition of diamond film.

Well-crystallized diamond film with grain size of 1-3 μm was produced in 1 hour deposition run, displaying the Raman spectrum with only minor content of amorphous component. The continuous film was grown on this composite without preliminary seeding the surface with diamond powder, which is normally required when other materials are used as substrates in this type of reactor [7]. The boron carbide-carbon composite supports a sufficiently high diamond nucleation density presumably due to the presence of ultrafine diamond particles in its composition, however a positive role of a-C and/or B_4C in nucleation process also can not be excluded.

3.3. SINTERED DIAMOND COMPACT

The sample of polycrystalline diamond sintered under static high pressure-high temperature conditions without a metallic binder showed a surface morphology with rounded grains without any sign of faceting (Fig. 7a). A homoepitaxial growth of diamond film during 2 hours rebuilds the grains to well-faceted crystals of 30-40 μm size (Fig. 7b). The facets carry numerous small crystallites born by secondary nucleation.

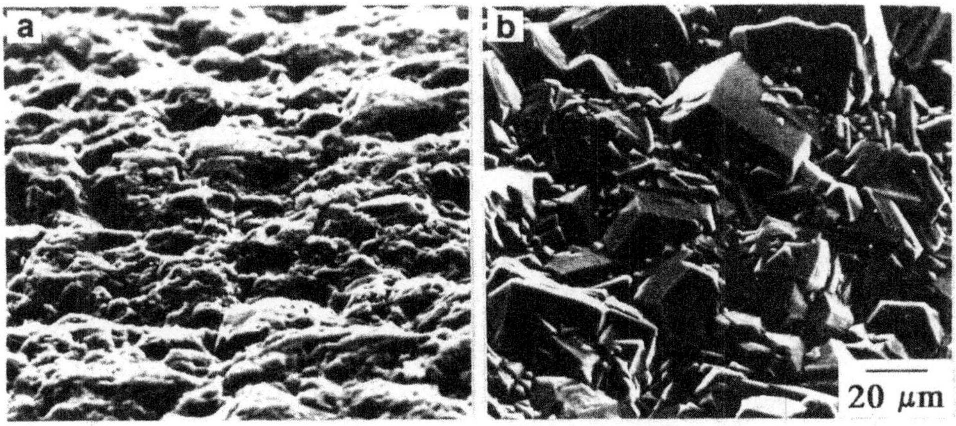

Fig. 7. SEM picture of sintered diamond compact before (a) and after (b) diamond film deposition with a DC plasma.

Wide-range Raman spectrum shown in Fig. 8a reveals a strong luminescence background for the compact, which is reduced after film deposition. On the other hand some quantity of amorphous carbon, which is absent in compact, appears in the film. Narrow-range Raman spectra taken before and after film growth (Fig. 8b) clearly show that the diamond peak width for film, Γ=10.8 cm^{-1}, is by a factor of 2 smaller than for compact ($\Gamma \approx 20$ cm^{-1}). Thus the coating is less defective and, perhaps, less strained material than the substrate.

Fig. 9 shows a corner of the prism-shaped compact with deposited diamond film. Important that the film completely covers the edges of the compact, therefore the DC plasma deposition technique can be used to modify or re-built the sintered diamond cutting inserts and other diamond-based tools.

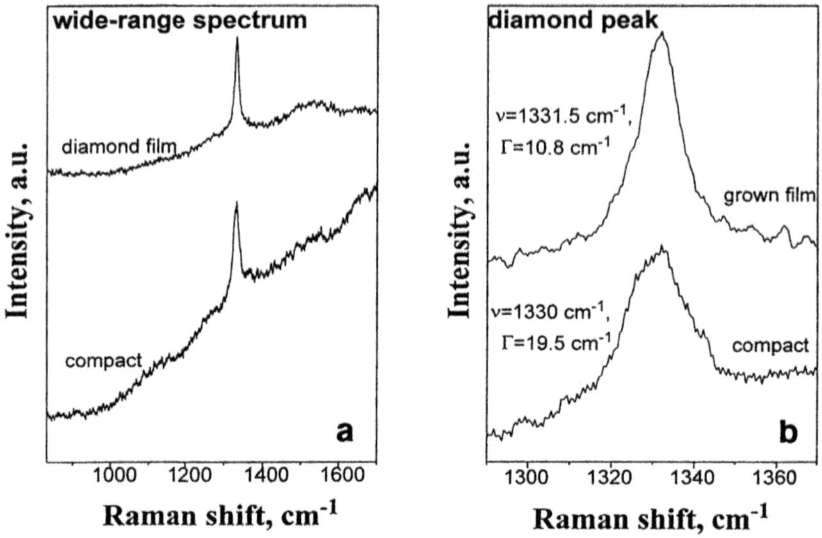

Fig. 8. Raman spectra of sintered diamond compact and CVD film taken in wide frequency range (a) and narrow range (b).

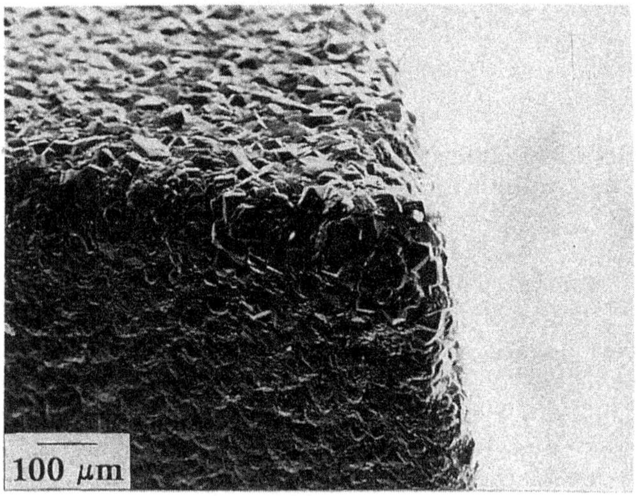

Fig. 9. Edge of sintered diamond compact covered with diamond film.

4. Conclusions

Surface morphology of diamond composites can be significantly modified by diamond film deposition. Diamond-copper and boron carbide-carbon composites produced by shock explosion method, and diamond ceramics

produced by static sintering, were used as substrates for CVD diamond growth. Polycrystalline diamond films were grown on different diamond composites using DC arc-jet plasma and DC plasma deposition techniques.
Continuous diamond coatings have been obtained on conductive diamond-Cu substrate. The quality of the epitaxially grown film remained high, while poorer diamond was deposited on copper islands, as evidenced from Raman spectra. No hexagonal diamond has been detected with Raman measurements in shock-loaded compact. The observed low-frequency shift of 1332 cm^{-1} Raman diamond peak was ascribed to laser-induced heating effect caused by low thermal conductivity of the weakly compressed compact. Deposition of diamond film significantly improved the surface thermal conductivity of the composite.
Diamond films easily nucleate on boron carbide-carbon substrates upon deposition in DC plasma.
Sides and edges of sintered diamond compact plate were covered by a continuous diamond film with higher quality. This may be useful for improvement of mechanical properties of PCD cutting tools.

5. References

1. Voronov, O.A. (1993) Diamond compacts, in *Proc. Third Int. Symp. on Diamond Materials*, The Electrochem. Soc., Pennington, NJ, Vol. 93-17, 1018.
2. Kondo, K. and Sawai, S. (1990) Diamond compact of nano-sized grains, in S. Saito, O. Fukunaga and M. Yoshikawa (eds), *Science and Technology of New Diamond*, KTK Scientific Publishers/Terra Scientific, pp.245-249.
3. Yoshikawa, M. and Tokura, H. (1990) Gas-phase synthesis of diamond film on a sintered diamond and its application to a cutting tool, in *New Diamond*, Japan New Diamond Forum, pp. 86-88.
4. Otake, N., Tokura, H., Yoshikawa, M. and Yang, C.F. (1990) Deposition of diamond film on a sintered diamond surface and its application to a cutting tool, in S. Saito, O. Fukunaga and M. Yoshikawa (eds), *Science and Technology of New Diamond*, KTK Scientific Publishers/Terra Scientific, pp. 139-143.
5. Lyamkin, A.I., Petrov, E.A., Ershov, A.P., Sakovich, G.V., Staver, A.M. and Titov, V.M (1988) Diamond production from explosives, *Sov. Phys.: Doklady*, **302**, 611-613 (in Russian).
6. Pereverzev, V.G., Smolin, A.A., Ralchenko, V.G., Vlasov, I.I., Obraztsova, E.D., Konov, V.I., Metev, S., Ozegowski, M., Sepold, G. and Khomich, A.V. (1996) Diamond films grown with DC arc-jet plasmatron on Mo and Si substrates, *paper presented at the 3rd Int. Symp. on Diamond Films*, 16-19 June, St. Petersburg, Russia.
7. Konov, V.I., Smolin, A.A., Ralchenko, V.G., Pimenov, S.M., Obraztsova, E.D., Loubnin, E.N., Metev, S.M. and Sepold, G. (1995) D.c. arc plasma deposition of smooth nanocrystalline diamond films", *Diamond Relat. Mater.* **4**, 1073-1078.

8. Nemanich, R.J., Glass, J.T., Luckovsky, G. and Shroder, R.E. (1988) Raman scattering characterization of carbon bonding in diamond and diamondlike thin films, *J. Vac. Sci. Technol. A* **6**, 1783-1787 .
9. Obraztsova, E.D., Korotushenko, K.G., Pimenov, S.M., Ralchenko, V.G., Smolin, A.A., Konov, V.I. and Loubnin, E.N. (1995) Raman and photoluminescence investigations of nanograined diamond films, *NanoStructured Materials* **6**, 827-'830.
10. Ralchenko, V.G., Obraztsova, E.D., Korotushenko, K.G., Smolin, A.A., Pimenov, S.M., and Pereverzev, V.G. (1995) Stress in thin diamond films on various materials measured with microRaman spectroscopy, in M.D. Drory, D.B. Bogy, M.S. Donley and J.E Field (eds.), *Mechanical Behavor of Diamond and Other Forms of Carbon,* Mat. Res. Soc. Symp. Proc. Vol. 383, pp. 153-158.
11. Wild, C., Kohl, R., Herres, N., Muller-Sebert, W. and Koidl, P. (1994) Oriented CVD diamond films: twin formation, structure and morphology, *Diamond Relat. Mater.* **3**, 373-381.
12. DeCarli, P.S. (1995) Shock wave synthesis of diamond and other phases, in M.D. Drory, D.B. Bogy, M.S. Donley and J.E Field (eds.), *Mechanical Behavor of Diamond and Other Forms of Carbon,* Mat. Res. Soc. Symp. Proc. Vol. 383, pp. 21-31.
13. Knight, D.S. and White, W.B. (1989) Characterization of diamond films by Raman spectroscopy, *J. Mater. Res.* **4**, 385-393.
14. Herchen, H. and Capelli, M.A. (1991) First-order Raman spectrum of diamond at high temperatures, *Phys. Rev. B*, **43**, 11740-11744.

THE DIAMOND AND HARD ALLOY-BASED COMPOSITE MATERIAL

V.S. URBANOVICH
Institute of Solid State and Semiconductor Physics
Academy of Sciences of Belarus
17, P. Brovka str., Minsk, 220 072, Belarus

1. Introduction

Composite diamond-containing materials posses certain properties that enable them to combine high abrasive durability of diamond with its unique stress-proof qualities. As a rule they represent composition of diamond particles evenly dispersed within a ceramic matrix, generally based upon tungsten carbide [1-3]. The most well-known representatives of that class are slavutich and tvesal [4]. Mining and drilling peculiarities of the above materials are those that the stone destruction is being performed by the diamond itself along with a hard-alloy mold. Diamond particles' depression within the hard-alloy mold leads to their higher resistance towards dynamic strains.

From the durability's point of view, it seems much more preferable to use the finest possible diamond grain in composite materials [5]. Those materials are products of the powder metallurgy prepared by simple sintering or hot pressing. It is widely accepted [6,7] that high pressures used in hard-alloy WC-Co sintering help to increase its hardness and durability by forming a fine-grain structure along with shorter sintering times. [1] informs of increased tvesal durability when manufactured under high pressures. Scientific literature, though, lacks any specific information on sintering properties of the above materials under high pressures as well as their production technologies.

That is the reason why we undertook this effort to investigate manufacture conditions of such materials by means of high pressure sintering and to evaluate its physical and mechanical properties.

2. Experimental Procedure

To perform sintering of composed diamond-containing materials by DO 137A pressing unit at 5 MN, the anvil-type stepped surface device of our own design, has been used

M. A. Prelas et al. (eds.), Diamond Based Composites, 53–62.

[8]. It consists of two hard-alloy WC-Co molds of 41 mm in diameter and 20 mm in height, with trapeziform depressions of 25 mm in diameter and 6.5 mm in height at their reversed ends. The surface around a depression is cone-stepped with 5° cone angle. The molds are being held together by a three-layer coat of steel supporting rings.

The container is manufactured of backelite-binded lithographic stone. Axis opening of the container holds a composed pressed-graphite heater made of 2 equal parts separated by a horizontal joint.

Pressure measurement inside the device has been carried out with 4% accuracy according to methodology [9] at room temperature regarding phase transitions of bismuth, thallium and ytterbium at 2,55; 2,69; 3,67 and 4 GPa respectively [10]. Figure 1 shows the results of unit pressure graduation in accordance with the container's height.

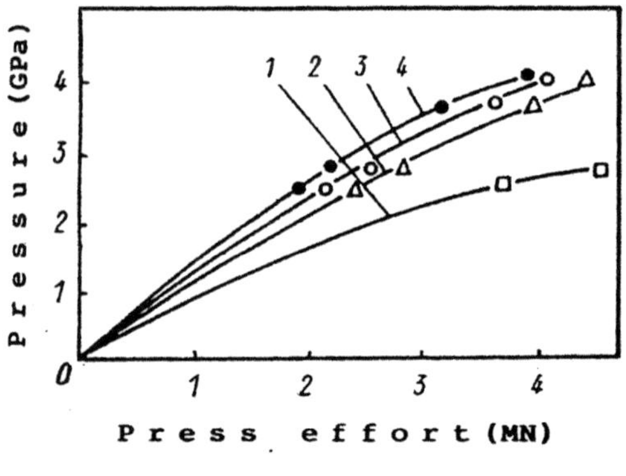

Figure 1. Dependencies of internal pressure on the pressing unit power at variable container heights, mm (1-17,0; 2-17,5; 3-18,0; 4-18,5)

Based upon data received, dependence tables has been drawn of the chamber pressure to the container's height for fixed efforts of the pressing unit. These permitted to work out the optimum height of the container to achieve wanted pressure levels.

Temperature control with 4% accuracy has been carried out by means of the chromel-alumel thermocouple entered through a gap between the molds, as well as by melting points of nickel and platinum at 4 Gpa. Special experiments determined time dependencies of cell temperature in the process of heating and cooling of a sample to different values of heating power (Figure 2 a,b).

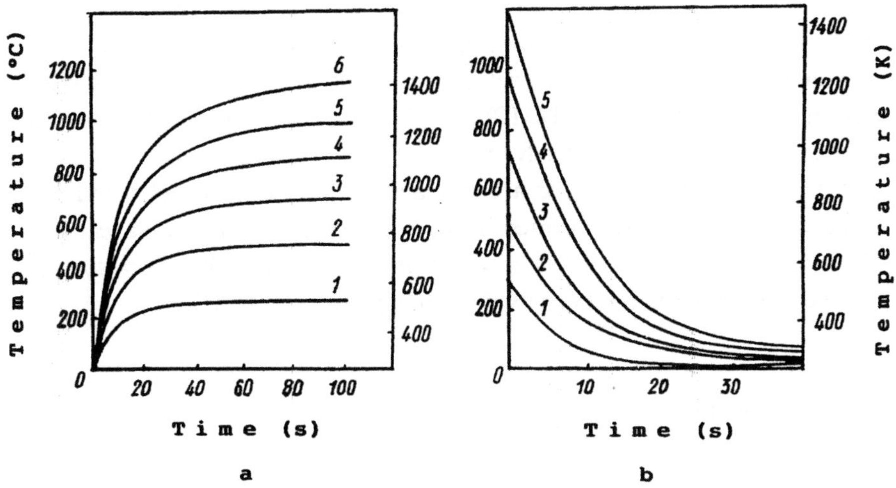

Figure 2. Dependence of cell temperature on heating times for variable current power values (a) and time at variable current powers being cut off (b)
(1-0,34 kW; 2-0,55; 3- 0,74; 4(a)-0,83; 4(b)-0,94; 5(a)-0,94; 5(b)-1,17; 6(a)-1,17 kW)

The above data shows that condition of equal temperature in the cell is being achieved within 10-30 seconds depending on heating current power; the sample's temperature decreases to 300 K for the same period of time after the current has been cut off. Average cooling value is estimated close to 50 K/s. Results of the research in temperature conditions inside the reaction chamber have been used to choose the best modes of composite sintering.

We used the above device to perform further series of experiments to determine influence of high pressure/high temperature treatment upon sintering qualities and other properties of composed diamond-containing materials, that resulted in absolutely new technology of extra-resistant ceramics production [4].

Fine-grained WC-Co alloy powders containing 6 and 8% of Co and those of AC 20 diamond with particle size of 80/100 have been used as basics in the experiments. Mixtures have been prepared by multiple sifting of the given basic components. After that, samples of 11 mm in diameter and 6 mm in height have been manufactured by pressing under 530 MPa, with further deployment in the heater and high pressure/high temperature treatment under 4 GPa at 1000-1500 °C. Prior to sintering some of the samples had been subjects to vacuum annealing at temperatures close to 700 °C. Density of those pressure-sintered samples has been measured by Archimed method in four-chlorous carbon.

The samples have been also studied by XRD using Cu-K_α-radiation. In order to evaluate working properties of a composite material meant for cement-based products, traditional methodology has been used for such materials now being used in drilling devices. Durability of a composite towards the abrasive weariness has been estimated by ratio value of the cutting element radial wear to the way made by the cutting device on the surface of rotating K3 63C12MCM2K10 abrasive circle of 350 mm in diameter, at line speed of V = 12 m/s., with advance of S = 0,1 mm/r., depth of cutting t = 0,125 mm. The cutting element, made as a disc plate of 10 mm in diameter and 5 mm in height, has been mechanically secured in the holder assuring front and rear angles of 10°. The circle under treatment has been put on 16K20-type lathe. Two-layer plate of $K02D_1$ superhard material has been used as an experimental sample.

3. Results and Discussion

Figure 3 presents dependence of sintered composite density on the temperature of sintering itself. It is quite obvious, that changes in the sample's density following those of the sintering temperature, are quite irregular. The density value decreases at 1200 °C, then starts to increase insignificantly, reaching its maximum at roughly 1300 °C.

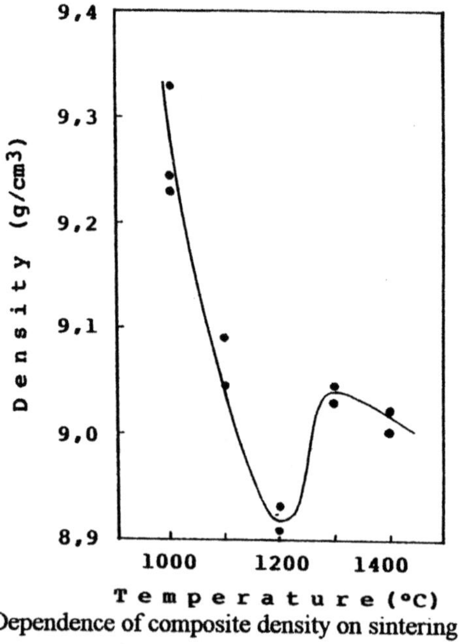

Figure 3. Dependence of composite density on sintering temperature

In order to determine specific reason to such a dependence, we have performed several additional experiments on matrix powder sintering of WC-8wt%Co hard alloy. The results (Figure 4) show different density values at temperatures of 1300 °C and above.

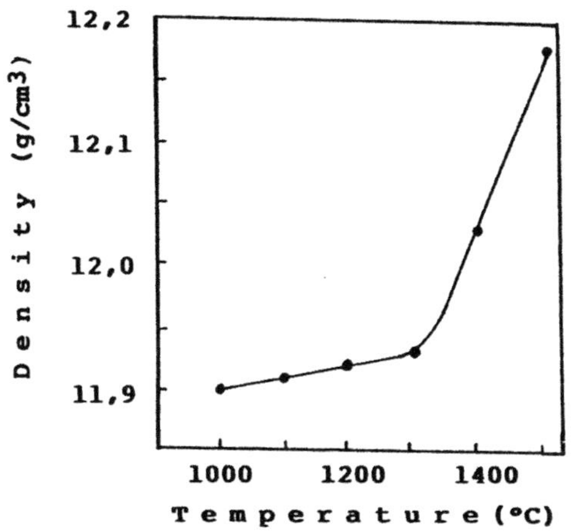

Figure 4. Dependence of WC-8wt%Co hard alloy density on sintering temperature

Presumably, changes in angle of the curve on the diagram "density - sintering temperature" (Figure 4) can be explained by the following: at temperatures of more than 1300 °C sintering continues in its liquid phase, triggered by eutectic γ + WC melting. On early melting stages particles of solid cobalt are being enriched by tungsten carbide, and at around 1300 °C liquid γ-phase appears. Thanks to its perfect moistening of tungsten carbide grains, density of a solid alloy sample under sintering increases. That is the reason why one cannot blame the mold behavior on density decreases of WC-8wt%Co — 20vol% diamond composite at sintering temperatures of 1000-1200 and 1300-1400 °C. The phenomenon can be explained by presence of oxygen particles, as well as those of other gases, adsorbed on powder particles of a solid alloy or within the sample's cells. According to [5] oxygen may play a catalyst in the process of diamond graphitization. Taking the above said into account, we presume, that the partial graphitization of diamond grains' surfaces could well be the reason of composite density decrease at higher sintering temperatures. When sintering temperature increases because of the liquid phase, density of sintered samples increases as well. There is a possibility of chemical interference on the diamond-hard alloy borderline, that could also influence certain properties of a composite. But, no doubt, this problem requires a separate investigation.

Table 1 presents specific durability results of composite samples received from sintering at different temperatures.

TABLE 1. Durability of diamond-containing ceramics to abrasive weariness

Plate type (Sintering t)	Number of tries	Total length (km)	Rear side wear (mm)	Radial wear (mm)	Dulling criteria (mm/km)
BK-19 (1100 °C)	10	4,457	4,6	0,79	0,18
BK-20 (1200 °C)	3	1,337	3,4	0,58	0,43
BK-21 (1300 °C)	5	2,229	3,7	0,63	0,28
K02D	7	3,12	2,0	0,34	0,109

Results of X-ray/structural analysis show, that in the process of sintering, metastable diamond modification of carbon is preserved. Formation of new phases has not been mentioned. Ratio analysis of diamond line intensity leads to the conclusion of strong diamond grains' anisotropy. Displacement of diamond line (111) points to their rather tense condition. Along with sintering temperature increase, hard alloy line (112) narrows, providing that its crystal grid is being modified.

The table shows that durability of different composite samples is comparable to that of cubic nitride boron biplates, received by straight phase transactions at higher pressure levels (approximately 7-8 GPa).

To examine the influence of diamond powder presence in the initial material on composite sintering abilities and the possibility of its simultaneous securing, we have made several two-layer samples. The basic layer consisted of WC-8wt%Co hard alloy, the upper layer represented a composite based on the former with variable diamond powder content of 25-50%. At that, total diamond powder content remained stable — 20%. This phenomenon has been achieved by reduction of the composite layer with further increase of diamond content in it, while the sample's height remained unchanged. The samples have been pressed in such a way, that its hard-alloy powder layer faced the mobile puncheon of the pressform.

The results are given in Table 2. It also contains some data received from sintered samples that have been pressed in the presence of composite layer placed at the side of the mobile puncheon.

TABLE 2. Dependence of sample density on diamond concentration in composite layer. Sintering temperature 1300 °C

Sample Number	Diamond Content (wt%)	Density (g/cm^3)
BK-12	25	8,96
BK-11	33	9,17
BK-9	50	9,212
BK-8*	50	9,354

* — under pressing procedure composite layer has been placed at the side of the mobile puncheon

The table shows, that, along with diamond content increase within composite layer, pressed samples' density increases as well and gets close to saturation for 50% concentration, because its values for the samples BK-9 and BK-11 differ insignificantly; i.e. sample's density value decreases while composite layer becomes thicker. What concerns the other sample pressing scheme, where composite layer is situated at the side of the pressform mobile puncheon, the density values of sintered samples (BK-8) are somewhat higher.

Comparing both density values for the samples BK-8 and BK-9 to the information of the Figure 3, we may conclude, that, by structural changes in the plates as well as in sample pressing scheme, significant improvement of sintered composed materials themselves and their properties is very much likely.

In order to avoid diamond graphitization, the molds to be sintered have been treated by vacuum annealing at 300 and 700 °C, being kept under each of the temperatures for 30 minutes. Table 3 presents the results. Problems of chemical interference on the diamond-hard alloy borderline under high pressure requires separate investigation and are of no subject to the work being presented.

TABLE 3. Influence of sintering temperature and vacuum annealing of composite samples on their density

Sample Number	Sintering Temperature °C	Density (g/cm^3)	Annealing
BK-17	1400	9,02	-
BK-57	1400	10,18	+
BK-58	1500	10,31	+

Table 3 shows the density of sintered composite samples (BK-57), being preliminary annealed, is approximately 13% higher. It grows further, reaching 10,3 g/cm^3 with temperature increase up to 1500 °C.

Same picture can be observed for the composite with larger ACP diamond powder grains and those of hard alloy WC-6wt%Co with the same content. Along with sintering temperature increases, its durability rises owing to increased hardness of the sintered components.

That is why, in order to receive a high-density material based on the above composite under 4 GPa, one would need sintering temperature on no less than 1500 °C and preliminary annealing of samples to get rid of gases adsorbed on the surface of powder particles.

Preliminary tests of the composite permit us to recommend it for use in stone-processing, mining/boring and ruling instruments.

4. Conclusions

The high pressure unit for diamond-based composites' sintering has been elaborated. Sintering conditions and structure of diamond and hard-alloy composites have been investigated under pressures up to 4 GPa and temperatures up to 1500 °C. It has been found, that, in order to achieve the highest possible mechanical properties of a composed material, sintering temperatures of no less than 1500 °C and samples' preliminary annealing are necessary. Conducted investigations enable to improve technology of stone-processing, dressing and mining tools production.

5. Acknowledgments

The author is thankful to Mr. V.M. Kuchinski for the X-ray diffraction analysis and to Mr. V.B. Nichipor for his help in wear-resistance determination of composites.

6. References

1. Tsypin, N.V., Simkin E.S (1986) Composed Diamond-Containing Materials for Boring Instruments, in N.V. Novikov (ed.) *Synthetic superhard materials* Vol. **2**. *Composed instrumental superhard materials*, Naukova Dumka, Kiev, (in Russian) pp. 119-130.
2. Acc. request 61-58432 Japan, MKI[4] C04 B 35/52 Sintered Materials for Instruments and Methods of their Production/ Sumitomo denki koge k.k. — Publ. 11.12.86.
3. Acc. request 3-6209 Japan, MKI[4] C22 C 1/05 Production Methods of Diamond-containing Composite/ ISM AS Ukraine — Publ. 29.01.91 (in Russian).
4. Shulzhenko A. A., Gargin V. G., Shishkin V. A., Bochechka A.A. (1989) Diamond-based Polycrystallic Materials, Naukova Dumka, Kiev (in Russian).
5. Vovchansky I.F., Bughaev A. A., Finkelstain E. M., (1986) Usage of Superhard Material Tools in Boring, in N.V. Novikov (ed.) *Synthetic superhard materials* Vol. **3**. *Usage of synthetic superhard materials*, Naukova Dumka, Kiev, (in Russian) pp. 233-256.
6. Dotsenko V.M., Kruchkova A.P., Ischuk V.L. (1978) Structural Properties and Pecularities of Hard Alloys Manufactured under High Pressures, in I.N. Frantsevich (ed.) *High pressures influence on a substance*, Naukova Dumka, Kiev (in Russian), pp. 91-94.
7. Dotsenko V.M., Ischuk V.L., Pugatch E.A. (1978) Investigation of High Pressures' Influence on Hard Alloy Resistance, in I.N. Frantsevich (ed.) *High pressures influence on a substance*, Naukova Dumka, Kiev (in Russian), pp. 94-95.
8. Mazurenko A.M., Urbanovich V.S., Kuchinski V.M. (1994) The High Pressure Apparatus for Sintering of Ceramics Based High-melting Compounds, Vestsi AS of Belarus (ser. Phys.-techn. Sci.) (in Russian), **1**, 42-45.
9. Samoilovich M.I., Zadneprovski B.I. (1987) *Diamond Synthesis Apparatus in Synthesis of Minerals*, 2 Vol., **1**, Nedra, Moscow, (in Russian), pp. 317-336.
10. Tonkov E.Y. (1988) The Phase Transformation of Compounds at High Pressure, 2 Vol.**1**, Metallurgiya, Moscow (in Russian)

11. Acc. request 93-047490, Russia, MKI[4] D 22 F 3/12, *Method of Abrasive-resistant Composed Material Production.* Mazurenko A.M., Urbanovich V.S., Bartashevich S.A., Lemesh E.A. — Prioritet 12.10.93, (in Russian).

GROWTH KINETICS OF CUBIC BORON NITRIDE FILMS AND COMPOSITES

CHARLES A. TAYLOR II and ROY CLARKE
Harrison M. Randall Laboratory of Physics
University of Michigan
Ann Arbor, MI 48109-1120, U.S.A.

ABSTRACT

We have developed a novel plasma-assisted physical deposition process for the growth of cubic boron nitride films. Our approach recognizes the central role of kinetics in the formation of this metastable phase and utilizes high substrate temperatures, a low flux of boron, and energetic nitrogen ions from a compact ECR source. We have been able to reduce the ion-energy needed to form the cubic phase to substantially below 100 eV with substantial improvements in the film crystallinity and orientation and an associated reduction in the film stress. Using this method we have demonstrated the growth of cubic BN films on silicon to thicknesses in excess of 1.9 μm. In-situ RHEED measurements show that the deposition process is optimized in a narrow range ion energies ($\Delta E \sim 15$ eV) controlled by a dc substrate bias potential. The optimum substrate bias potential is inversely proportional to the nitrogen ion flux. Additional insight into the evolution of thin-film stress during growth is provided by polarized FTIR spectroscopy measurements which also shed light on the initial nucleation mechanism of the cubic phase.

1. Introduction

Interest in III-V nitride materials for optoelectronic and high-performance electronic devices has been increasing steadily over the past several years [1-5]. Their bandgaps, which range from 1.9 eV for wurtzitic InN to >6.3 eV for cubic BN, make these materials well suited for optical applications throughout the visible and into the near-UV region. With very favorable mechanical and thermal transport properties, some of these nitrides are also very promising candidates for hard coatings on a variety of surfaces [6].

In order to attain the material quality required for such applications it is necessary to gain a much improved basic understanding of the physics of the growth mechanisms, particularly the *kinetic* processes which lead to the formation of metastable phases. In many ways the growth issues being addressed in the III-V nitrides are similar to those that are pertinent to diamond and related composites and thus have broad relevance to a wide class of materials.

M. A. Prelas et al. (eds.), Diamond Based Composites, 63–113.

The research program at the University of Michigan described in this article is aimed specifically at the cubic phase of boron nitride. This metastable structure is a paradigm for the importance of kinetic processes in the growth of wide-bandgap materials. We show that the dominance of kinetics in these systems, as opposed to processes occuring strictly in thermodynamic equilibrium, can be utilized to great effect in tuning the growth conditions to achieve a desired structural form.

Cubic BN shows great promise with material properties in many ways superior to those of other III-V nitrides. A comparison of selected properties is shown in Table 1, including those of SiC and diamond.

TABLE 1. Comparison of selected properties of the III-V nitrides, SiC, and diamond. w and c refer to the wurtzitic and cubic crystal structures, respectively. Values have been obtained from recent review articles [1,3,7,11] and Landolt- Börnstein [12].

	Bandgap (eV)	Dielectric Constant	Breakdown Voltage (10^7 V/cm)	Thermal Conductivity (W/cm K)
InN (w)	1.89	?	?	~0.8
GaN (w)	3.39	e(0) 9 e(∞) 5.35	~0.5	1.3
AlN (w)	6.28	e(0) 8.5±0.2 e(∞) 4.84	?	2.8
BN (c)	6.4 ± 0.2	e(0) 7.1 e(∞) 4.5	1.0	~13
SiC 6H (SiC 3C)	2.9 (2.2)	e(0) 9.7	0.5 (0.4)	3.5 (5)
Diamond	5.5	e(0) 5.5	2.0	~20

As shown in the table, cubic BN has the highest thermal conductivity, dielectric breakdown strength, and perhaps largest bandgap of the III-V nitrides. Together with the existence of active n- and p-type dopants [8], these properties pave the way for high power/temperature device development.

1.1 BORON NITRIDE OVERVIEW

Boron nitride is a synthetic compound semiconductor which crystallizes in at least four known phases, each with similar characteristics to analogous structures of carbon. The hexagonal structure, shown in Figure 1(a), is the most common form of boron nitride and is the thermodynamically stable form at room temperature and pressure. First synthesized more than a century ago[13], it is a microcrystalline layered material with a soft flaky texture often compared to graphite. The intralayer bonds are strongly covalent (sp^2) whereas the interlayer bonds are weak (van der Waals-type).

Whereas the electronic properties of graphite have been extensively studied and agreement between theory and experiment is considered quite good, there is continued disagreement in this area for boron nitride. For example, values for the hexagonal BN bandgap range from 3.2-5.8 eV [14-21] (experimental) and 2.45-12.7 eV [17,19,22-31] (calculated) with continued disagreement as to whether the bandgap is direct or indirect.

The many similarities of hexagonal boron nitride to graphite led to a series of attempts to prepare the structural analog of diamond, a BN compound with tetrahedral sp^3 bonding and the cubic zinc blende lattice, as shown in Figure 1(b). Bulk samples of cubic BN were first achieved by Wentorf in 1957 [32] using the same high pressure and high temperature techniques developed for synthetic diamond growth.

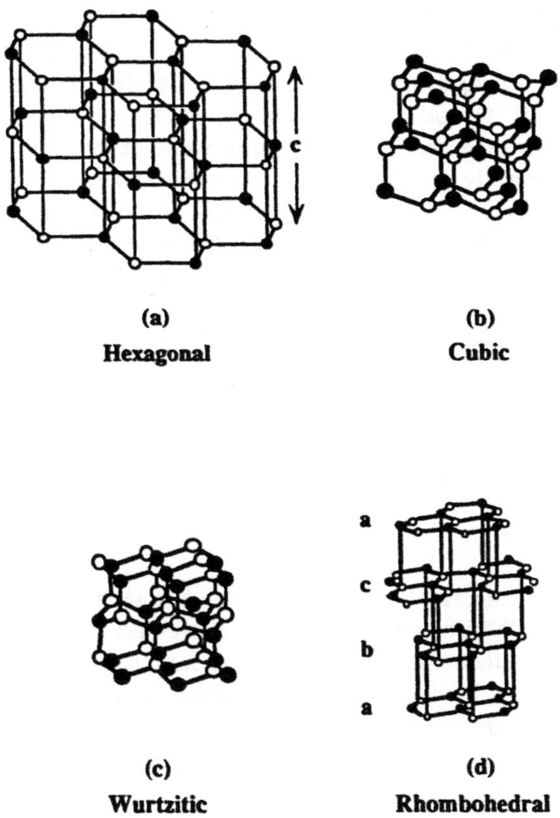

Figure 1. Crystal structures of the boron nitride allotropes.

Now commercially available in millimeter size crystals, cubic BN is an extremely inert, electrically resistive material with high thermal conductivity and a hardness second only to diamond.

The high pressures and temperatures needed for synthesis indicate that cubic BN is a metastable compound - in the same manner that diamond is the metastable form of carbon. High pressure and temperature are needed to overcome the activation energy of formation for the cubic phase. Once formed, the potential well is sufficiently deep that the cubic phase remains stable at room temperature and pressure. The stability of cubic BN at high temperatures and pressures has been studied in detail [33]. A recent measurement [34], using a Raman microprobe, indicates that reconversion to the hexagonal BN phase occurs at 1840±60 K, which is in good agreement with the previously published value[33].

As is the case with hexagonal BN, the exact value of the bandgap for cubic BN is not yet established, but it is generally accepted that the bandgap is indirect. Values in the literature fall in the ranges of 6.0-8.8 eV [35-38] (experimental) and 2.9-11.3 eV [39-52] (calculated). Recent absorption measurements on bulk single crystals indicate the optical absorption edge is ~6.3 eV.[53]

Two other hexagonal modifications of BN are occasionally reported in the literature. A wurtzitic structure, corresponding to Lonsdalite (hexagonal diamond), was synthesized in 1963[54] and is shown in Figure 1(c). This structure is also metastable and is typically obtained by static high pressure or dynamic shock methods. The crystallites are generally small, highly defective, and contaminated by other phases. An even less common form of graphitic BN is the rhombohedral structure, with a stacking sequence ABCABC.[55] For rhombohedral BN, the hexagonal rings are no longer occluded, but are translated parallel to each other from one layer to the next. The sequence of N to B to N between layers is maintained as shown in Figure 1(d). The space group for this structure is R3m. While little is known about the physical properties of either the wurtzitic or rhombohedral phases. and the limits of their stability are not well defined, their existence underlines again the structural similarities between the boron nitride and carbon allotropes.

1.1.1 *Thin-film cubic BN*

The growth of cubic BN films has been of technological interest for many years. The first evidence for a cubic BN film was reported by Sokolowski [56] in 1979 who used a pulsed plasma method. Since then, many other techniques have been used in the attempt to grow cubic films. Reviews of these techniques have been given by Arya and D'Amico [57], and Pouch and Alterovitz [58].

It is apparent from a review of the literature that the growth of high quality BN films, either hexagonal or cubic, is extremely difficult. In the past, forming the metastable cubic phase of BN was a challenge in itself which was met with only limited success. Early films were typically characterized as nanocrystalline materials containing mainly sp^2-bonded phases, with only a small fraction of the sp^3-bonded cubic phase. The common link to successful deposition processes is the use of an energetic plasma to promote the formation of the cubic phase. From these investigations it is clear that the interaction of the plasma with the growing film plays an important role in the synthesis of cubic BN.

The approach we have taken for the deposition of cubic BN is to develop a high temperature growth process which utilizes a very low boron flux combined with a high-density, low-energy nitrogen plasma. In particular, we have developed an RF-magnetron sputtering process which combines an electron cyclotron resonance (ECR) ion-source and negative substrate potentials to precisely control the nitrogen ion-energy at the substrate surface. Using this novel process we have been able to significantly reduce the ion-energy needed to form the cubic phase, with values now substantially less than 100 eV. Through a better understanding of the growth kinetics, we have made improvements in film crystallinity and orientation, with an associated reduction of the high film stress which has severely limited film thickness in the past. Significant improvements continue to be made which give promise for the technological development of this material.

1.1.2 Layout and Overview of this Article

A large part of this research project has been the development of a deposition process for high quality cubic BN films. In the following section, the details of our plasma-assisted RF-magnetron sputtering process are presented, along with structural characterization of films grown by this technique. We show that through a combination of high-temperature growth, a stable boron flux, and a high flux of low-energy nitrogen ions, predominately cubic BN films can be grown on silicon substrates to thicknesses previously unobtainable by any other deposition process. Furthermore, through the use of in-*situ* reflection high energy electron diffraction (RHEED) and transmission electron microscopy (TEM), we show that the films are well crystallized and grow in a layered morphology: a thin amorphous layer at the substrate interface, followed by an oriented hexagonal layer, upon which the cubic phase nucleates. Scanning force microscopy measurements show that the cubic BN nucleates as well aligned triangular crystallites, indicative of (111) film growth.

Section 3 details the use of Fourier transform infrared (FTIR) spectroscopy for the characterization of BN films. IR spectroscopy supports many of the structural results and, furthermore, provides a means to study the films quantitatively via phonon-related features. Through the use of lattice dynamics calculations performed by Fahy[59] we have been able to estimate film strain based on shifts in the transverse optic (TO) phonon frequencies in thick films. We show that the use of low energy ions and high temperature growth minimizes compressive film stress, allowing the growth of thick cubic BN films. Furthermore, we use IR spectroscopy to probe the initial nucleation and growth of the cubic BN layer. We find that the behavior of the TO phonons is strongly affected by geometrical factors resulting from the island nucleation and coalescence. A phenomenological model, based on depolarization fields resulting from the island growth mode, is presented to explain the phonon behavior at the initial nucleation stage; the results are in excellent agreement with our experimental findings.

2. Film Growth and Structural Characterization

2.1 INTRODUCTION

As discussed in Section 1, stabilizing the cubic phase of BN in thin-film form is a challenge in itself which has eluded many researchers. Processes which have shown limited success in the past have all utilized some type of energetic-ion or neutral-particle irradiation of the growing film to promote growth of the metastable cubic phase.

This section contains a description of a novel plasma-assisted sputter deposition process we have developed for the growth of cubic BN. The structural characterization results for films grown by this technique demonstrate that a combination of high temperature growth, a stable boron flux, and a high flux of nitrogen ions with precisely controlled energy, enables the growth of predominately cubic BN films on silicon substrates. Using this technique we have grown cubic BN films to thicknesses previously unobtainable by any other growth technique.

The structural characterization techniques we have used include: in-*situ* reflection high energy electron diffraction (RHEED), high resolution transmission electron microscopy (TEM), and scanning force microscopy (SFM). As a result of the distinct differences between hexagonal and cubic diffraction patterns, RHEED provides a clear indication when cubic BN nucleation occurs. Results indicate that cubic BN growth is optimized within a narrow "window" of ion energy which, for a fixed boron rate, is strongly dependent on the nitrogen ion flux.

2.2 NATURE AND ENERGY DISTRIBUTION OF SPUTTERED PARTICLES

The interaction of a high energy bombarding ion with the surface of a target material can result in a variety of elastic and inelastic collisional effects as illustrated in Figure 2. Analysis shows that under typical deposition conditions most sputtered particles are in a neutral state. The fraction of charged sputtered species from clean metal or semiconductor surfaces is on the order of 10^{-4}.[60-62] Mass spectrometry shows also that the majority of sputtered species are atomic. For example, an analysis of the emission from a GaAs target shows mainly (99.4%) neutral Ga and As atoms.[63] However, the fraction of molecular species sputtered from ionic crystals can be considerably higher.[64] There is evidence that a large fraction of the molecular species originate from recombination above the target surface.[65]

The most notable feature associated with sputter deposition is that sputtered species have energies considerably greater than thermal. For a thermal source operating at 2000 K, the average energy of the emitted atoms is ~ 0.2 eV. Sputtered atoms, depending on the energy of bombarding ions, typically have energy

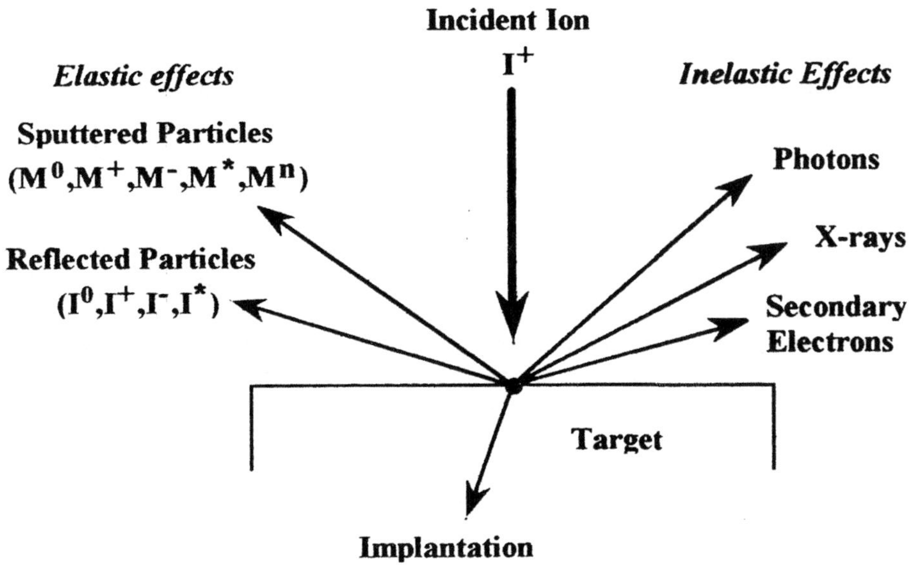

Figure 2. Schematic diagram of ion-surface interactions.
[G. M. McCracken, Rep. Prog. Phys. **38**, 241 (1975)]

distributions strongly peaked around 5 to 20 eV. The sputtered ion energies can be thermalized by collisions but at the sputtering pressures at which we operate (<1 mTorr) this would be suppressed. Our work shows that ion energies on a somewhat larger scale are required for stabilization of cubic BN. For this reason we have employed an ECR nitrogen ion source to provide a high flux of nitrogen ions with a peak energy of ~30 eV and energy spread ~20 eV. The energy of ions from this source can be tuned with the addition of a dc substrate bias potential up to 100 eV and beyond.

The role of ion/surface interactions and photo-induced reactions during film growth have been reviewed extensively by Greene *et al.*.[66] A considerable amount of research has been performed on nucleation and film growth using both ion and energetic neutral irradiation during deposition. Although both energetic ion and neutral particle bombardment is an integral part of sputter deposition processes, many experiments are carried out by other means including evaporation in which a portion of the evaporant is ionized, evaporation in the presence of a separate ion source providing energetic particles, and bias sputtering in which the substrate is biased to attract and accelerate ions from the plasma discharge. The point being that energetic particle bombardment of the film surface is not limited to sputtering processes and is in many cases added to thermal processes for the advantageous effects seen in nucleation and growth kinetics. Ion bombardment of the substrate and growing film can modify the stoichiometry, microstructure, state of stress, defect concentration, and hence the physical properties of deposited films.

In order to understand the role of low energy particle/surface interactions in controlling film properties, it is important to realize that the nature of the effects observed is strongly dependent on the incident energy and ratio of the accelerated to thermal deposition fluxes, as well as the growth temperature and deposition rate. Failure to pay attention to these details explains conflicting reports in the literature which variously claim energetic-particle irradiation is both beneficial and detrimental for a given application. The effects of bombardment are in fact quite different when applied to growth at low temperature as opposed to growth at high temperature. Low-energy bombardment of films during deposition at low temperatures, where radiation-induced defects cannot be annealed out, generally leads to the formation of compressive stress[67-69], a smaller average grain size[70], and a higher dislocation density.[70-72] On the other hand, irradiation can be much more beneficial when coupled with high temperature growth. At high temperatures, defects which are produced in the near surface region are rapidly annealed out during film growth. The primary use of low-energy bombardment becomes one of coupling energy directly to the growth surface. At high growth temperatures, low-energy ion or neutral particle irradiation has been shown to increase grain size, decrease defect concentration. [73-77] We will return to this issue in Section 3.4.2.

2.3 EXPERIMENTAL DETAILS

2.3.1 Chamber Configuration and Deposition Process

The cubic BN growth studies were conducted using a custom designed ultrahigh-vacuum-compatible (UHV) deposition system with an *in-situ* reflection high energy electron diffraction (RHEED) system for analysis of film structure. The deposition system is detailed schematically in Figure 3. Substrates are introduced individually through a turbo-pumped load-lock chamber which typically reaches a pressure of ~1 x 10^{-8} Torr before the substrate is introduced to the UHV growth chamber. The main growth chamber is cryopumped (APD Cryogenics, Model APD-8) and has a base pressure of ~8.0 x 10^{-11} Torr after baking. Typical residual gas content at base pressure, examined by a quadrupole mass spectrometer with an electron multiplier, consists mainly of H_2 with H_2O and N_2/CO content at least an order of magnitude below that of H_2, and an O_2 partial pressure below the detection limits of the spectrometer (~2 x 10^{-12} Torr). The base pressure routinely returns to ~1.0 x 10^{-10} Torr after film growth at ~1.0 mTorr total chamber pressure. The UHV environment allows for thermal desorption of the silicon substrate oxide prior to growth and also ensures a clean environment for film deposition.

The reactive sputtering of a hexagonal BN target, 99.99% purity, is performed using a modified UHV 3-inch planar magnetron source (US Guns) mounted in a source-up configuration on a custom five-port base flange assembly. The source has been modified for low pressure operation by the addition of high-field $Nd_2Fe_{14}B$ magnets in a custom field geometry. The high-field magnets provide additional electron confinement to enable source operation at pressures down to ~8 x 10^{-4} Torr in

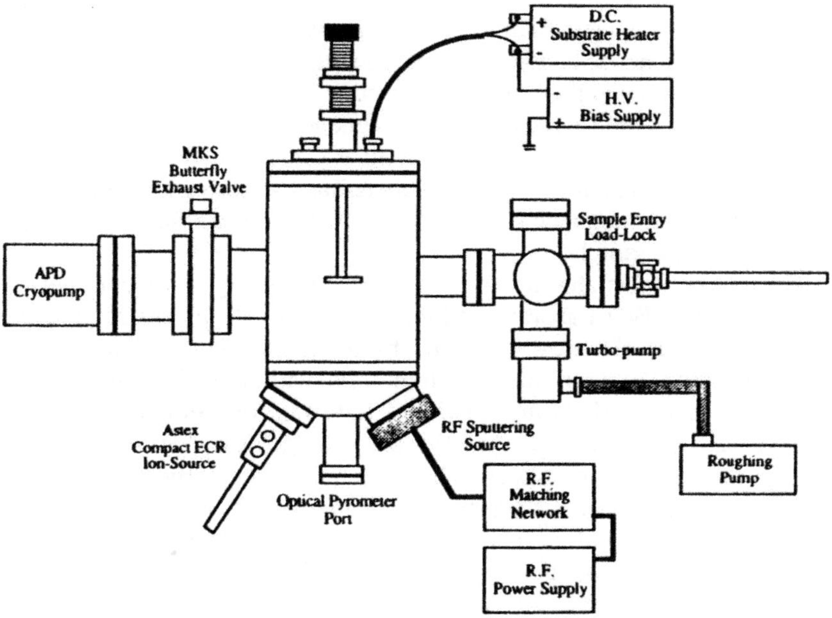

Figure 3. Schematic drawing of the ion-assisted RF sputtering deposition system

an Ar/N_2 environment. The low pressure operation of the magnetron is needed to provide a suitable mean free path and high ion-current operation for the primary nitrogen plasma source (Astex compact ECR) in the growth process. The magnetron source is mounted at angle of 45° to the substrate normal with a source-to-substrate distance of 20 cm. Radio frequency (RF) power at 13.56 Mhz is supplied by an Advanced Energy 650 watt RF power supply (Model RFX-600) to an autotuning network (Advanced Energy, Model ATX-600) which minimizes the reflected power from the magnetron source. Under typical operating conditions of 600 watts forward power, the reflected power is approximately 4 watts and the DC bias of the target, a measure of the surface charge of the insulating BN sputter target, is approximately -60 V.

An Astex compact electron cyclotron resonance (ECR) plasma-source is mounted in one of the base ports at a 45° angle to the substrate normal with a source-to-substrate distance of 8 cm. In this configuration, the substrate surface is completely immersed in the ECR plasma but is well removed from the plasma confinement region of the magnetron sputtering source. Ion current measurements, performed at the substrate position with a commercially available Faraday probe[78], confirm that the substrate is not exposed to charged species from the magnetron source. Microwave power (2.45 GHz) is supplied to the ECR source by a 250 watt Astex power supply through a manual tuning network to minimize reflected power. The ion energy distribution produced by the Astex source is Gaussian, centered at approximately 30 eV with a full width at half maximum (FWHM) of approximately 20 eV.[79] The ion current produced by the source is a function of both the microwave power applied to the source cavity and the magnetic field which establishes the electron cyclotron resonance conditions. Many resonance conditions can be found by varying the microwave power and magnet current. Since there are no extraction grids at the output of the source, ions simply drift out of the source along the magnetic field lines which run parallel to the source axis. The ions quickly diverge along the magnetic field lines at the output of the source. The ion flux at the sample is therefore a sensitive function of source-to-sample distance and the magnetic field strength. Increasing the field strength provides a higher ion flux at the sample.

The silicon substrates are heated resistively using a custom heater stage originally designed at Bell Laboratories.[80] The heater stage consists of a tantalum foil assembly which is electrically isolated from the chamber ground. The silicon substrate is mounted between the tantalum foils with minimal force so that the substrate is free to expand during heating. Substrate temperature is continuously monitored through the center port of the base flange using an optical pyrometer optimized for measurement of silicon substrate temperature (wavelength = 1.0 ± 0.1 μm). Resistive heating of silicon substrates is very efficient, requiring as little as 120 watts to reach 1100 °C, and extremely clean since minimal heating of heater stage components occurs. Pressures in the mid 10^{-10} Torr range can be routinely maintained with the substrate at 1100 °C. Electrical isolation of the sample provides a means to bias the sample with respect to ground during growth. Negative substrate bias provides an ideal method to control the incident ion energy during growth. The magnitude of the bias is limited to ~150 V due to the isolation limit of the DC current supply. Higher substrate bias values (not needed for the growth of cubic BN) can be achieved by isolating the DC current supply from case ground. A further advantage of resistive heating is that substrate charging (a significant problem with radiatively heated substrates in plasma-assisted processes) is virtually eliminated.

2.3.2 Structural Characterization Techniques

The sample growth chamber is equipped with a Vieetech (Model VE-052) 30 keV electron gun and phosphor screen for in-*situ* reflection high energy electron diffraction (RHEED) analysis of the film surface structure. Since the growth process is performed at ~1 mTorr chamber pressure, RHEED analysis is limited to examination of the substrate prior to growth and to the film surface immediately after growth. A CCD detection and analysis system (k-Space Associates, Model KSA 300) is used for diffraction pattern image acquisition and line profile analysis.

Since RHEED is a glancing angle scattering technique it is extremely sensitive to surface structure and morphology. A thorough review of RHEED theory and examples of the basic fundamentals of RHEED analysis are provided by Arrott.[81] Essentially, the surface of a crystalline sample can be thought of as a two dimensional diffraction grating. When the electron beam is incident along high symmetry directions of the crystal surface, a spot (or streak) pattern of diffraction maxima, which satisfy the Laue condition, appear on the phosphor screen. The position, spacing, and intensity of the features within a RHEED pattern provide a wealth of information about the sample surface. Typical types of information obtained include; the quality of surface preparation, the orientation of the crystal or a selected grain of a mosaic crystal, the crystal structure and morphology of the sample surface, and the number of atomic layers deposited during layer-by-layer growth.

Plan-view transmission electron microscopy (TEM) was performed using a JEOL 2000FX Analytical Microscope with an acceleration voltage of 200 kV. High resolution TEM was performed on sample cross sections using a JEOL 4000, 400 kV electron microscope. The point-to-point resolution of the latter instrument is 1.7 Å.

We have used scanning force microscopy (SFM) for imaging a series of BN films grown to varying thicknesses to investigate the nucleation and growth of the cubic BN layer. A Digital Instruments NanoScope III SFM was used for imaging under ambient (in air) conditions. The microscope was operated in "tapping" mode, a contact-SFM method where a very sharp stylus mounted on the tip of a cantilever is in constant contact with the surface of the sample. The sample is mounted on a piezoelectric transducer (PZT) and the cantilever tip is scanned in a raster fashion over a selected area of the sample. The force sensor is integrated into an optical readout system for detecting bending motions due to changes in surface topography. These motions are measured with angstrom accuracy via the reflection of a laser beam focused on the rear of the cantilever to a split photodiode. In the constant force (topographic) mode, an electronic feadback control compensates for changes in the topography by modifying the *z*-extension of the PZT, i.e., by adjusting the height of the sample. Microfabricated Si tips integrated into triangular cantilevers (Ultralevers, Park Scientific. Sunnyvale , CA) with force constants between 0.03 and 0.06 N/m were used. Scanning was with a J-tube PZT with 9.5 x 9.5 μm (*x*,*y*) and 5.4 μm (*z*) scan range. The scan rate was 4 lines/second and only the standard planefit (surface-leveling) correction was applied to the data. The image size was 512x512 pixels so that the acquisition time for an image was approximately 2.5 minutes. All images were checked for rotational and scaling variability to insure that the observed features were not simply artifacts of the triangular cantilever geometry.

2.4 STRUCTURAL CHARACTERIZATION : RESULTS

2.4.1 In-Situ RHEED

We have carefully investigated the growth of cubic BN on silicon (001) substrates as a function of substrate temperature, nitrogen ion-current, and ion-energy. As described in the previous section, silicon substrates are individually transferred into the growth chamber from a turbo-pumped loadlock chamber. The substrate oxide is thermally desorbed at a base pressure of ~ 1.0 x 10^{-10} Torr and is monitored by RHEED until a sharp 2x1 surface reconstruction is observed.

A series of BN samples was grown under identical conditions except for substrate bias in order to investigate the effect of ion energy on the formation of cubic BN. The sample growth parameters for the series are listed in Table 2. The samples were grown for the same length of time (180 minutes, corresponding to ~1000 Å total film thickness) and were characterized using RHEED immediately after growth.

TABLE 2. Growth parameters for the series of BN films presented in Figure 4.

RF Sputtering Gun Power: 600 W
Gas Flow Rate: 10 sccm Ar into chamber
10 sccm N_2 into ECR source
Total Chamber Pressure: 1 mTorr
ECR Microwave Power: 200 W ECR Magnet Current: 23.0 A
Nitrogen Ion Current at the Sample: ~ 0.65 mA/cm^{-2}
Substrate Temperature: 850 °C Growth Time: 3 hours

Each of the diffraction patterns, shown in Figure 4, were found to consist of a series of diffuse rings which are indicative of a polycrystalline, randomly oriented surface. Of particular interest is the evolution of the diffraction patterns as the negative bias voltage is increased from -100 V to -120 V. Initially (-100 V and -105 V) the diffraction patterns show two distinct rings. Near -106 V, a transition begins in which the two-ring pattern changes to a distinct three-ring pattern. This three-ring pattern becomes noticeably sharper and more intense just a few volts beyond the transition (-108 V). As the bias voltage is increased further (-110 V, -120 V), the three-ring pattern becomes more diffuse and less intense.

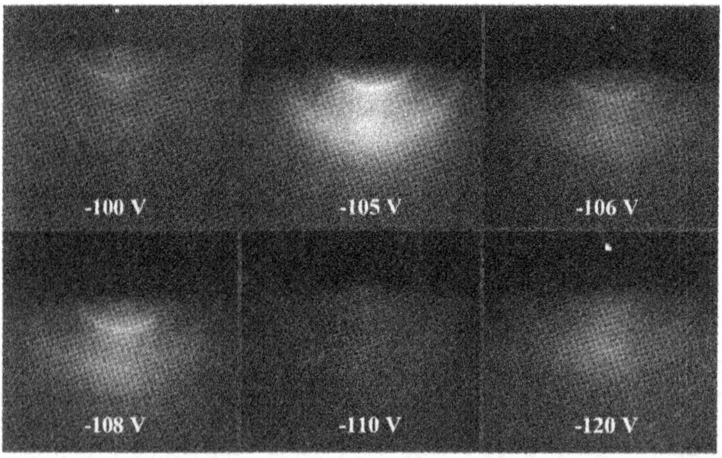

Figure 4. RHEED diffraction patterns from a series of BN films grown with substrate bias voltages ranging from -100 V to -120 V. The deposition parameters are listed in Table 2.1. The patterns are seen to evolve from a two-ring pattern to a three-ring pattern near -105 V.

To assign and index the patterns conclusively as either hexagonal or cubic BN we examined the RHEED pattern from commercially available cubic BN powder[82] (dispersed on a silicon substrate) as a standard. We found that the cubic BN powder exhibited a three-ring pattern with positions and intensities similar to the patterns observed from the -106 V to -120 V samples. Furthermore, TEM analysis (discussed in following paragraphs) and infrared spectroscopy results (see Section 3) confirm that the two-ring RHEED pattern is associated with hexagonal BN films while the three-ring pattern is indicative of cubic BN. Having confirmed the origin of the ring patterns using several analysis techniques, the indexing can be determined by measuring the ring spacings from the central electron spot which is visible at very shallow incident angles. The two-ring pattern is indexed as the hexagonal (0002) and $(10\bar{1}2)$ while the three-ring BN pattern is indexed with the cubic (111), (220), and (311) reflections. Linescan profiles of the diffraction intensity, taken along the shadow edge of the RHEED images, are shown in Figure 5. The linescans clearly reveal the distinct changes in the number and spacing of the rings associated with the hexagonal and

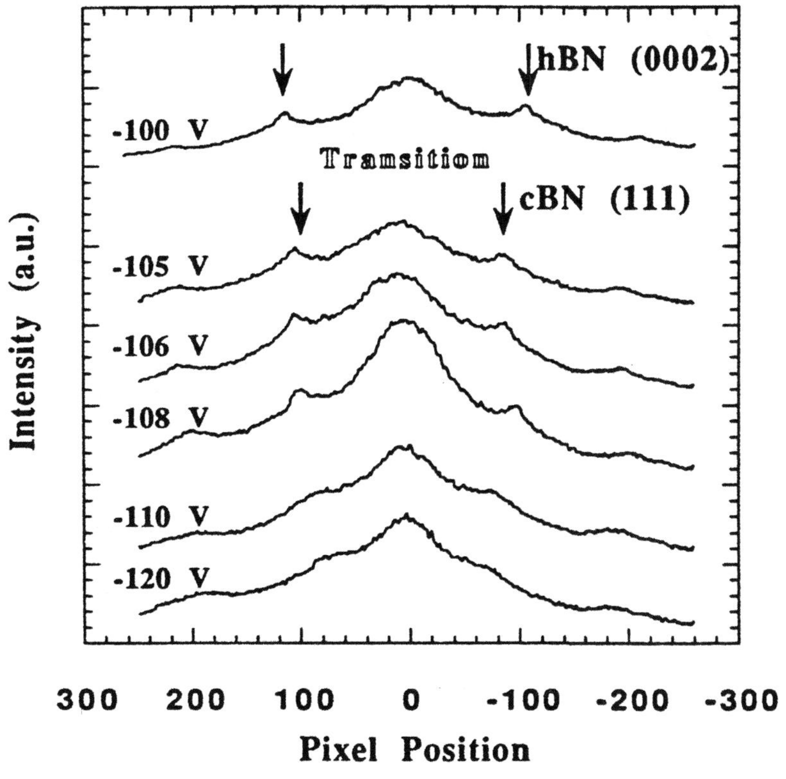

Figure 5. Intensity profiles of the RHEED patterns shown in Figure 4. The linescan profiles were taken along the shadow edge of the RHEED patterns and clearly show the distinctions between the two-ring hexagonal and three-ring cubic patterns.

cubic BN patterns. Furthermore, the structural quality of the films, as monitored by the ring width of the RHEED patterns, is optimized in a narrow range of substrate bias voltage (see Table 3). In the region of -105 to -108 V the rings in the cubic RHEED pattern are much sharper and brighter, indicating larger grain sizes and a generally superior structure. Beyond -108 V the rings begin to noticeably broaden indicating a decrease in structural quality, presumably caused by damage induced by excess ion-energy.

Significant differences in the hexagonal to cubic transition voltage were observed when the nitrogen ion-current was increased from 0.65 mA/ cm^2 to 0.90 and 1.1 mA/cm^2. In each case, cubic BN growth was optimized in a narrow range of ion-energy, but the transition voltage dropped significantly with increased ion-current. Growth of cubic BN was optimized near -80 V with a nitrogen ion-current of 0.90 mA/cm^2 and near -60 V with an ion-current of 1.1 mA/cm^2. This inverse relationship between the ion-energy needed to form the cubic BN phase and the nitrogen-ion:boron flux ratio was discussed in a recent publication by Reinke *et al.*.[83] The authors[83] analyzed data published in the literature for a variety of plasma assisted physical and chemical deposition processes and observed that the optimum ion-energy for cubic BN growth decreases dramatically as the ion:boron flux ratio increases. Our results are in agreement with this general observation. Infrared spectroscopy measurements, which are sensitive to the distinct bonding differences of hexagonal and cubic BN, are presented in Section 3 and support the results obtained and conclusions drawn from the RHEED images.

TABLE 3. RHEED ring width vs. substrate bias voltage for samples grown near the transition voltage. Diffraction ring widths are obtained from fitting the intensity profile to Gaussian line shapes.

Substrate Bias (V)	Diffraction Ring Width (Pixels)	Structure and Ring Index
-100	51.6	hBN (0002)
-105	27.5	cBN (111)
-106	25.2	cBN (111)
-108	23.7	cBN (111)
-110	38.1	cBN (111)
-120	57.2	cBN (111)

2.4.2 TEM

Transmission electron microscopy was performed on many of our early samples grown with low ion-current (~ 0.65 mA/cm^2) and substrate bias optimized near -110 V. The results, which are presented in detail by Kidner [84], have contributed significantly to our understanding of BN film growth and will be summarized in this section.

Figure 6 shows a plan-view TEM micrograph from a cubic BN film grown with low ion-current (0.65 mA/cm^2) and bias voltage of -110 V. Seen in the inset are two diffuse arcs, labeled 1 and 3, and a series of spotty rings (labeled 2 and 4-12). The diffraction spots correspond to the [110] zone axis of the silicon substrate. The two arcs (1 and 3) index to the hexagonal BN (0002) and (0004) while the series of spotty rings, in both the inset pattern and the over-exposed image, all index to cubic BN.

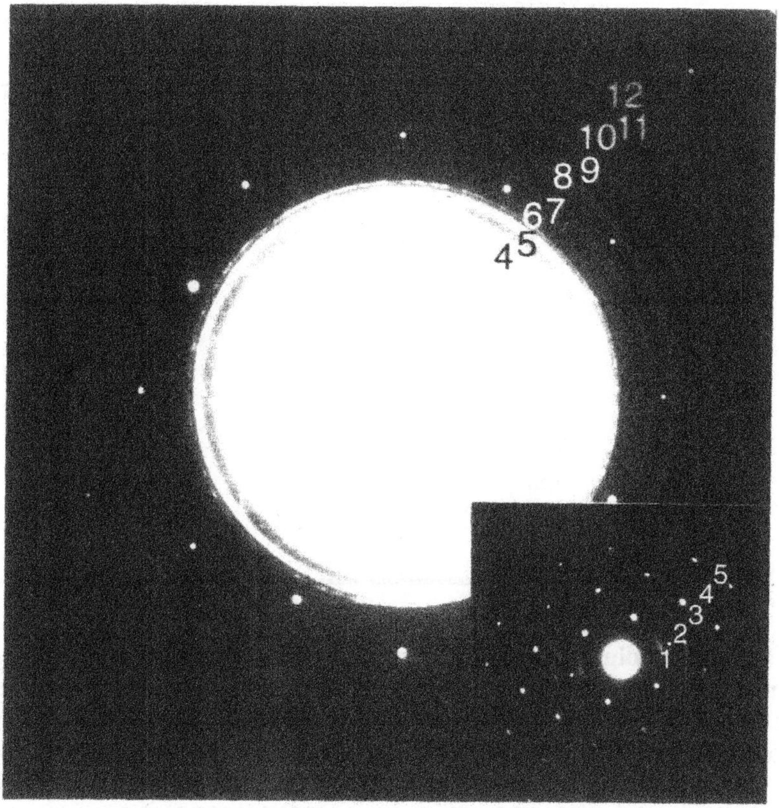

Figure 6. Plan-view TEM micrograph of a cubic BN film grown with low ion-current (0.65 mA/cm^2) and bias voltage of -110 V. The silicon spot pattern is from the [110] zone axis. The ring indexings are listed in Table 5.

The d-spacings and indexing are listed in Table 4. From the inset micrograph we see that (for this sample) the hexagonal BN [0002] (*c*-axis) is oriented about the silicon [002]. Interestingly, we have also observed, in samples grown under nearly identical conditions, the hexagonal BN [0002] is oriented about the silicon [220]. When tilting away from the silicon [110] zone axis to the [100] zone axis (not shown), the hexagonal BN features become complete rings while the cubic BN rings retain their appearance. This indicates that the BN film contains randomly oriented cubic BN crystallites and textured hexagonal BN. Before we discuss the plan-view diffraction micrograph further, a cross-sectional TEM image of the silicon/film interface region will provide considerable insight regarding the presence of both hexagonal and cubic BN in the film, as well as the orientation of the hexagonal BN within the sample.

TABLE 4. Indexing of the plan-view electron diffraction pattern shown in Figure 6.

Ring Label	Measured d-spacing (Å)	c-BN d-spacing (Å) JCPDS 35-1365	c-BN (hkl)	h-BN d-spacing (Å) JCPDS 35-421	h-BN (hkl)
1	3.34			3.33	0002
2	2.10	2.09	111		
3	1.67			1.67	0004
4	1.28	1.28	220		
5	1.10	1.09	311		
6	1.05	1.05	222		
7	0.90	0.904	400		
8	0.83	0.830	331		
9	0.74	0.738	422		
10	0.70	0.696	333		
11	0.64	0.639	440		
12	0.61	0.611	531		

A high resolution cross-sectional TEM micrograph of a thin BN sample is shown in Figure 7. The micrograph reveals that the film consists of three distinct regions: an amorphous region at the silicon substrate interface (~40 Å), followed by a region of ordered parallel fringes (~300 Å), oriented nearly perpendicular to substrate surface, upon which a crystallite, with very small fringe spacing, has nucleated. Similar findings were reported by Kester *et al.*[85] in which they reported BN films grown by an ion-assisted process which exhibited a layered morphology. They observed an amorphous region near the silicon substrate interface, followed by a central region which they identified as hexagonal BN, oriented with the *c*-axis parallel to the Si substrate surface, and a top layer of polycrystalline cubic BN. This film morphology is by now well documented for BN films grown by ion-assisted processes on silicon[84,86,87] as well as other[88,89] substrates.

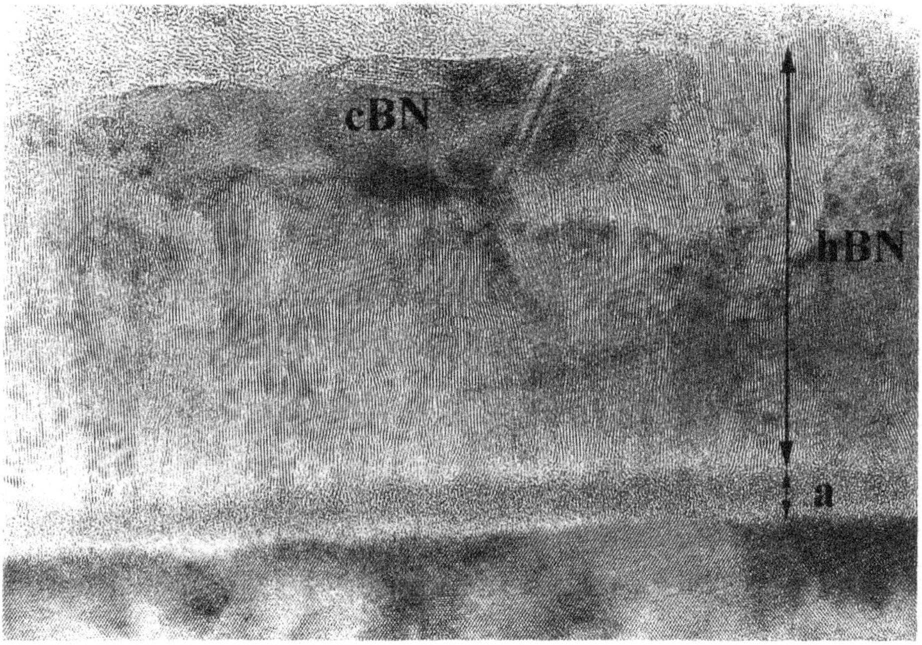

Figure 7. High resolution cross-sectional TEM micrograph of a thin BN sample near the initial nucleation stage of cubic BN.

An analysis of the high resolution micrographs from our samples indicate that the fringe spacing of the central region is 3.34 Å, which corresponds to the hexagonal BN (0002) plane spacing. This is consistent with the *c*-axis of the hexagonal BN growing parallel to the silicon substrate surface. Selected area diffraction patterns taken from the cross-section of a thick (~4500 Å) BN sample are shown in Figure 8. The first diffraction micrograph (a) is of the interface region along the [110] zone axis of the silicon substrate and includes information from all three (amorphous/hexagonal/cubic) strata of the BN film. Similar to the plan-view micrograph presented previously, this pattern indicates that the hexagonal BN layer is oriented with the *c*-axis parallel to the silicon substrate surface and that the cubic BN is randomly oriented. Note that in this sample the hexagonal BN [0002] (*c*-axis) is

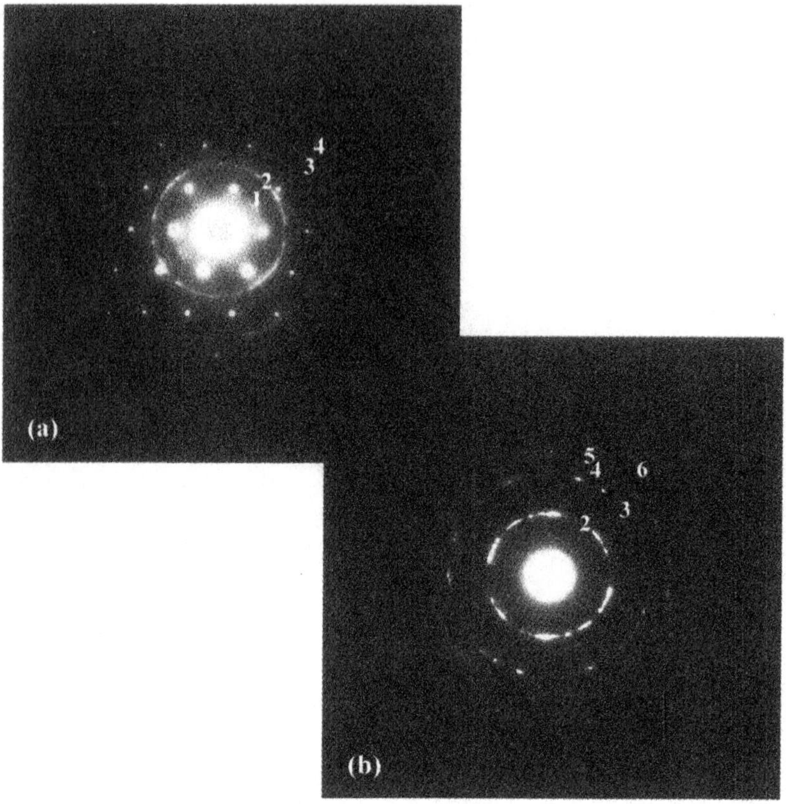

Figure 8. Selected area diffraction patterns taken from the cross section of a thick (~4500 Å) BN film. The first micrograph (a) is from the interface region along the [110] zone axis of the silicon substrate and includes information from all three (amorphous/hexagonal/cubic) layer of the film. Pattern (b) is from *only* the cubic layer of the film as evidenced by the absence of hexagonal BN diffraction features.

oriented along the silicon [220]. The *c*-axis texturing of the hexagonal layer will be discussed shortly. The second selected area pattern (b) is from *only* the top cubic layer of the film. The hexagonal BN (0002) and (0004) features are absent from this pattern and all rings index to cubic BN. This is an important result which shows that once the cubic layer is formed, the film continues to grow as single phase cubic BN. This result is supported by infrared absorptance measurements (presented in Section 3), which can easily distinguish the sp^2 and sp^3 bonding configurations of hexagonal and cubic BN, respectively. Indexing and spacings for the two diffraction patterns are listed in Table 5.

TABLE 5. Indexing of the selected area diffraction patterns shown in Figure 8.

Ring Label	Measured d-spacing (Å)	c-BN d-spacing (Å) JCPDS 35-1365	c-BN (hkl)	h-BN d-spacing (Å) JCPDS 35-421	h-BN (hkl)
1	3.35			3.33	0002
2	2.09	2.09	111		
3	1.27	1.28	220		
4	1.08	1.09	311		
5	1.05	1.05	222		
6	0.90	0.904	400		

McKenzie *et al.*[90,91] have proposed that the layered film morphology results from the buildup of biaxial strain caused by ion-induced damage. They argue that the biaxial strain field drives the amorphous to hexagonal transition, and show that free energy minimization requires that the hexagonal BN *c*-axis orients in the substrate plane. Since the compressibilities of the *c*-axis direction and the *a-b* plane of hexagonal BN are highly anisotropic, the free energy of the film will be minimized when the highly compressible *c*-axis orients in the direction of highest strain, i.e., in the plane of the film. They support this theory with in-*situ* stress measurements which show that their BN films are under high compressive stress during initial growth. Strong evidence supporting this theory is also provided by Kuhr *et al.*[92] who show, using infrared spectroscopy measurements, that hexagonal BN films can be grown on silicon (001) and (111) substrates with the c-axis normal to the substrate when using a very low energy plasma process. When ion energy is increased by biasing the film by as little as -10 V, the hexagonal BN reverts to the *c*-axis in-plane orientation, presumably to minimize strain caused by ion radiation-induced defects.

In a uniform biaxial strain field there is no thermodynamically favorable in-plane orientation for the hexagonal BN *c*-axis and random orientation is predicted.[90] What we observe however, in our samples grown at 850 to 1100 °C, is that the *c*-axis is highly textured along either the silicon [220] or [200]. The [220] orientation has been observed by at least two other research groups[86,93]. Exactly what determines this in-plane orientation, from sample to sample, is not yet understood. One could speculate that the direction of in-plane texturing is influenced by the direction of the nitrogen ion irradiation during film growth. Such effects can be particularly relevant at low growth temperatures where ion-induced defects can accumulate during growth. (The melting temperature of hexagonal BN is estimated at 3240 ± 200 K [94].) We tend to dismiss the directional influence of the ion-irradiation since the orientation of the substrate axes during growth (with respect to the nitrogen ion source and sputtering source) have been similar for all samples we have grown. Considering that the [220] orientation has been observed for growth on silicon (001) by several groups, it is likely that the orientation is substrate influenced. Lattice matching to the substrate may occur through some type of short range ordering within the amorphous layer. We have examined high resolution cross-section micrographs for evidence of ordering extending through the amorphous layer but have found no evidence to support this conclusion. In any case, the in-plane orientation of the hexagonal layer is an interesting observation which may provide insight to the evolution of the film morphology and requires further investigation.

The compressive strain that develops during the ion-assisted growth of BN films may also play an important role in the unusual transformation from the equilibrium hexagonal structure to the metastable cubic structure. McKenzie *et al.*[91] have suggested that the continued buildup of biaxial in-plane strain brings the film into the pressure-temperature regime where cubic BN is thermodynamically stable, thus driving the hexagonal to cubic transition. They propose a mechanism in which an ion-induced thermal spike locally melts a small volume of the film that resolidifies, under pressure, to form cubic BN.

From the experimental point of view, high pressure conversion processes are relatively well established for bulk BN and reasonably accurate pressure-temperature phase diagrams are available.[95-98] The conversion processes are of two types, static compression or shock-loading. Corrigan and Bundy[99] have performed experiments involving both static compression and shock-loading of well crystallized hexagonal BN at low temperatures. In both cases, the resulting material is predominately wurtzitic BN. The pressure to induce the transformation is approximately 11 Gpa[95,100] at room temperature and drops to approximately 7 GPa[100] at 1000 °C. Sato *et al.*[101,102] have performed shock-loading experiments on both hexagonal and rhombohedral BN at pressures from 17 to over 40 GPa. At high pressure (>40 GPa) they found that hexagonal BN transformed to wurtzitic BN, while under identical conditions rhombohedral BN transformed to cubic. At lower pressures they observed that rhombohedral BN could be transformed to a mixture containing both wurtzitic and cubic BN. Similar results (a mixture of wurtzitic and cubic) were obtained by Onodera *et al.*[103] in the low-temperature static compression of rhombohedral BN, indicating that alternative pathways exist for the rhombohedral to cubic transformation.

From the microscopic point of view, the nature of such transformations is still largely a mystery. As indicated by the experimental data, the formation of the high density phases (wurtzitic or cubic) seems to depend not only on temperature and pressure, but also on the structural orientation of the starting material. The structural transformations of the layered forms of BN into their dense modifications (hexagonal to wurtzitic and rhombohedral to cubic) have been studied theoretically by Wentzcovitch *et al.*[104]. They argue that results from the bulk conversion experiments support models in which the transformations proceed by diffusionless mechanisms with well defined epitaxial relationships between the parent and newly formed phases. They show (see Figure 9) that both diffusionless transformations can be achieved by decreasing the interlayer bond lengths (*c*-axis compression), buckling the honeycomb layers into "chair" configurations, and simultaneously stretching the intralayer bond lengths. The transformation pathways are such that the hexagonal BN *c*-axis becomes the *c*-axis of the wurtzitic structure, whereas the rhombohedral *c*-axis becomes the [111] direction of the cubic structure. Both transformations are found to have similar activation energies, approximately 0.39 eV/pair. The authors note that the hexagonal to wurtzitic transformation that they consider may not be the lowest energy pathway.

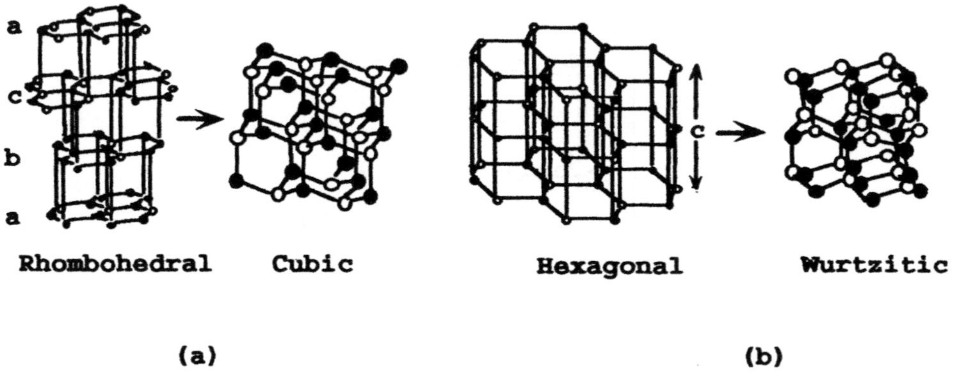

Figure 9. Structural relationships considered in the theoretical energy calculations of Wentzcovitch *et al.*[104]. (a) Rhombohedral to zinc-blende transition. (b) Hexagonal BN to wurtzitic transition.

An alternative pathway proposed by Bundy and Kasper[105] for the hexagonal to wurtzitic transformation of carbon is such that the layers of the initial phase buckle into "boat" instead of "chair" configurations. A distortion of this kind, combined with intralayer bond stretching and interlayer bond compression, results in a wurtzitic phase with *c*-axis perpendicular to the hexagonal *c*-axis, as observed experimentally.[105,106] The possibility of a similar mechanism being operative in the hexagonal to wurtzitic transformation of BN was raised by Riter[107]. He argues that although this pathway requires a lateral displacement of the layers in the initial hexagonal BN phase, it should involve a smaller activation energy based on the total deformation energies of individual layers. In support of this alternative pathway, Johnson and Michell[108] reported results of flash x-ray diffraction measurements, taken during shock-wave compression of highly oriented hexagonal BN, in which the hexagonal *c*-axis became the [100] direction of the wurtzitic phase. There has, however, also been a report[109] in which the hexagonal BN (0002) planes are textured parallel to the wurtzitic (0002) planes.

Interestingly, Wheeler and Lewis[110] have proposed a mechanism for the conversion of graphite to diamond via a wurtzitic carbon (Lonsdalite) intermediate as shown in Figure 9 . The epitaxial relationships are such that the wurtzitic phase is oriented with *c*-axis perpendicular to the *c*-axis of graphite and the cubic $[11\bar{1}]$ is then oriented parallel to the wurtzitic *c*-axis. More precisely,

$[0001]$ wurtzitic || $[10\bar{1}0]$ hexagonal, $[\bar{1}2\bar{1}0]$ wurtzitic || $[\bar{1}2\bar{1}0]$ hexagonal
and
$[11\bar{1}]$ cubic || $[0001]$ wurtzitic, $[\bar{1}10]$ cubic || $[\bar{1}2\bar{1}0]$ wurtzitic.

The wurtzitic/cubic structural relationship has actually been observed in bulk BN samples which have undergone high pressure conversion synthesis.[99] Thus, the existence of an intermediate hexagonal phase offers yet another plausible diffusionless mechanism, in this case for a hexagonal to cubic transformation.

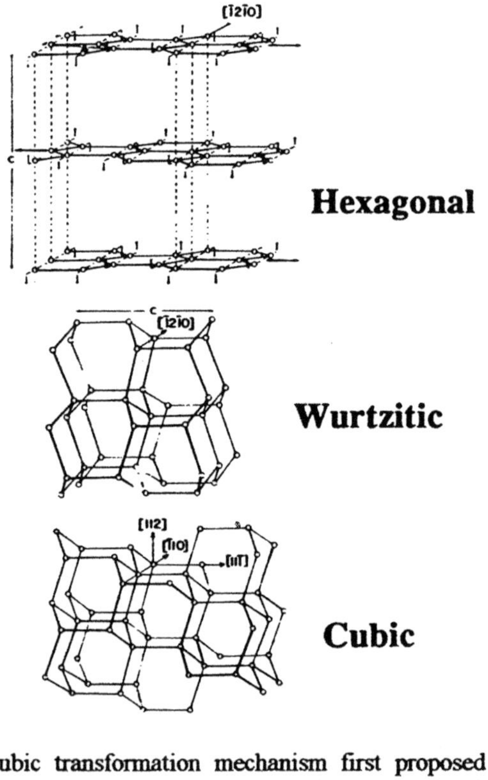

Figure 10. Hexagonal to cubic transformation mechanism first proposed by Wheeler and Lewis[110]. This transformation mechanism provides an epitaxial relationship, via a wurtzitic intermediate, between the edges of hexagonal (0002) basal planes and the cubic structure, such that the cubic (111) planes are normal to the hexagonal (0002) planes.

In contrast to diffusionless mechanisms, Corrigan and Bundy[99] suggest that reconstructive pathways exist, but only at high temperature (>1500 °C), for the direct conversion of hexagonal BN to cubic. In addition to a change in the chemical nature of the bonds from sp^2- to sp^3-hybridization, this type of transformation requires bond breaking. Their experimental data indicates that the activation energy for the transformation is extremely high, a barrier of approximately 200 kcal/mole (~4 eV/atom). They found a similarly high activation energy (~220 kcal/mole) for the wurtzitic to cubic transformation at high temperature. The principal conclusions, from experimental measurements and theoretical estimates, are: (a) that diffusionless mechanisms are energetically favored in bulk materials at the low temperatures (<1200 °C) at which thin film growth is typically performed, and (b) that the stacking sequence of the parent phase plays an important role in determining the newly formed phase.

Returning to the transformation we observe in BN films, it remains unclear whether strain is responsible for the hexagonal to cubic transition. From the foregoing discussion one could speculate that the formation of cubic BN is facilitated by either stress-induced transformations or some type of lattice matching to the vertically arranged hexagonal layers. In the latter case, strain may still be needed for creation of the textured hexagonal layer. Several recently published studies have reported evidence to support both scenarios.[111-113]

Medlin *et al.*[111] have observed, by high resolution TEM, domains within the hexagonal/turbostratic layer that exhibit a three-layer stacking sequence. They show that the spacing and stacking of the fringes is consistent with, although not conclusive proof of, the rhombohedral BN phase. Although the rhombohedrally configured domains have limited spatial extent, ~ 30 Å, and do not lead into regions with a spacing and stacking sequence consistent with cubic BN, the result shows that sufficient disorder exists within the hexagonal layer to initiate a diffusionless solid-state transformation pathway to cubic BN. Another interesting report was made by Reinke *et al.*[112] in which they observed local regions where lattice fringes with 3.1 Å spacing continue directly into fringes with 2.1 Å spacing. They interpret the result as compressed hexagonal BN (0002) planes transforming to cubic BN (111) planes and note that the observed spacings are close to a 2:3 lattice matching. Medlin *et al.* have confirmed this finding and report[113] that their films are preferentially oriented such that individual crystallites have at least one cubic BN [111] direction lying in the plane of the film but are otherwise randomly oriented about (1) the substrate normal and (2) the in-plane cubic BN [111] axis.

2.4.3 Scanning Force Microscopy

We have recently grown a series of thin BN films in order to investigate the nucleation of the cubic phase. Our initial interest was focused on the investigation of large shifts in the cubic BN TO phonon frequencies observed during initial nucleation. We have shown that the shift in phonon frequencies results from polarization of small isolated islands by the long wavelength infrared radiation, (see Section 3). To further study the island nucleation process, this same series of films was examined using scanning force microscopy (SFM). Shown in Figure 11 are 2x2 μm images of four films at different growth stages of the cubic BN layer. The films are identified by total deposition time,

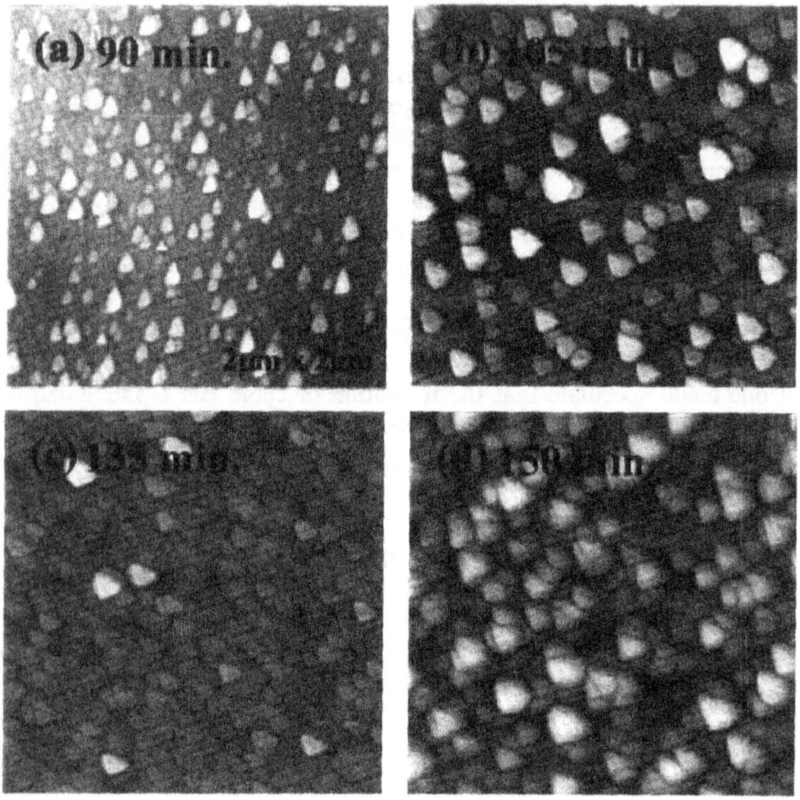

Figure 11. Scanning force microscopy images of a series of BN films grown near the initial nucleation stages of the cubic BN layer. The triangular crystallites, which are indicative of cubic (111) growth, are seen in panel (a) and (b) to grow in size to approximately 700 Å. The grains then coalesce, panel (c), to form a smooth continuous layer. Thicker films, panel (d), show a more random orientation as growth continues.

which includes the time (~35 min.) to deposit the amorphous and hexagonal layers of the film. (The growth rate for the cubic BN layer, measured by cross-sectional TEM from thick cubic BN films, is ~4.5 Å/minute.) Image (a) shows the first cubic BN islands nucleating on the hexagonal BN film surface. The islands are clearly triangular, indicative of (111) cubic growth, and are very well aligned with one another, suggesting some type of epitaxial relationship with the underlying hexagonal BN planes. A slightly thicker film, panel (b), shows that further nucleation occurs and the average grain size grows to approximately 700 Å. The film grown for 135 minutes, panel (c), is relatively smooth (RMS roughness is on the order of 20 Å), indicating that the islands have coalesced to form a continuous layer. The islands remain triangular with an average grain size of approximately 700 Å. The sample grown for 150 minutes, panel (d), shows crystallites of similar size and shape but appears more randomly oriented compared with the thinner films.

Although the SFM results do not provide conclusive evidence of crystallographic orientation, the images strongly suggest that the cubic BN initially nucleates in an ordered [111] growth direction on the edges of the (0002) planes of the hexagonal BN layer. More specifically, the cubic BN [111] is normal to the hexagonal BN [0002] (c-axis). The well aligned triangular crystallites, most obvious in panels (a) and (b), suggest an underlying epitaxial relationship between the hexagonal and cubic layers such as that suggested by Wheeler and Lewis[110]. We are currently performing transmission electron microscopy on this series of samples to provide a more detailed and conclusive study of the crystallographic orientation suggested by the SFM images.

This indication of oriented cubic BN growth on silicon substrates not only helps define the mechanism for the hexagonal to cubic transition but also gives promise for the technological development of this material. The results suggest a pathway to "compliant" oriented growth on a variety of substrates in which the hexagonal BN planes serve to both absorb in-plane strain and provide a nucleation mechanism for cubic BN.

In the following section, infrared absorptance spectroscopy is used to study the phonon frequencies of the BN films. All of the results obtained from the structural characterization techniques, including the island nucleation and coalescence of the cubic BN crystallites, are independently confirmed by this powerful spectroscopic technique.

3. Infrared Spectroscopy Results

3.1 INTRODUCTION

Boron nitride films were characterized optically using infrared (IR) absorptance spectroscopy and cathodoluminescence spectroscopy. Our cathodoluminescence results were reported in a previous publication [114] and we will not elaborate on them here.

At this stage in the development of cubic BN films we have found that IR spectroscopy is an extremely valuable characterization technique and so we will focus on the FTIR results in this section. Since the different phases of BN have different bonding configurations and distinct infrared-active phonon frequencies, it is straightforward to determine film bonding based on the presence of well separated IR-absorption peaks. For example, it is immediately apparent whether the film is predominately sp^3-bonded, characteristic of the cubic structure, or whether its bonding is sp^2-like, which is characteristic of hexagonal structures.

The relative heights of specific absorption peaks provide qualitative information about phase concentrations which verify that our growth process for cubic BN is optimized within a narrow window of ion energy. We find that within this range of optimized growth conditions, the strength of the sp^3-bonding feature of cubic BN is maximized and the width of this absorptance peak decreases to a minimum. This indicates larger cubic BN grain sizes and increased structural quality when correlated with TEM diffraction results.

Through the use of theoretical calculations we have used IR spectroscopy to estimate film strain based on shifts in the transverse optic phonon frequencies. We show that the use of low energy ions and high temperature growth minimizes compressive stress in cubic BN films, allowing us to grow cubic films to thicknesses previously unobtainable by any other growth process. Furthermore, we have recently used IR spectroscopy to probe the initial nucleation and growth of the cubic BN layer. We find that the behavior of transverse optic (TO) phonons is strongly affected by geometrical factors resulting from island nucleation and coalescence. A phenomenological model is presented to explain the phonon behavior at the initial nucleation stage of cubic BN and is shown to be in excellent agreement with our experimental findings.

3.2 FTIR

3.2.1 Hexagonal BN

Bulk samples of hexagonal BN were investigated by Geick *et al.* using polarized IR spectroscopy in both reflection and transmission geometry.[115] As discussed by the authors, the structure of hexagonal BN as identified by Pease[116] has uniaxial symmetry with point group D_{6h} and four atoms per unit cell. The irreducible representations for the zone center normal modes are:

3 acoustical modes	$A_{2u} + E_{1u}$ (doubly degenerate)
3 infrared active optical modes	$A_{2u} + E_{1u}$ (doubly degenerate)
4 Raman active optical modes	$2E_{2g}$ (doubly degenerate)
2 inactive optical modes	$2B_{1g}$.

Two transverse optic modes were identified; the first, at 783 cm^{-1}, has A_{2u} symmetry and corresponds to out-of-plane B-N-B bond bending. The second transverse mode, at 1367 cm^{-1}, corresponds to in-plane B-N bond stretching and has E_{1u} symmetry. An illustration of the vibrational modes in the hexagonal BN lattice is shown in Figure 12. Two infrared-active longitudinal modes were also identified at 828 cm^{-1} and 1610 cm^{-1} corresponding to out-of-plane and in-plane vibrational modes respectively.

Hoffman *et al.* also performed a careful study of the normal modes of hexagonal BN using reflectance spectroscopy.[117] They observed the A_{2u} mode at 770 ± 3 cm^{-1} and the E_{1u} mode at 1383 ± 5 cm^{-1}. A comparison of the work of Hoffman and Geick is provided in Table 6. The differences in phonon frequencies observed, especially in the case of the in-plane E_{1u} phonon, were attributed to differences in the samples.

In thin film hexagonal BN materials the A_{1u} mode is reported typically between 770 cm^{-1} and 810 cm^{-1}, while the E_{1u} mode is reported between 1360 cm^{-1} and 1400 cm^{-1}. The wide range of observed values is most certainly due to differences in sample quality, e.g., grain size, degree of orientation, etc..

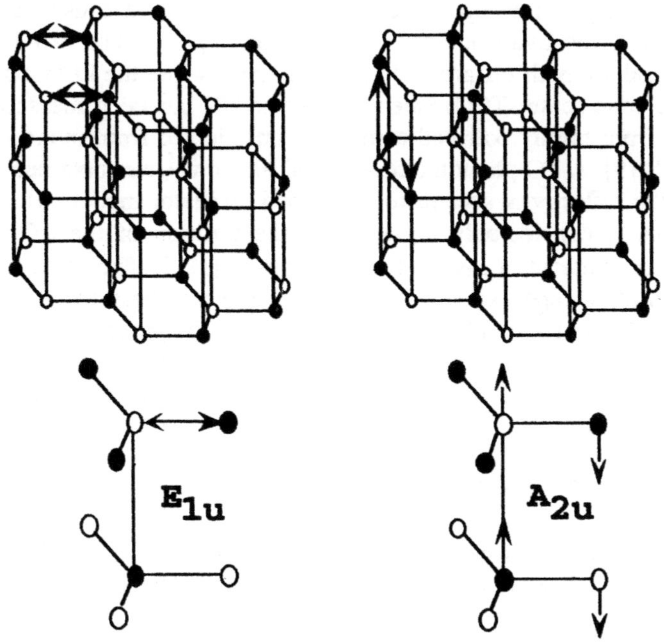

Figure 12. The atomic displacements associated with the two infrared-active normal modes in hexagonal BN. The E_{1u} and A_{2u} modes are associated with the in-plane B-N bond stretch and the out-of-plane B-N-B bond bend, respectively.

TABLE 6. Measured values for the infrared-active TO phonon modes of hexagonal BN. The frequency of the mode is given by ω and the half-width by Γ.

	A_{2u} (out-of-plane)		E_{1u} (in-plane)	
	ω_1 (cm^{-1})	Γ_1 (cm^{-1})	ω_2 (cm^{-1})	Γ_2 (cm^{-1})
Geick *et al.* (Ref. 115)	767	35	1367	29
Hoffman *et al.* (Ref. 117)	770 ± 3	18 ± 4	1383 ± 5	30 ± 4

3.2.2 Cubic BN

Cubic BN is a zinc-blende structure with two atoms per unit cell and T_2 symmetry. The six branches of the phonon dispersion are therefore separated into three acoustic branches and three optical branches. The optical modes are degenerate at $k=0$ but immediately away from the zone center the degeneracy is lifted as a result of the induced electric field that arises from the partially ionic character of the B-N bond. The coupling of the longitudinal optic mode to the electric field causes an LO-TO splitting as predicted by the Lyddane-Sachs-Teller relation.[118] The two TO modes remain degenerate and are both infrared and Raman active while the LO mode is only Raman active. The zone center TO frequency of cubic BN has been measured using infrared reflectivity from highly polished compacts consolidated from monocrystalline cubic boron nitride.[119] A TO phonon frequency of 1065 cm^{-1} was obtained for these samples by fitting the reflectivity spectra to a damped oscillator model. It should be noted that this value is considerably higher than the position of the TO mode as determined by Raman scattering ($\omega_{TO} = 1056 \pm 1$ cm^{-1}).[120] The value of the TO phonon frequency is often reported as evidence for the cubic phase in BN films and reports range from 1050 cm^{-1} to 1100 cm^{-1}. As will be discussed in the following sections, the phonon frequencies are extremely sensitive to finite size effects, geometrical effects, and strain, which can account for the widely varying values reported in cubic BN films.

3.3 EXPERIMENTAL DETAILS: FTIR

Infrared absorption measurements were performed using a Nicolet Magna 550 Series II Fourier Transform Infrared (FTIR) spectrometer. An infrared spectrometer using an interferometer has several advantages over a conventional dispersive spectrometer. The interferometer does not separate (disperse) light into individual frequencies before measurement. The information from all wavelengths present in the incident light are collected simultaneously (multiplex advantage) rather than one at a time. The resolution of an FTIR spectrometer is increased by lengthening the path of the moving mirror (resolution and throughput advantage), rather than narrowing slits and decreasing energy throughput, as with a dispersive instrument. Therefore as the wavelength resolution is increased, the advantage of interferometric measurement over dispersive measurement increases. The combination of these advantages is what allows an FTIR spectrometer to obtain a high quality infrared spectrum in a fraction of the time required by a dispersive instrument.

The Nicolet spectrometer used for these measurements was configured for mid-IR operation with a deuterated tri-glycine sulphate (DTGS) detector and KBr beam splitter. Samples were mounted on a 2-circle rotation stage so that measurements could be performed at off-normal incidence. The spectrometer was purged continuously with dry nitrogen to remove water and carbon dioxide absorptions from the spectra. Background spectra were collected using a bare silicon substrate from the same wafer as the sample so that the effects of the silicon substrate could be removed from the sample spectra. Both background and sample spectra were collected using 2 cm^{-1} resolution over the spectral region 400 cm^{-1} to 2000 cm^{-1}. Four hundred scans were typically averaged to obtain the final spectrum

3.4 RESULTS AND DISCUSSION: FTIR

3.4.1 Determination of Optimum Bias Voltage

We have found that with our ion-assisted sputter deposition process the formation of the cubic BN phase is a complex function of boron deposition rate , ion current at the sample, ion energy, and substrate temperature. We first observed, using *in-situ* RHEED, that with a fixed boron deposition rate and nitrogen ion-current at the sample, the formation of cubic BN could be optimized in a narrow window of ion energy.[121] As discussed in Section 2, the onset of cubic BN formation is marked by a distinct change in the RHEED pattern from a hexagonal to a cubic BN diffraction pattern. As the ion energy at the sample is varied by controlling the substrate bias, the formation of the cubic phase is optimized. This is evidenced by an increase in intensity and narrowing of the linewidth of the cubic diffraction pattern. The cubic BN diffraction pattern is optimized in a narrow (approximately 10 eV) range of substrate bias voltages.

Infrared spectroscopy can also be used to determine the optimum bias voltage (or ion energy) for cubic BN formation under fixed deposition conditions. Shown in Figure 13 are infrared absorptance spectra from a series of BN films grown under identical deposition conditions except for substrate bias. This series of films was grown with a nitrogen ion-current of 0.65 mA/cm^2. The ion-current was measured at the sample with a commercially available Faraday probe.[122] The growth time for each sample was three hours and the complete deposition conditions are listed in Table 7. The peaks in the absorption spectra centered near 767 cm^{-1} and 1390 cm^{-1} correspond to the hexagonal BN A_{2u} and E_{1u} transverse optic phonon modes, respectively. The peak at 1076 cm^{-1} corresponds to the cubic BN zone center TO phonon absorption. At a substrate bias of -108 V the intensity of the sp^2-bonding features from hexagonal BN are approximately equal in intensity to the sp^3-bonding feature from the cubic BN. As the substrate bias becomes more negative, thus increasing the ion energy at the sample, the hexagonal sp^2-bonding features decrease in absorptance intensity while the cubic sp^3-feature at 1076 cm^{-1} shows a corresponding increase.

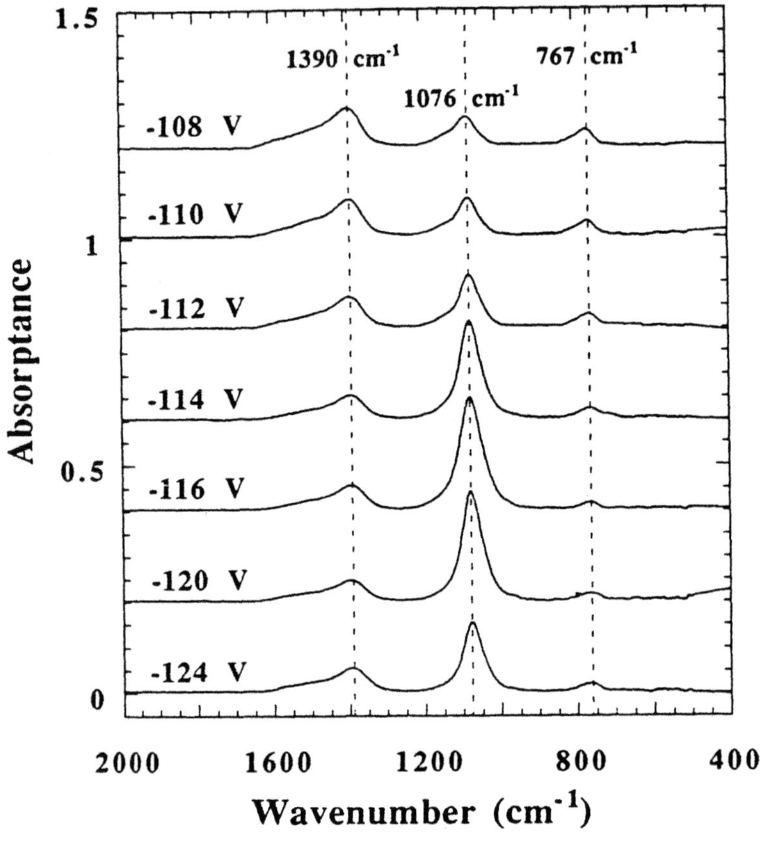

Figure 13. Infrared absorptance spectra for a series of BN films grown near optimum bias voltage. The cubic BN TO phonon absorptance (1076 cm^{-1}) is seen to increase as the substrate bias voltage approaches the optimum voltage, between -116 V and -120 V, for cubic BN formation. The cubic BN TO phonon absorptance decreases again as the bias voltage is increased past the optimum "window" of formation.

TABLE 7. Growth parameters for the series of BN films presented in Figure 13.

RF Sputtering Gun Power: 600 W	
Gas Flow Rate: 10 sccm Ar into chamber	
10 sccm N_2 into ECR source	
Total Chamber Pressure: 1 mTorr	
ECR Microwave Power: 200 W	ECR Magnet Current: 23.0 A
Nitrogen Ion Current at the Sample: 0.65 mA/cm^{-2}	
Growth Temperature: 1100 °C	Growth Time: 3 hours

It has become fairly common in the BN related literature to report the fraction of cubic BN present in a sample by ratioing the intensities of the sp^3 to sp^2 infrared bonding features.[123-129] In particular, the percentage of cubic BN is estimated by taking the ratio of the peak height of the cubic BN mode near 1076 cm^{-1} to the sum of the peak heights of the hexagonal BN mode near 1380 cm^{-1} and the cubic BN mode. Figure 14 is a plot of the percentage of cubic BN (sp^3-bonding) as a function of substrate bias for the complete series of films grown with parameters detailed above (Table 7). The line plotted in the Figure is a polynomial fit to the data points. The data clearly show that the percentage of sp^3-bonding is maximized near a substrate bias voltage of -118 V and that this bonding percentage is maximized within a very narrow window. The infrared spectroscopy results are in very good agreement with, and support the conclusions drawn from, the in-*situ* RHEED data presented in Section 2. Furthermore, we see that as the ion-energy is increased beyond the optimum window, the sp^3-fraction begins to decrease rapidly. This indicates that damage is occurring from excessively high ion-energy and that further increases in ion energy can inhibit film growth. We have observed that film growth is completely inhibited at energies as little as 20 eV beyond the peak in the sp^3-bonding curve.

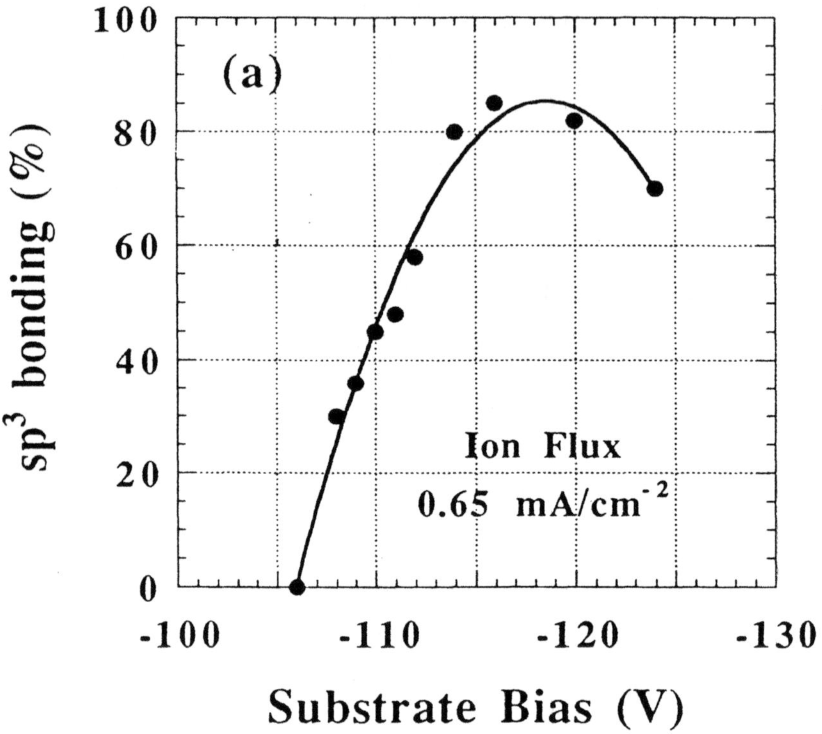

Figure 14. Percent of sp^3-bonding calculated for the series of absorptance spectra presented in Figure 13. The solid line is a polynomial fit to the data and indicates that the maximum sp^3-bonding percentage can be achieved at an optimum bias voltage near -119 V.

We have investigated cubic BN growth using a range of nitrogen ion-currents in the growth process. When one of the process parameters such as nitrogen ion-current is changed, a complete series of samples must be grown to identify the new substrate bias "operating point" which optimizes the growth process for cubic BN. A list of the nitrogen ion-currents we have investigated and the resulting optimum substrate-bias is provided in Table 8. All other growth parameters were identical to those listed in Table 7. The highest nitrogen ion-current we have investigated is 1.1 mA/cm^2. The sp^3-bonding percentage as a function of substrate bias voltage is shown in Figure 15. At this higher ion-current, cubic BN growth is optimized near a substrate bias of -59 V.

TABLE 8. The nitrogen ion currents we have investigated and resulting optimum substrate bias potentials for cubic BN formation.

Nitrogen Ion Current (mA/cm^{-2})	Optimum Substrate Bias (V)
0.65	-119
0.90	-80
1.10	-59

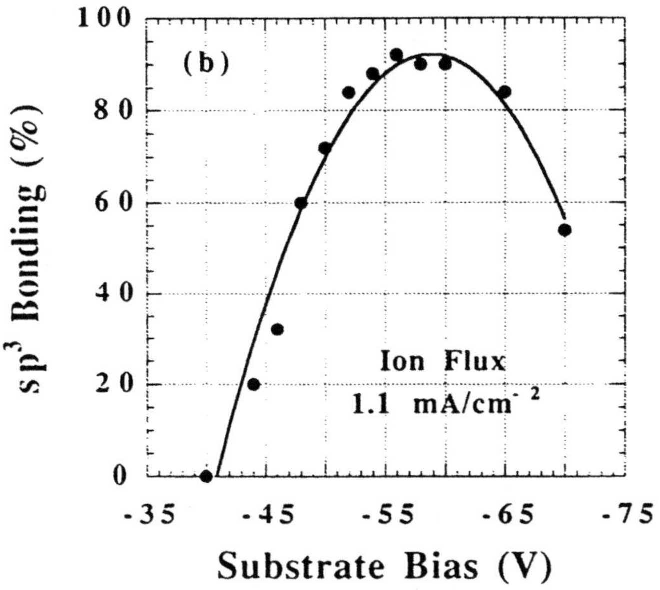

Figure 15. Percent of sp^3-bonding calculated from a series of absorptance spectra from BN films grown with a nitrogen ion current of 1.1 mA/cm^{-2}. The solid line is a polynomial fit to the data and indicates that the maximum sp^3 bonding percentage can be achieved at an optimum bias voltage near -59 V at this ion current.

Comparing the data shown in Figures 14 and 15, one could conclude that higher ion-current and lower ion-energy leads to a slightly higher fraction of cubic BN (sp^3-bonding) in our films. While this may in fact be correct, it is not a conclusion that can be drawn solely from IR data. We have found that the method of ratioing IR peak heights is very useful for comparing films of identical thickness, grown under similar conditions, when the morphology of the films is well understood from structural characterization techniques. It is *very misleading* to use this method to compare random samples grown under different conditions or by different processes, as is often done in the literature. As discussed in Section 2, BN films which are primarily cubic typically grow in a layered structure (amorphous/hexagonal/cubic). Selected area diffraction patterns from sample cross sections show that the cubic layer does not contain a detectable fraction of hexagonal BN. Once the cubic BN layer has formed, the fraction of sp^3-bonding becomes simply a function of film thickness. Infrared absorptance measurements on a series of cubic BN films grown under identical growth conditions (see Figure 16) demonstrates the effect of film thickness on the ratio of sp^3 to sp^2-bonding features. The cubic BN zone center TO phonon mode is seen to

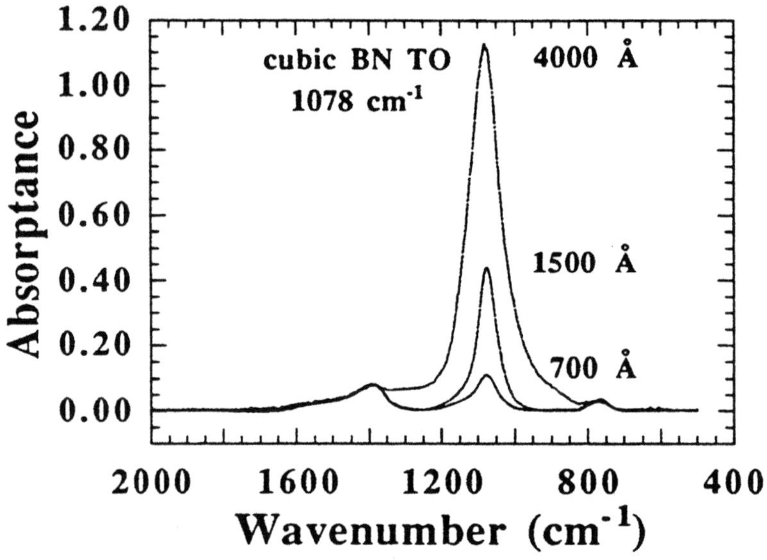

Figure 16. Infrared absorptance spectra from a series of cubic BN films of different thickness. The cubic BN TO phonon absorptance increases with film thickness while the hexagonal BN features are of similar intensity in each film.

continually increase with film thickness while there is no noticeable increase in the sp^2-bonding features associated with hexagonal BN. This confirms the conclusions drawn in Section 2 from the selected area diffraction results. The amorphous and hexagonal sp^2-bonded BN is isolated to the layers near the substrate and the cubic layer does not contain a detectable fraction of hexagonal BN. In terms of sp^3-bonding percentage, the thickest film presented in Figure 16 indicates 96% sp^3-bonding. Clearly any comparisons of sp^3-bonding fraction between films should only be made on samples of identical thickness, grown under similar conditions.

To conclude this section, we have found that our growth process has enabled us to stabilize the cubic phase of BN at much lower energies than is typically reported in the literature. Many researchers report cubic BN growth with ion energies from 300-600 eV[127, 128, 130-2] and several reports have detailed growth over 1000 eV[124-5] Our results represent a significant advancement in the growth of cubic BN films since the realization of high quality material will require a low energy process. The structural evidence we have seen and discussed in Section 2 clearly indicates that the sample quality is increasing as the ion energy becomes lower. We have seen dramatic increases in grain size and degree of orientation as we have developed the growth process to stabilize cubic BN with lower energy ions and higher temperature growth. In general, reports of cubic BN growth with low ion energies, on the order of 100 eV or less, utilize a high flux of ions in comparison to the boron deposition rate. As Reinke *et al.* have pointed out[123], one of the lowest ion energies reported (~40 eV) for successful cubic BN deposition was with an ECR assisted chemical vapor deposition process[133] in which the ion:deposited-boron-atom ratio was as high as 300:1. Our results are in strong agreement with their general conclusion, that cubic BN can be stabilized at lower ion energies by increasing the ion:boron flux ratio. We believe that continued efforts in this direction will lead to significant advances in the growth of cubic BN films.

3.4.2 Analysis of Film Strain and Growth Temperature Studies

Experimental and theoretical studies concerning the strain shifts of the bulk LO and TO phonon frequencies and elastic properties of cubic BN allow us to estimate the effects of film strain on the absorptance peaks in a continuous film of uniform thickness. Sanjurjo *et al.* studied the isotropic pressure dependence of the LO and TO phonons in bulk cubic BN samples by means of Raman spectroscopy.[134] The following isotropic pressure dependencies were found for the bulk LO and TO phonon frequencies:

$$\omega_{LO} = (1305 \pm 1) + (3.45 \pm 0.07)p \quad (1)$$
$$\omega_{TO} = (1054.7 \pm 0.6) + (3.39 \pm 0.08)p \quad (2)$$

Recently, several theoretical papers have focused on the effect of strain on the zone-center TO phonon frequencies[21] as well as the elastic properties of both single crystal and isotropic polycrystalline cubic BN.[136-7] The elastic constants of cubic BN are not well known experimentally due to limitations in available sample size and the values reported in the literature vary widely.[138-9] An estimate for the strain induced frequency shift of the cubic BN TO mode would be beneficial for determining residual compressive stress in BN films.

Fahy[135] notes that the symmetry of cubic BN allows one to characterize completely the first-order effects of strain on the zone-center TO mode frequencies by specifying three fundamental types of frequency shift: (1) the shift of all TO modes when the sample is subjected to isotropic strain; (2) the shift of modes with polarization parallel to the strain axis when the sample is under uniaxial strain; (3) the shift of modes with polarization perpendicular to the strain axis when the sample is under uniaxial strain. The calculated values of the linear strain induced frequency shifts (in cm^{-1} per 1% strain) were updated in a recent erratum[21] and are listed below:

Isotropic strain (type 1)	$-36\ cm^{-1}$
Uniaxial strain (type 2)	$-29\ cm^{-1}$
Uniaxial strain (type 3)	$-13\ cm^{-1}$

Note that the uniaxial strain results, listed above, are for strain in which the dimensions perpendicular to the strain axis are kept fixed. A comparison of the calculated isotropic strain shift to the measurements of Sanjurjo *et al.* can be made by using the measured value[26] of 369 GPa for the bulk modulus of cubic BN. Using this value, the measurements of reference [134] give a TO phonon shift of -37.2 cm^{-1} per 1% isotropic strain which is in good agreement with the calculated value.

As indicated in the erratum[135], uniaxial strain shifts depend on the direction of strain, which is in general true for cubic systems. The phonon calculations give shifts of -5 cm^{-1} per 1% uniaxial, *traceless* strain along the [001] crystal direction (E irreducible representation) and a -26 cm^{-1} shift per 1% uniaxial, *traceless* strain along the [111] direction (T_2 irreducible representation). These results are for modes with polarization along the strain axis. For both E and T_2 strains, Fahy finds that the shift of modes with polarization perpendicular to the strain axis is -1/2 the shift of modes with polarization along the strain axis.

Fahy has recently made additional calculations to estimate the elastic properties of cubic BN.[137] He reports a bulk modulus B = 375 GPa which is in good agreement with the calculations of Rodrigues-Hernandez *et al.*[136] (386 GPa) and with the experiments of Yakovenko *et al.*[140] on single crystal cubic BN (369 GPa). In calculating the Poisson ratio for bulk cubic BN, Fahy found that a change of 4% in the x-y lattice constant causes a change in the relaxed z dimension of 2.53%, giving a Poisson ratio of 0.316. The shear modulus $C_s = (C_{11}-C_{12})/2 = 326$ GPa and $C_{44} = 486$ GPa also compare well with the calculations of Rodrigues-Hernandez *et al.* ($C_s = 326$ GPa, $C_{44} = 483$ GPa).

Assuming a polycrystalline material, in which the orientation of the cubic grains is isotropic, Fahy estimates[137] a shear modulus for the isotropic material, $C_s = C_{44} = 422$ GPa. This estimate is made by spherically averaging the cubic single crystal shear moduli as $(3C_{44} + 2C_s)/5$. This gives elastic constants for the polycrystalline isotropically averaged material, $C_{11} = (3B + 4C_s)/3 = 938$ GPa and $C_{12} = (3B - 2C_s)/3 = 93.7$ GPa, which corresponds to a Poisson ratio $\nu = C_{12}/(C_{11} + C_{12}) = 0.0908$ and a Young modulus $E = C_{11} - 2\nu C_{12} = 921$ GPa. Fahy notes that the Poisson ratio of the isotropic, polycrystalline material is anomalously small compared to that of many materials (including single crystal cubic BN, calculated Poisson ratio = 0.316) which is usually ~ 0.3.

A similar averaging of the phonon shifts for the isotropic, polycrystalline material gives a shift of -17.6 cm^{-1} of modes with polarization along the strain per 1% uniaxial, *traceless* strain and +8.8 cm^{-1} for modes with polarization perpendicular to the strain. (The isotropic averaged shift = [3(T_2 shift) + 2(E shift)]/5). Using the elastic properties and the isotropically averaged phonon calculations, Fahy estimates a shift of +29 cm^{-1} per 1% compressive film strain (assuming that the film surface is free to relax) for modes with polarization parallel to the film surface - the modes observed in IR absorptance measurements at normal incidence. For modes with polarization perpendicular to the film surface, the estimate is a +7 cm^{-1} shift per 1% compressive strain of the film. Using these values we can estimate the residual compressive strain typically observed in cubic BN films using IR spectroscopy.

We have investigated the growth of cubic BN films over the temperature range 800 °C to 1200 °C. All of the samples were characterized using infrared absorptance spectroscopy and the spectra were analyzed to determine the peak position of the cubic zone center TO phonon mode. The temperature dependent position of the cubic TO phonon mode is shown in Figure 17. The infrared spectra show, within the experimental error, a linear dependence of peak position with temperature in the measurement range. The solid line in the figure is a least squares fit to the data using either equation

$$\omega_{TO} = 1122.10 - 0.0412T \tag{3}$$

$$\text{or strain } (\%) = -1.9679 + 1.4199 \times 10^{-3}T. \tag{4}$$

The fit has been extrapolated (dashed line) to the bulk TO phonon frequency (1065 cm^{-1}) as measured by IR reflectivity. We attribute the shift of the TO phonon frequencies, *in continuous films of uniform thickness,* to residual strain due to unannealed defects. (As is discussed in the following section, shifts in TO phonon

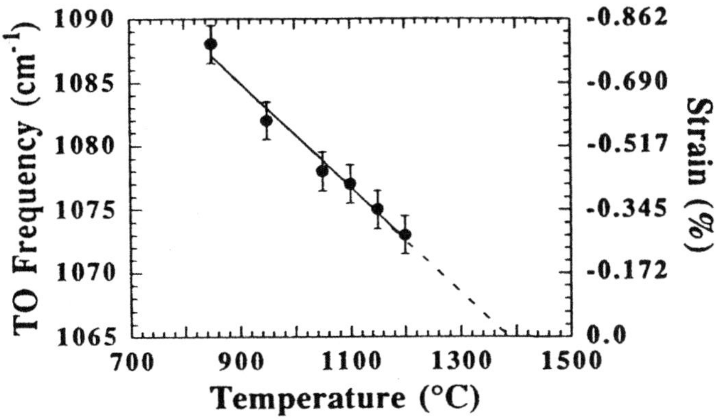

Figure 17. Growth temperature dependence of the cubic BN zone center TO phonon frequency measured by infrared absorptance. The line is a linear fit to the data using equation (3).

frequencies can be caused by a variety of factors including strain, finite crystallite size, and other experimental artifacts.) Compressive strain in cubic BN films is commonly reported in the literature[124, 140-2] and is expected due to the ion-assisted processes needed to stabilize the cubic phase. We suspect that a large portion of the film strain is relieved through the substrate since the growth temperatures (and extrapolated data) are approaching the melting temperature, T_m, of silicon (T_m = 1420°C). The principal conclusion is that high growth temperatures are needed to provide the necessary mobility to anneal defects generated during growth. An empirical rule, established from years of experimental epitaxial growth studies of covalently bonded materials, is that growth temperatures need to be ~1/2 to 5/8 T_m (the melting point of the material being deposited) to achieve reasonable mobility and grain growth.

Although 1200 °C is quite low for cubic BN deposition by the 1/2 to 5/8 T_m standard (cubic BN T_m>2973 °C), we have seen significant improvements in film quality at these temperatures. We have found that a growth temperature above 1000°C reduces the film strain, probably through a combination of defect annealing in the film and strain relief through the substrate, so that the films remain completely adherent to the substrate. Film delamination, which is often reported for cubic BN films grown at lower temperatures (<1000 °C)[141,143], is not seen in films grown at high temperature. We have successfully grown cubic BN films to 1.9 μm thickness at 1100 °C which remain completely adherent. That is not to say that film strain is not a significant problem. The residual film strain, even for growth at 1200 °C, could cause considerable problems for any type of device development. Further development is needed toward reducing ion energies and reducing trapped defects during growth in order to further reduce film strain.

3.4.3 Cubic BN Nucleation Studies

As mentioned previously, recent efforts in the growth of cubic BN films have concentrated on deposition processes which use energetic plasmas to promote the growth of the cubic phase. BN films grown under these highly non-equilibrium conditions undergo an unusual series of transitions leading to an amorphous/hexagonal/cubic layer-sequence. Interestingly, the hexagonal layer is found to orient with the c-axis parallel to the substrate surface; perhaps so that the highly compressible c-axis of hexagonal BN can accommodate in-plane film strain.[30] Although the mechanism is not yet understood, cubic BN nucleates on this oriented hexagonal layer. It has been suggested[142] that the buildup of unannealed defects generates sufficient internal film stress to reach the regime in the phase diagram where cubic BN is stable. Several researchers have since reported that cubic BN formation either scales with[124] or is linked to[142,145-6] compressive stress. Yet there is little evidence to differentiate between cubic BN forming *as a consequence* of compressive stress or simply forming *in conjunction* with it. Adding to the controversy over nucleation mechanisms is a recent publication[125] reporting a large shift in the zone center cubic TO phonon frequency during the initial nucleation of the cubic phase, which decreases to an equilibrium value as growth continues. The authors[125] proposed that the shift is consistent with high compressive stress during initial nucleation and that it supports a stress induced nucleation mechanism. It is this report which has led us to investigate the mechanism for phonon frequency shifts during the initial nucleation and growth of cubic BN.

The approach we have taken is to investigate the transverse optic phonon modes at different stages of initial nucleation and growth, using infrared absorptance spectroscopy at normal and off-normal incidence. The series of films used in this study were exactly the same as those investigated with scanning force microscopy (SFM) in

Section 2. The films were grown at 1100 °C and the nitrogen ion current, measured at the sample using a Faraday cup, was 0.9 mA/cm^{-2}. Optimum cubic BN growth conditions were found with a substrate bias of -80.0 V. The SFM images indicated (see section 2.4.2) that the cubic BN nucleates as triangular crystallites, which are initially isolated and well separated. As growth continues, the crystallites remain triangular, reaching an average size of approximately 700 Å. The crystallites coalesce to form a relatively smooth and continuous layer with an RMS roughness on the order of 20 Å. We show that the observed behavior of the TO phonon frequencies, *at initial nucleation*, is consistent with small isolated crystallites which coalesce to form a smooth, continuous layer, as opposed to the strain induced shifts suggested by Friedmann *et al.*[125].

In Figure 18 we show infrared absorption spectra collected at normal incidence for this series of samples. The cubic BN zone center TO phonon absorption (with mode polarization parallel to the film surface) is seen to increase and shift with growth time. Peak positions for the cubic and hexagonal modes are listed in Table 9. We observe that the cubic TO mode is shifted to higher frequency at the initial nucleation stage and rapidly decreases to a minimum value, slightly above that of bulk cubic BN. Our interpretation of this behavior is that the large frequency shift is based on geometrical factors resulting from island nucleation. Specifically, at the initial nucleation stage the long wavelength infrared radiation polarizes small separated cubic BN crystallites leading to a long range depolarization field. The effect of this field is seen as a shift in the TO phonon frequency. As islands coalesce to form a continuous film, the long range field parallel to the substrate surface rapidly vanishes and the TO mode returns to an equilibrium value. The remaining small shift in the TO mode is a measure of the strain of the cubic BN layer. We will show that this interpretation is compatible with measurements of infrared absorption at off-normal incidence, whereas the expected effects of frequency shifts due to strain alone are not.

In the previous section (3.4.2) we have shown that the calculated strain shifts of the bulk TO phonon frequencies and elastic properties of cubic BN allow us to estimate the effects of film strain *alone* on the absorption peaks in a continuous film of uniform thickness. Assuming the cubic BN crystallites are randomly oriented (i.e., taking an isotropic average of the single crystal strain shifts and elastic properties), Fahy[137] found that the frequency of modes with polarization *normal* to the film changes by approximately +7 cm^{-1} for 1% strain parallel to the film surface. The mode with polarization *parallel* to the film (the only cubic mode seen at normal incidence) has a much larger frequency shift of +29 cm^{-1} per 1% strain. Our measurements of peak shifts in thick, continuous films, where the peak near the bulk TO position shows a small systematic shift to higher frequencies with increasing film thickness and the mode near the bulk LO position (i.e., the mode with polarization normal to the film surface) shows no discernible shift, are in agreement with these estimates. The principal conclusion, from theoretical estimates and the experimental measurements, is that the strain shift for modes with polarization normal to the surface is *much less* than that for modes with polarization parallel to the surface.

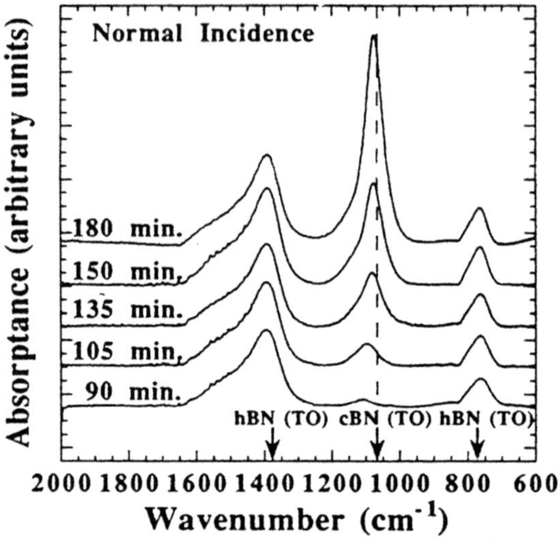

Figure 18. Infrared absorptance spectra from BN films collected at normal incidence. The films are identified by total growth time which includes the growth time for the amorphous and hexagonal BN layers.

In contrast to strain effects, simple models[147] of the effects of islanding on the mode frequencies predict that the shift of modes with polarization normal to the surface should be *substantially greater* than (approximately twice) the shifts for modes with polarization parallel to the surface. Moreover, the shifts are expected to be large in the initial nucleation stage of the cubic BN layer and to fall off rapidly as the crystallites coalesce to form an approximately uniform film. These features are well reproduced in absorption spectra taken at off-normal incidence for films in the initial stages of cubic BN growth (see Table 9). [148]

For the purposes of the present analysis, Fahy[147] has adopted a simple, phenomenological model of the initial stage of cubic BN nucleation and growth:

(1) It is assumed that the cubic BN islands are spheres of radius r during their initial stage of growth. For a constant deposition rate the island radius r is assumed to be proportional to the cube-root of the time since initial nucleation.

(2) At the initial growth stage, the cubic BN islands are embedded in the surrounding hexagonal BN film whose dielectric constant is assumed to be approximately frequency-independent and equal to the optical dielectric constant of cubic BN. This film morphology is in agreement with high resolution TEM analysis of the films (see Section 2, Figure 7) which shows cubic BN grains surrounded (but

not covered) by hexagonal BN during the initial stage of nucleation and growth.

(3) The depolarization electric field within a cubic BN island is the sum of two contributions:

(a) the field due to induced surface charges on the island itself, and
(b) the field due to the other polarized cubic BN islands in the film.

When the layer coverage is small, the contribution of (b) is negligible. When the islands have coalesced to form a continuous layer, the contribution of (a) and (b) together is that of a uniform layer of cubic BN. In the intermediate regime, contribution (b) is modeled as that of a distribution of polarized spheres of radius r whose centers are randomly and uniformly distributed in the plane of the film, except that no sphere has its center within a radius of 2r of the cubic BN island in question. The typical separation between the centers of islands is assumed to be constant throughout the initial stage of growth and the film is assumed to have coalesced into a continuous layer when the island diameter 2r equals the typical separation distance d.

The cubic BN spheres are assumed to be uniformly polarized, either normal or parallel to the substrate, depending on which mode is excited. For polarization normal to the substrate, the contribution of (b) to the depolarization electric field within each grain is of opposite sign and exactly twice the magnitude of that for polarization parallel to the substrate. At any stage of initial growth, this contribution is proportional to r^2. The contribution of (a) to the depolarization field is independent of time and of polarization direction and (due to the spherical geometry) is equal to 1/3 the depolarization field in a uniform film polarized normal to its surface. Fahy finds the following frequency dependencies for modes with polarization normal and parallel to the surface:

$$\omega^2 = (\omega_{LO}^2 + 2\omega_{TO}^2)/3 + 2(2r/d)^2(\omega_{LO}^2 - \omega_{TO}^2)/3 \qquad \text{normal} \qquad (5)$$

$$\omega^2 = (\omega_{LO}^2 + 2\omega_{TO}^2)/3 - (2r/d)^2(\omega_{LO}^2 - \omega_{TO}^2)/3 \qquad \text{parallel} \qquad (6)$$

where r^3 is proportional to time since the start of cubic BN growth and ω_{LO} and ω_{TO} are the bulk LO and TO frequencies respectively.

The only fitting parameter in the equations is the factor (2r/d). We have estimated 2r by assuming the growth rate = 4.5 Å/min., as measured directly from cross-sectional TEM micrographs from thick cubic samples. The experimental data indicates that the cubic layer coalesces after ~115 minutes of cubic BN growth (150 minutes total growth time including amorphous and hexagonal BN layers), which limits 2r = d = (115 min.)(4.5 Å/min.) = 520 Å. Using this value as our first guess, we have plotted the frequency dependence for modes with polarization parallel to the surface, against the normal incidence data as a function of growth time in Figure 19. During the initial nucleation stages the predicted behavior of the phonon

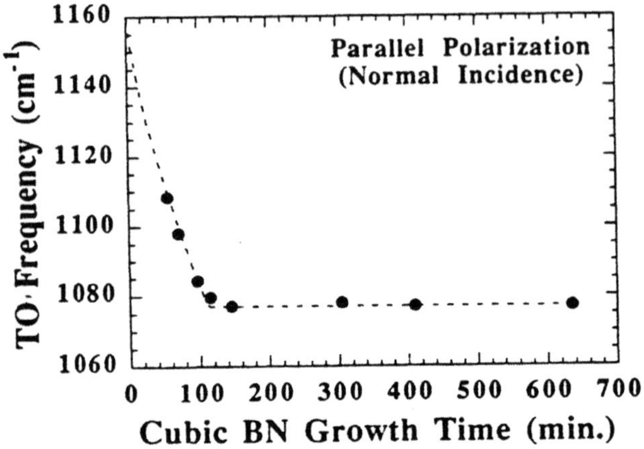

Figure 19. Frequency dependence of the cubic BN zone center TO mode as a function of growth time. The plotted line is the frequency dependence for modes with polarization parallel to the film surface as derived from the island nucleation model. A 12 cm^{-1} offset has been added to the results of the model to account for the residual strain observed in experimental data.

modes is in excellent agreement with the data. Fahy notes[147] that while the above model is a relatively crude representation of the real depolarization fields affecting the mode frequencies during the initial stages of growth, it illustrates the fact that the magnitude of such effects is large for inhomogeneous films. In particular it leads one to expect large shifts in the frequency of modes polarized normal to the surface, in contrast to the relatively small shift expected in strained homogeneous films. An important prediction of the model is that the shift $(\omega_{LO}^2 - \omega^2)$ for modes with polarization normal to the film is exactly twice the shift $(\omega^2 - \omega_{TO}^2)$ for modes with polarization parallel to the film. Although the specific form of the time-dependence of the frequencies assumes spherical islands, this assumption is not essential for this result; Fahy has found that it also holds for spheroidal particles with symmetry axis normal to the film, which is not an unreasonable approximation to the shape of actual crystallites. However, the result is relatively sensitive to the dielectric constant of the surrounding medium, which is assumed to be the same as the optical dielectric constant of cubic BN. In the range of frequencies of the absorption peaks of cubic BN, the dielectric constant of hexagonal BN is frequency dependent. We find that this assumption, however, does not affect the correlation between the model and our experimental data. The ratios of the (negative) shift of the ω_{LO}^2 to the (positive) shift of the ω_{TO}^2 at the initial stages of growth, as found in the off-normal absorption experiments, are listed in Table 9. We assume that the islands have some intrinsic strain at early nucleation, or become strained from differential thermal contraction upon cooling after growth (i.e., the TO frequency shift due to islanding equals the measured TO frequency minus the equilibrium frequency of 1077 cm^{-1}). The 2:1 ratio, which is only expected for samples at the very early stages of nucleation where the cubic BN is most likely to consist of small, well separated islands, is seen consistently in the first few samples. Considering the complications of the system geometry which occur in the real growth process, the experimental results are again in remarkably good agreement with the model.

TABLE 9. Peak positions from the infrared absorptance spectra taken at normal incidence (top of table) and off-normal incidence (bottom half). The peak-shift ratio calculated for the off-normal incidence spectra assumes that the cubic TO frequency is affected by residual strain (i.e. ω_{TO} = measured position - 1077 cm^{-1}).

Growth Time (min.)	hex BN "LO" mode (w.n.)	hex BN E_{1u} TO mode (w.n.)	cubic BN "LO" mode (w.n.)	cubic BN TO mode (w.n.)	hex BN A_{2u} TO mode (w.n.)	peak-shift ratio $\frac{(\omega^2_{LObulk}-\omega^2_{LO})}{(\omega^2_{TO}-\omega^2_{TObulk})}$
90	1619	1376	1249	1113	767	2.1
105	1619	1376	1279	1093	766	2.0
135	1619	1376	1293	1084	766	2.1
150	1617	1376	1299	1078	767	7.3
180	1617	1376	1304	1077	767	n.a.

4. Conclusions

We have developed a novel plasma assisted physical deposition process for the growth of cubic BN films. The approach we have taken is to develop a high temperature growth process (~1100 °C) which utilizes a very low boron flux combined with a high-density, low-energy nitrogen plasma. Specifically, we have developed an RF-magnetron sputtering process which combines the high density nitrogen plasma of an electron cyclotron resonance (ECR) ion-source and negative substrate bias potentials to enable us to precisely control the nitrogen ion-energy at the substrate surface. Using this novel process we have been able to significantly reduce the ion-energy needed to form the cubic phase, with values now substantially less than 100 eV. Through a better understanding of the growth kinetics, we have made improvements in film crystallinity and orientation, with an associated reduction of the high film stress which has severely limited film thickness in the past. Cubic BN films have been grown on silicon substrates to thicknesses in excess of 1.9 µm which are stable and show no signs of film delamination.

In-*situ* reflection high energy electron diffraction (RHEED) measurements show that for our deposition process cubic BN growth is optimized in a narrow range of ion-energy which, for a fixed boron rate, is strongly dependent on the nitrogen ion flux. We find that the optimum substrate bias (ion-energy) for cubic BN formation shifts from -119 V to -58 V as the ion flux is increased from 0.65 mA/cm^2 to 1.1 mA/cm^2. In all cases the range or "window" of ion-energy for cubic BN formation is approximately ±15 eV centered about the optimum energy. Below the window of formation only hexagonal BN diffraction features are observed, while only several eV above the optimum energy an increase in structural damage is observed. Our results, which are in strong agreement with recent reports in the literature, demonstrate that nitrogen ion energy can be decreased substantially by increasing the ion: boron flux ratio and furthermore, attest that precise control of the nitrogen ion energy is essential for the optimization of cubic BN film growth.

High resolution transmission electron microscopy (HRTEM) reveals that the films grow in a layered structure: a thin (20-50 Å) amorphous region near the silicon substrate, followed by a 300-500 Å oriented hexagonal BN layer upon which the cubic BN nucleates. Interestingly, the hexagonal BN layer is found to orient such that the (0002) planes are perpendicular to the substrate surface, perhaps to accommodate in-plane film strain generated by the ion-assisted deposition process. Selected area diffraction patterns from thick cubic BN films demonstrate that once the cubic layer forms, growth continues as single phase cubic BN - an important result for technological applications of this material.

Scanning force microscopy (SFM) was performed on a series of films grown to investigate the initial nucleation stage of the cubic BN layer. The SFM images reveal that the cubic BN nucleates as triangular crystallites, indicative of (111) cubic growth. More specifically, the cubic BN appears to be nucleating on the edges of perpendicularly oriented hexagonal planes such that the cubic BN [111] is normal to the hexagonal BN [0002] (c-axis). The crystallites are seen to coalesce, as growth continues, to a relatively smooth film (RMS roughness ~20 Å) with an average grain size of approximately 700 Å. This indication of oriented cubic BN growth on silicon substrates not only helps define the mechanism for the hexagonal to cubic transition but also gives promise for the technological development of this material. The results suggest a pathway to "compliant" oriented growth on a variety of substrates in which the hexagonal BN planes serve to both absorb in-plane strain and provide a nucleation mechanism for cubic BN.

Boron nitride films were characterized using infrared (IR) absorptance spectroscopy. The IR data support many of the structural results and provide a means to study the films quantitatively via phonon-related absorptance features. Since the different phases of BN have different bonding configurations and distinct infrared-active phonon frequencies, it is straight-forward to determine film bonding based on the presence of well separated IR-absorption peaks. For example, it is immediately apparent whether the film is predominately sp^3-bonded, characteristic of the cubic structure, or whether its bonding is sp^2-like, which is characteristic of hexagonal structures.

The relative heights of specific absorption peaks provide qualitative information about phase concentrations which verify that our growth process for cubic BN is optimized within a narrow range of ion energy. We find that within this range of optimized growth conditions, the strength of the sp^3-bonding feature of cubic BN is maximized and the width of this absorptance peak decreases to a minimum. This indicates larger cubic BN grain sizes and increased structural quality when correlated with TEM diffraction results.

Through the use of theoretical calculations we have used IR spectroscopy to estimate film strain based on shifts in the transverse optic phonon frequencies. We have investigated cubic BN film growth over the temperature range 800 °C to 1200 °C and show that the use of low energy ions and high temperature growth minimizes compressive stress, allowing us to grow cubic films to thicknesses previously unobtainable by any other growth process. Furthermore, we have recently used IR spectroscopy to probe the initial nucleation and growth of the cubic BN layer. We find that the behavior of transverse optic (TO) phonons is strongly affected by geometrical factors resulting from island nucleation and coalescence. A phenomenological model is presented to explain the phonon behavior at the initial nucleation stage of cubic BN and is shown to be in excellent agreement with our experimental findings. More importantly, we find that understanding the behavior of the phonon frequencies can provide important information about the nucleation process. The very large frequency shifts we observe during early growth indicate that cubic BN island nucleation begins in our process when the hexagonal BN film is approximately 30 nm thick. Moreover, the model predicts, based on our measured growth rate, that the islands will coalesce at a growth time of ~110 minutes, with an average island size of approximately 520 Å. These values are also in excellent agreement with the scanning force microscopy data which indicate an average grain size of ~700 Å.

Although cubic BN has proven to be an extremely difficult material to grow in either bulk or thin film form, great strides have been made in just a few short years which provide encouragement that applications of thin films and composites will be possible in the near future. The many attractive tribological and electronic properties of cubic BN make it a very important semiconductor material for high power/temperature devices and perhaps UV-optical applications.

Acknowledgments - This work was supported by Office of Naval Research Grants N00014-91-J-1398 and N00014-94-J-0763.

REFERENCES

1. Wide Band Gap Semiconductors, eds. T. D. Moustakas, J. I. Pankove, and Y. Hamakawa, MRS Symposium Proceedings **242**, Pittsburgh, PA (1992).
2. S. Strite and H. Morkoç, J. Vac. Sci. Technol. B **10**, 1237 (1992).
3. Silicon Carbide and Related Materials, eds. M. G. Spencer, R. P. Devaty, J. A. Edmond, M. Asif Khan, R. Kaplan, and M. Rahman, Institute of Physics Conference Series **137**, (1994).
4. Diamond, SiC and Nitride Wide-bandgap Semiconductors, eds. C. H. Carter, G. Gildenblat, S. Nakamura, and R. J. Nemanich, MRS Symposium Proceedings **339**, Pittsburg, PA (1994).
5. III-Nitride, SiC, and Diamond Materials for Electronic Devices, eds. D. K. Gaskill, C. Brandt, and R. J. Nemanich, MRS Symposium Proceedings **423**, Pittsburgh, PA (1996).
6. G.L. Doll, A.K.Ballal, L. Salamanca-Riba, C.A. Taylor, S. Kidner, and R. Clarke, in *Beam Processing of Advanced Materials*, J. Singh, S.M. Copley, eds., Proc. Matls. Soc., pp. 419-434 (1993).
   Kiyoku, Y. Sugimoto, Jpn. J. Appl. Phys. **35**, L74 (1996).
7. H. Morkoç, S. Strite, G. B. Gao, M. E. Lin, B. Sverdlov, and M. Burns, J. Appl. Phys. **76**, 1363 (1994).
8. R. H. Wentorf, Jr., J. Chem. Phys. **36**, 1990 (1962).
9. O. Mishima, J. Tanaka, S. Yamaoka, and O. Fukunaga, Science **238**, 181 (1987).
10. O. Mishima, K. Era, J. Tanaka, and S. Yamaoka, Appl. Phys. Lett. **53**, 962 (1988).
11. *Properties of Group III Nitrides*, EMIS Datareviews Series No.11, edited by James H. Edgar, (INSPEC, London, (1994).
12. *Landolt-Börnstein Numerical Data and Functional Relationships in Science and Technology-Semiconductors: Physics of Group IV and III-V Compounds*, Volume 17, Subvolume a, edited by O. Madelung, (Springer-Verlag, Berlin, 1982), pp. 148-149.
13. W. Balmain, Phil. Mag. **21**, 170 (1842).
14. S. Larach and R. E. Shrader, Phys. Rev. **104**, 68 (1956).
15. J. Cazaux, C. R. Acad. Sci. B **270**, 700 (1970).
16. R. Vilanove, C. R. Acad. Sci. B **272**, 1066 (1972).
17. J. Zupan and D. Kolar, J. Phys. C **5**, 3097 (1972).
18. W. Baronian, Mater. Res. Bull. **7**, 119 (1972).

19. A. Zunger, A. Katzir, and A. Halperin, Phys. Rev. B **13**, 5560 (1976).
20. R. Mamy, J.Thomas, G. Jezequel, and J. C. Lemonnier, J. Phys. (Paris) Lett. **42**, 473 (1981).
21. V. A. Fomichev, Fiz. Tverd. Tela (Leningrad) **13**, 907 (1971). [Sov. Phys.- Solid State **13**, 754 (1971)].
22. E. Tegeler, N. Kosuch, G. Wiech, and A. Faessler, Phys. Status Solidi B **91**, 223 (1979).
23. D. M. Hoffman, G. L. Doll, and P. C. Eklund, Phys. Rev. B **30**, 6051 (1984).
24. A. I. Lukomskii, V. B. Shipilo, and L. M. Gameza, Zhur. Prikl. Spektr. **57**, 100 (1992).
25. E. Doni and G. P. Parravicini, Nuovo Cimento **64B**, 117 (1969).
26. M. S. Nakhmanson and V. P. Smirnov, Fiz. Tverd. Tela (Leningrad) **13**, 905 (1971).
27. J. Zupan, Phys. Rev. B **6**, 2477 (1972).
28. J. Robertson, Phys. Rev. B **29**, 2131 (1984).
29. R. Taylor and C. A. Coulson, Proc. Phys. Soc. London, Sect. A **65**, 834 (1952).
30. A. Zunger, J. Phys. C **7**, 76 (1974); **7**, 96 (1974).
31. R. Dovesi, C. Pisani, and C. Roetti, Int. J. Quantum Chem. **17**, 517 (1980).32 R. H.
32. Wentorf, Jr., J. Chem. Phys. **26**, 956 (1957).
33. L. Vel, G. Demazeau, and J. Etourneau, Mater. Sci. Eng. **B10**, 149 (1991).
34. H. Herchen and M. A. Cappelli, Phys. Rev. B **47**, 14193 (1993).
35. P. J. Gielisse, S. S. Mitra, J. N. Plendl, R. D. Griffis, L. C. Mansur, R. Marshall, and E. A. Pascoe, Phys. Rev. **155**, 1039 (1967).
36. R. M. Chrenko, Solid State Commun. **14**, 511 (1974).
37. V. A. Fomichev and M. A. Rumsh, J. Phys. Chem. Sol. **29**, 1015 (1968).
38. N. Miyata, K. Moriki, O. Mishima, M. Fujisawa, and T. Hattori, Phys. Rev. B **40**, 12028 (1989).
39. L. Kleinman and J, C. Phillips, Phys. Rev. **117**, 460 (1960).
40. F. Bassani and M. Yoshimine, Phys. Rev. **130**, 20 (1963).
41. D. R. Wiff and R. Keown, J, Chem. Phys. **47**, 3113 (1967).
42. L. A. Hamstreet Jr. and C. Y. Fong, Phys. Rev. B **6**, 1464 (1972).
43. H. C. Hwang and J. Henkel, Phys. Rev. B **17**, 4100 (1978).
44. Y. F. Tsay, A. Vaidyanathan, and S. S. Mitra, Phys. Rev. B **19**, 5422 (1979).
45 C. Prasad and J. D. Dubey, Proc. Nucl. Phys. Solid State, 386 (1982).
46. C. Prasad and J. D. Dubey, Phys. Status Solidi B **125**, 629 (1984).
47. M. Z. Huang and W. Y. Ching, J. Phys. Chem. Solids **46**, 977 (1985).
48. R. M. Wentzcovitch, K. J. Chang, and M. L. Cohen, Phys. Rev. B **34**(2), 1071 (1986).
49. K. T. Park, T. Terakura, and H. Hamada, J. Phys. C: Solid State Phys. **20**, 1241 (1987).
50. A. Zunger and A. J. Freeman, Phys. Rev. B **17**, 2030 (1978).
51. R. Dovesi, C. Pisani, C. Roetti, and P. Dellarole, Phys. Rev. B **24**, 4170 (1981).
52. P. E. Van Camp, V. E. Doren, and J. T. Devreese, Phys. Status Solidi B **146**, 573 (1988).

53. A. Onodera, M. Nakatani, M. Kobayashi, Y. Nisida, and O. Mishima, Phys. Rev. B **48**, 2777 (1993).
54. F. P. Bundy and R. H. Wentorf Jr., J. Chem. Phys. **38**, 1144 (1963).
55. T. Ishii, T. Sato, Y. Sekikawa, M. Iwata, J. Cryst. Growth **52**, 285 (1981).
56. M. Sokolowski, J. Cryst. Growth **46**, 136 (1979).
57. S. P. S. Arya and A. D'Amico, Thin Solid Films **157**, 267 (1988).
58 J. J. Pouch and S. A. Alterovitz (eds.), *Synthesis and Properties of Boron Nitride*, Materials Science Forum, Vols. 54 and 55, (Trans Tech Publications, Brookfield, NY, 1990).
59. S. Fahy, University College Cork, Ireland, private communication (to be published).
60. C. A. Anderson and J. R. Hinthorne, Science **175**, 853 (1972).
61. A. Benninghoven and A. Mueller, Phys. Lett. A **40**, 169 (1972).
62. H. W. Werner, Surf. Sci. **47**, 301 (1975).
63. J. Comas and C. B. Cooper, J. Appl. Phys. **38**, 2956 (1967).
64. A. B. Campbell and C. B. Cooper, J. Appl. Phys. **43**, 863 (1972).
65. B. J. Garrison, N. Winograd, and D. E. Harrison, J. Vac. Sci. Technol. **16**, 789 (1979).
66. J.E. Greene, in CRC Critical Reviews of Solid State and Materials Science **11**, issues 1 and 2 1984.
67. J. M. E. Harper, J. J. Cuomo, R. J. Cambino, and H. R. Kaufman, in *Ion Bombardment of Surfaces*, eds., O. Auciello and R. Kelly (Elsevier, Amsterdam, 1984).
68. J. A. Thornton and D. W. Hoffman, J. Vac. Sci. Technol. **18**, 203 (1981).
69 A. Pan and J. E. Greene, Thin Solid Films **78**, 25 (1981).
70. T. C. Huang, G. Lim, F. Parmiagiani, and E. Kay, J. Vac. Sci. Technol. **A3**, 216 (1985).
71. D. Henderson, M. H. Brodsky, and P. Chaudhari, Appl. Phys. Lett. **25**, 641 (1974).
72. A. G. Dirks and H. J. Leamy, Thin Solid Films **47**, 219 (1977).
73. L. Hultman, U. Helmersson, S. A. Barnett, J. -E. Sundgren, and J. E. Greene, J. Appl. Phys. **61**, 552 (1987).
74. M. Marinov, Thin Solid Films **46**, 267 (1977).
75. T. Narusawa, S. Shimizu, and S. Komiya, J. Vac. Sci. Technol. **16**, 366 (1979).
76. K. Yagi, S. Tamura, and K. Tokuyama, Jap. J. Appl. Phys. **16**, 245 (1977).
77. K. Tokuyama, K. Yagi, K. Miyaki, M. Yamura, N. Natsuaki, and S. Tachi, Nucl. Instr. and Meth. **182/183**, 241 (1981).
78. Kimball Physics Inc., 311 Kimball Hill Road, Wilton, NH 03086-9742.
79. Technical data, Astex Compact ECR Source, Applied Science and Technology Inc., Woburn MA.
80. The original design of the substrate heater is credited to Ray Tung of the former AT&T Bell Laboratories (presently Lucent Technologies).
81. A. S. Arrott, in *Ultrathin Magnetic Structures I*, eds. J. A. C. Bland and B. Heinrich, (Springer-Verlag, Berlin, 1994).
82. "Borazon", Cubic Boron Nitride Powder Trademark of General Electric Company.

83. S. Reinke, M. Kuhr, W. Kulisch, and R. Kassing, Diamond Rel. Mater. **4**, 272 (1995).
84. S. Kidner, Ph. D. Thesis, University of Michigan (1994).
85. D. J. Kester, K. S. Ailey, and R. F. Davis, J. Mater. Res. **8**, 1213 (1993).
86. D. L. Medlin, T. A. Friedmann, P. B. Mirkarimi, P. Rez, M. J. Mills, and K. F. McCarty, J. Appl. Phys. **76**, 295 (1994).
87. S. Reinke, M. Kuhr, and W. Kulisch, in *Proc. of The 4th International Symposium on Diamond Materials*, (Reno, Nevada, May 1995).
88. D. J. Kester, K. S. Ailey, and D. J. Lichtenwalner, J. Vac. Sci. Technol. A **12**, 3074 (1994).
89. P. B. Mirkarimi, D. L. Medlin, K. F. McCarty, and J. C. Barbour, Appl. Phys. Lett. **66**, 2813 (1995).
90. D. R. McKenzie, J. Vac. Sci. Technol. B **11**, 1928 (1993).
91. D. R. McKenzie, W. D. McFall, W. G. Sainty, C. A. Davis, and R. E. Collins, Diamond Relat. Mater. **2**, 970 (1993).
92. M. Kuhr, S. Reinke, W. Kulisch, Diamond Rel. Mater. **4**, 375 (1995).
93. H. Hofsass, C. Ronning, U. Griesmeier, M. Gross, S. Reinke, and M. Kuhr, submitted to Appl. Phys. Lett.
94. R. H. Wentorf, J. Phys. Chem. **63**, 1934 (1959).
95. F. P. Bundy and R. H. Wentorf, Jr., J. Chem. Phys. **38**, 1144 (1963).
96. M. Wakatsuki, K. Ichinose, and T. Aoki, Mater. Res. Bull. **7**, 999 (1972).
97. I. N. Dulin, L. V. Al'tshuler, V. Ya Vashchenko, and V. N. Zubarev, Fiz. Tverd. Tela (Leningrad) **11**, 1252 (1962) [Sov. Phys. Solid State **11**, 1016 (1969)].
98. T. Soma, A. Sawaoka, and S. Saito, Mater. Res. Bull. **7**, 755 (1974).
99. F. R. Corrigan and F. P. Bundy, J. Chem. Phys. **63**, 3812 (1975).
100. S. Nakano and O. Kukunaga, Diamond Relat. Mater. **2**, 1409 (1993).
101. T. Sato, T. ishii and N. Setaka, Commun. Am. Ceram. Soc. **65**, C162 (1982).
102. T. Ishii, T. Sato, Y. Sekikawa, and M. Iwata, J. Cryst. Growth **52**, 285 (1981).
103. A. Onodera *et al.*, J. Mater. Sci. **25**, 4279 (1990).
104. R. M. Wentzcovitch, S. Fahy, M. L. Cohen, and S. G. Louie, Phys. Rev. B **38**, 6191 (1988).
105. F. P. Bundy and J. S. Kasper, J. Chem. Phys. **46**, 3437 (1967).
106. A. Kurdyumov, V. Slesarev, N. Ostrovskaya, A. Pilyankevich, and I. Frankevich, Dok. Akad. Nauk SSSR **239**, 1337 (1978) [Sov. Phys.-Dokl. **23**, 278 (1978)].
107. J. R. Riter, Jr., J. Chem. Phys. **59**, 1538 (1973).
108. Q. Johnson and A. C. Mitchell, Phys. Rev. Lett. **29**, 1369 (1972).
109. V. A. Pesin, M. I. Sokhor, and L. I. Fel'dgun, Russ. J. Phys. Chem. **53**, 908 (1979).
110. E. J. Wheeler and D. Lewis, Mat. Res. Bull. **10**, 687 (1975).
111. D. L. Medlin, T. A. Friedmann, P. B. Mirkarimi, M. J. Mills, and K. F. McCarty, Phys. Rev. B **50**, 7884 (1994).
112. S. Reinke, M. Kuhr, and W. Kulisch, in *Proceedings of the 187th Meeting of the Electrochemical Society*, 1995 (unpublished).

113. D. L. Medlin, T. A. Friedmann, P. B. Mirkarimi, G. F. Cardinale, and K. F. McCarty, J. Appl. Phys. **79**, 3567 (1996).
114. C.A. Taylor, II, S.W. Brown, V. Subramaniam, S. Kidner, S.C. Rand, and R. Clarke Appl. Phys. Lett. **65**, 1251 (1994).
115. R. Geick, C. H. Perry, and G. Rupprecht, Phys. Rev. **146**, 543 (1966).
116. R. S. Pease, Acta Cryst. **5**, 536 (1952).
117. D. M. Hoffman, G. L. Doll, and P. C. Eklund, Phys. Rev. B **30**, 6051 (1984).
118. R. H. Lyddane, R. G. Sachs, and E. Teller, Phys. Rev. **59**, 673 (1941).
119. P. J. Gielisse S. S. Mitra, J. N. Plendl, R. D. Griffis, L. C. Mansur, R. Marshall, and E. A. Pascoe, Phys. Rev. **155**, 1039 (1967).
120. O. Brafman, G. Lengyel, S. S. Mitra, P. J. Gielisse, J. N. Plendl, and L. C. Mansur, Solid State Commun. **6**, 523 (1968).
121. S. Kidner, C. A. Taylor II, and R. Clarke, Appl. Phys. Lett. **64**, 1859 (1994).
122. Kimball Physics Inc., 311 Kimball Hill Road, Wilton, NH.
123. S. Reinke, M. Kuhr, W. Kulisch, and R. Kassing, Diamond Relat. Mater. **4**, 272 (1995).
124. P. B. Mirkarimi, K. F. McCarty, D. L. Medlin, W. G. Wolfer, T. A. Friedmann, and E. J.Klaus, J. Mater. Res. **9**, 2925 (1994).
125. T. A. Friedmann, P. B. Mirkarimi, D. L. Medlin, K. F. McCarty, E. J. Klaus, D. R. Boehme, H. A. Johnsen, M. J. Mills, and D. J. Otteson, J. Appl. Phys. **76**, 3088 (1994).
126. T. A. Friedmann, W. M. Clift, H. A. Johnsen, E. J. Klaus, K. F. McCarty, D. L. Medlin, M. J. Mills, and D. K. Ottesen, In Laser Ablation in Materials Processing: Fundamentals and Applications, ed. D. Braren, J. J. Dubowski, and D. P. Norton, Materials Research Society Proceedings Vol. 285, p. 507 (1993).
127. T. Wada and N. Yamashita, J. Vac. Sci. Technol. A **10**, 515 (1992).
128. D. J. Kester and R. Messier, J. Appl. Phys. **72**, 504 (1992).
129. K. Inagawa, K. Watanabe, H. Ohsone, K. Saitoh, and A. Itoh, J. Vac. Sci. Technol. A **5**, 2696 (1987).
130. N. Tanabe, T. Hayashi, and M. Iwaki, Diamond Relat. Mater. **1**, 883 (1992).
131. W. Dworschak, K. Jung, and H. Erhardt, Thin Solid Films **254**, 65 (1995).
132. D. Bouchier, M. A. Sene', Djouadi and P. Moller, Nucl. Instrum. Methods B **89**, 369 (1994).
133. H. Yokoyama, M. Okamoto, and Y. Osaka, Jpn. J. Appl. Phys. **30**, 344 (1991).
134. J. A. Sanjurjo, E. Lo'pez-Cruz, P. Vogl, and M. Cardona, Phys. Rev. B **28**, 4579 (1983).
135 S. Fahy, Phys. Rev. B **51**, 12873 (1995); erratum (in press).
136. P. Rodriguez-Hernandez *et al.*, Phys. Rev. B **51**, 14705 (1995).
137. S. Fahy, private communication (to be published).
138. N. N. Sirota and A. F. Revinskii, Vesti Akad, Nauk BSSR, Ser. Fiz.-Mat. Navuk **6**, 64 (1981).
139. T. D. Sokolovskii, Inorg. Mater. **19**, 1311 (1983).
140. Y. Yakovenko et al.
141. S. Reinke, M. Kuhr, and W. Kulisch, in *Proc. of 4th International Symposium on*

Diamond Materials, Reno, Nevada (1995).

142. D. R. McKenzie, W. D. McFall, W. G. Sainty, C. A. Davis, and R. E. Collins, Diamond Relat. Mater. **2**, 970 (1993).

143. D. J. Kester, K. S. Ailey, R. F. Davis, K. L. More, J. Mater. Res. **8**, 1213 (1993).

144.. D. R. McKenzie, J. Vac. Sci. Technol. B **11**, 1928 (1993).

145. D. L. Medlin, T. A. Friedmann, P. B. Mirkarimi, P. Rez, M. J. Mills, and K. F. McCarty, J. Appl. Phys. **76**, 295 (1994).

146. D. L. Medlin, T. A. Friedmann, P. B. Mirkarimi, M. J. Mills, and K. F. McCarty, Phys. Rev. B **50**, 7884 (1994).

147. S. Fahy (University College, Cork, Ireland) has developed several island nucleation models which predict the behavior of phonon frequencies observed in BN films, private communication. To be published with C. A. Taylor II and Roy Clarke.

148. C. A. Taylor, PhD Thesis, University of Michigan, Ann Arbor, 1996 (published by University Microfilms Inc., Ann Arbor, MI 48103).

PHASE TRANSFORMATIONS IN ULTRADISPERSED BORON NITRIDE UNDER THE CONDITIONS OF HIGH PRESSURES AND TEMPERATURES

N.I. POLOUSHIN
Dr, Prof.
High Temperature Materials Department, Moscow State Steel and Alloys University
Leninsky pr., 4, Moscow, 117936, Russia
Tel: 7 (095) 236-70-85
Fax: 7 (095) 236-21-05
E-mail: htm@diamond.misa.ac.ru

I.I. BAIRAMGOULOV
Research Scientist
NPO "SOYUZ"

V.I. CHOUKALIN
Dr, Senior Research Scientist
Institute of Modern Chemical Problems at Russian Academy of Sciences

V.N. TROITSKIY
DSc., Director of
Institute of Modern Chemical Problems at Russian Academy of Sciences

Phase transformation in boron nitride was first reported by F.P. Bandy and R.H. Wentorf [1]. According to their data, the pressure, necessary for the formation of the cubic modification (cBN) from the graphite-like modification (hBN), exceeds 10 GPa. Further investigations showed that phase transformations in BN could be realized at lower parameters as well [2-5]. What is more, the pressure and the temperature are not the only factors, which determine the mechanism and kinetics of the process. Physicochemical properties of the starting layered structure of BN, such as the degree of crystal perfection, the degree of dispersion, the existence of additions are of great significance, as well as the scheme of the high-pressure apparatus cell assembly (HPA) and some other factors.

In this paper the results of the research of the phase transformation in the ultradispersed boron nitride (u/d BN), obtained by the method of plasmochemical synthesis [6], are presented. Depending on the synthesis parameters, it is possible to

M. A. Prelas et al. (eds.), Diamond Based Composites, 115–119.

obtain the ultradispersed boron nitride powders with the specific surface from 700 to 100 m^2/g, that corresponds to an average particle size from 4 to 15 nm. The diffractograms of the powders contain two diffusion reflections (002) and (10), reflection (10) being asymmetric. The interplanar interval is d_{002}=0.350 nm.

The powders of the ultradispersed boron nitride consist of the separate particles of irregular form (the electron-microscope research was carried out using the equipment TESLA BS 540).

Phase transformation in u/d BN was studied in the pressure interval of 5.5-7.7 GPa and in the temperature interval of 1070-1670 K. The experiments were carried out using a "toroid" type high-pressure apparatus (HPA). The desired temperature was set up in accordance with a preplotted calibration curve. The pressure in the HPA was determined directly during each experiment in accordance with the points of polymorphic transformation in Ba (5.5 GPa) and Bi (7.7 GPa). Phase composition (qualitative and quantitative) of the test pieces after the thermobaric treatment was detected using the diffractometer DRON-2. Two types of HPA assembly were used (fig.1).

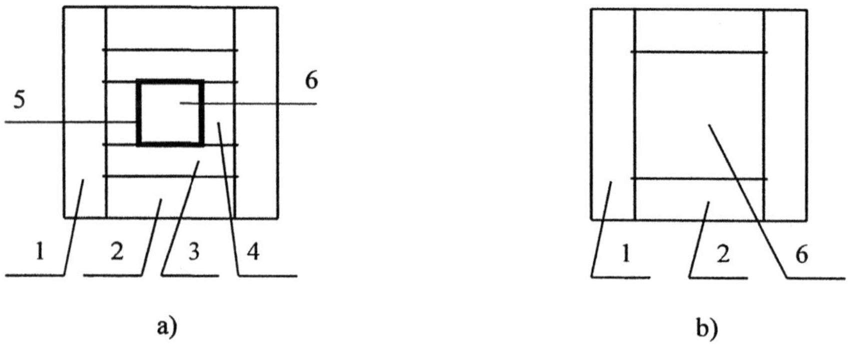

Figure 1. Scheme of the assembly of the HPA reaction cell

1 - graphite heater
2 - graphite plates
3 - plates of hBN
4 - tube of hBN
5 - ampoule of Ta foil
6 - tablet of u/d BN

Accordingly to the data of the spectrum analysis, the assembly (fig.1a) allows to seal hermetically the u/d BN tablet and prevent it from the penetration of the container and heater material.

To avoid the effect of air humidity and oxygen on the ultradispersed boron nitride, the work with the powder and the assembly of the containers were conducted in a dry cell (drying agent P_2O_5) in argon atmosphere. The u/d BN tablets were produced in a press mold of an appropriate diameter.

While the thermobaric treatment of the u/d BN test pieces (P=7.7 GPa), the intensive recrystallization of boron nitride starts at the temperatures above 1270 K. In the test pieces diffractograms obtained under these conditions, the peak (002) corresponds to the highly ordered hBN. The peaks (100) and (101) fully separate.

Table 1 shows the results of examining the phase transformations in u/d BN while thermobaric treatment in relation to the average sizes of the particles. The experiments were carried out at the following parameters: P=7.7 GPa, T=1670 K, τ=5 min. The columns 3,5 present the data on the analysis of the phase transformations in the u/d BN test pieces, prepared in a dry cell. The columns 4,6 present the analogous data for the test pieces of the ultradispersed boron nitride kept in the air for 24 hours before the thermobaric treatment.

TABLE 1. The dependence of cBN output upon the average sizes of the starting u/d BN particles and on the parameters of the experiment

#	Average size of u/d BN particles, nm	cBN output, % mass			
		Assembly 1a		Assembly 1b	
		unoxidized	oxidized	unoxidized	oxidized
1	4	20	100	-	100
2	5	20	-	60	100
3	6	-	100	-	100
4	7	30	-	-	-
5	8	28	85	66	93
6	9	26	-	-	-
7	10	26	-	65	85
8	11	24	57	-	-
9	12	-	-	64	-
10	15	18	40	60	85

As it can be seen from the presented data, the degree of phase transformation in the case of the assembly 1b is sufficiently higher than in the case of assembly 1a, all other conditions being equal. Obviously, that happens because of protecting the reaction space from the penetration of the additions from the container and the heater, which catalyze the phase transformation hBN→cBN.

In the case of using the "unoxidized" u/d BN powder (columns 3,5), the effect of the particles sizes on the phase transformation degree is weakly distinct. The obtained dependence needs the further research.

"Pre-oxidation" of the starting powder (columns 4,6), other conditions being equal, leads to the sufficient increase of cBN output. The degree of the phase transformation increases with the decrease of the average sizes of u/d BN particles. In our opinion, the cause is a higher degree of "oxidation" of smaller-sized u/d BN powders with larger specific surface. It is known [7] that the presence of B_2O_3 stimulates the formation of BN cubic modification from the graphite-like modification under conditions of high pressures and temperatures.

The effect of the u/d BN "oxidation" process on the tendency to the phase transformation in cBN was examined by the authors. The experimental data are presented in the table 2. The u/d BN powder with the average particle size of 8 nm was used. The experiments in HPA were carried out at the following parameters: P=7.7 GPa, T=1670 K, τ=5 min.

TABLE 2. The dependence of cBN output upon the period of the preliminary keeping the u/d BN in the air

#	Time of pre-oxidation, days (24 hours)	cBN output, % mass
1	unoxidized	28
2	1	90
3	2	80
4	4	80
5	8	80
6	30	80

The results presented in the Table 2 show that the increase of the period of keeping the u/d BN powders in the air leads to the increase of the cBN phase output in the case of a lasting keeping up to 1-2 days. The further increase of the period of keeping u/d BN powders in the air does not result in increasing the cBN phase output in the process of thermobaric treatment. Obviously, a fine protective film of B_2O_3 is forming on the surface of u/d BN particles and protects those from the further oxidation, or the process of oxygen and moisture vapours adsorption is coming to an end, while oxygen and moisture vapours form B_2O_3 or H_3BO_3 in the process of thermobaric treatment, activating the phase transformation in BN.

In comparison with the completely crystallized hBN, anomalously low P,T-parameters of cBN formation were observed. In accordance with the data of [7], where the effect of the starting graphite-like phase dispersity on the process of cubic boron nitride formation was examined, the following dependency exists: the smaller the granulars of the staring phase are, the more intensive are the processes of phase transformations. The authors' explanation is that the cBN phase growth begins on the edges of the granulars and hBN twins and spreads in the range of its individual particles. That is why the tendency to the direct phase transformation in cBN increases with the increase of the starting hBN dispersity.

In our opinion, the formation of boron nitride cubic modification at low P,T-parameters when using u/d BN can be additionally explained as follows. In [6] is shown that an effective pressure of approximately 4.5 GPa has an effect upon the ultradispersed BN particle, and this pressure causes the intralayer reduction. Under these conditions, the formation of the fragments with the tetrahedral boron and nitrogen links coordination is possible in the structure of u/d BN particles, which is characteristic for the close modifications of BN. Apparently, these fragments in the ultradispersed boron nitride structure at high pressures and temperatures can be the nuclei of the cBN phase. The existence of the boron nitride cubic modification nuclei in u/d BN can have an additional effect on the decrease of the pressure and temperature values, necessary for the phase transformation realization.

References

1. Bundy, F.P. and Wentorf, R.H. (1963) Direct transformation of hexagonal boron nitride to denser forms, *Journal of Chemical Physics*, **38**, **5**, 1144-1149.
2. Wakatsuki, M., Ichinose, K. and Aoki, T. (1972) Synthesis of polycrystalline cubic boron nitride, *Mater. Res. Bull.*, **7**, **9**, 999-1004.
3. Ichinose, K., Wakatsuki, M. and Aoki, T. (1976)) Synthesis of polycrystalline cubic boron nitride, *Pressure Eng. Jap.*, **14**, **1**, 26-35.
4. Corrigan, F.R. and Bundy, F.P. (1975) Direct transitions among the allotropic forms of boron nitride at high pressures and temperatures, *Journal of Chemical Physics*, **63**, **9**, 3812-3820.
5. Sumiga, H., Iseki, T. and Onodera, A. (1973) High pressure synthesis of cubic BN from amorphous state, *Mater. Res. Bull.*, **18**, **10**, 1203-1207.
6. Tchukalin, V.I., Tchukanov, N.V. and Gurov, S.V. (1988) Structure peculiarities and ultradispersed boron nitride infra-red spectra of absorption, *Powder Metallurgy*, **1**, 85-91.
7. Kurdiumov, A.V., Oleinik, G.S. and Ostrovskaya, N.F. (1979) The influence of the starting graphite-like phase dispersity on the structure of cubic boron nitride polycrystalls, *Powder Metallurgy*, **9**, 90-95.

COMPOSITE MATERIALS BASED ON CUBIC BORON NITRIDE: STRUCTURE AND PROPERTIES

V.B.SHIPILO, N.G.ANICHENKO, I.M.STARCHENKO, AND E.M.SHISHONOK

Institute of Solid State & Semiconductor Physics Academy of Sciences of Belarus, Minsk, Belarus

Physico-mechanical properties of the composites based on cubic boron nitride (cBN) depend essentially on a phase composition, thermal expansion coefficients, degree of dispersion, and interaction of their mixture components in the region of thermodynamic cBN stability. The presence of high- melting compounds in composite materials makes it possible to increase composite plasticity, to decrease the processes of recrystallisation and inverse phase transformation cBN-gBN (graphite-like boron nitride)as well as cBN grinding upon mixture compression [1-3].

The present authors have sintered composite materials for 15 sec in containers, made of lithographic stone, at the pressure of 5 GPa and at the temperatures of 1100, 1250. 1400, 1550, 1630, 1700, 1780 and 1850 ^{0}C. A diameter and a height of the sintered samples were 9 and 5 mm, respectively. The initial burden consisted of the mixture of cBN powders (60-95 wt, %) based on finely dispersed powders of aluminium and titanium nitrides.

X-ray studies of the samples were conducted using a diffractometer with CuK_α radiation. Vickers microhardness (H_V) was determined with the aid of PMT-3 microhardness gauge at the indenter load of 1.5N, a crack viscosity coefficient (K_{IC}) was evaluated by the length of radial cracks due to identation of Vickers indentor at the load $2H$: $K_{IC} = 0.16 \cdot H_V \cdot \sqrt{a}\,(c/a)^{-3/2}$, where H_V is the Vickers hardness (kg/mm^2), a is the identation amplitude (mm), c is the crack length (mm) [4]. IR spectra of reflection from the samples polished up to the roughness of grade 11 were recorded in the spectrum 400 - 1600 cm^{-1} using a " Perkin-Elmer 180 " spectrophotometer with the 2 cm^{-1} resolution. The measurement accuracy did not exceed 0,5 cm^{-1} for frequency and 1% for intensity.

The obtained *X*-ray data on a phase composition of the investigated samples are indicative of the absence of interaction between high-melting components (cBN, AlN,TiN) and of the formation of the mixture at the above sintering temperatures (T_S). No changes have been observed in the cBN lattice parameter determined by a diffraction peak (331). However, as far as sintering temperature increases, the spectra of all samples, irrespectively of their composition, show a specific change in the *X*-ray

M. A. Prelas et al. (eds.), Diamond Based Composites, 121–126.

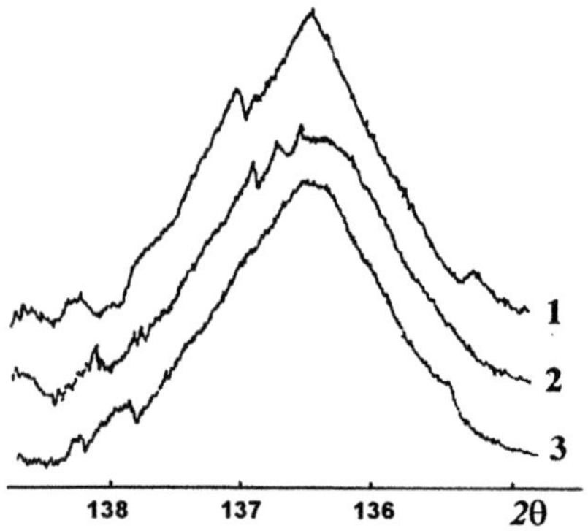

Figure 1. Profile of the X-ray line (331) of cubic boron nitride in the composite samples sintered at the pressure of 5 GPa and different temperatures: 1 -1100, 2 - 1630, 3 - 1850 °C

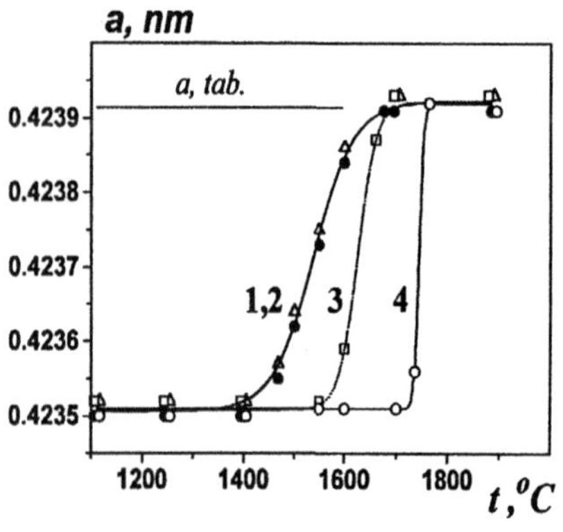

Figure 2. Lattice parameter of titanium nitride versus the sintering temperature at the pressure of 5 GPa in the composite samples of different compositions (wt.% cBN): • - 60; Δ - 70; ◻ - 80; o -95

diffraction profile (331) of cBN (Figure 1).Over the low-temperature sintering section (1100 - 1600 °C) this profile is asymmetrical (spectrum 1,2), while over the high-temperature section (1600 - 1850 °C) it is more symmetrical (spectrum 3). An intensity and halfwidth of the diffraction peaks in the cBN spectra almost do not change with a sintering temperature. *X*-ray analysis has not detected graphite-like boron nitride. For aluminium nitride, no pronounced changes in the angular position and intensities of *X*-ray lines have been revealed.

Unlike cubic boron nitride, the lattice parameter (*a*) of titanium nitride (Figure 2), determined by the diffraction peak (422), has changed on different sintering temperature sections in a different manner. Though it does not depend on the composition of the samples and the sintering temperature over a low-temperature section, however its value equal to 0,4235 nm is lower than the tabulated value (a_0 = 0,42395 nm). On the high-temperature section, the TiN lattice parameter increases with T_S and attains its tabulated value. With an increase of the cBN amount in the composites, the lattice parameter of TiN is found to more drastically increase due to composite sintering at higher temperatures. Thus, for the composites with a minimum cBN concentration (60 and 70 wt.%) the lattice parameter of TiN starts to increase at T_S =1400 °C and attains its maximum at T_S =1700 °C, while for the compositions with a maximum cBN concentration (80 and 95 wt.%) the limiting values of this temperature are 1550 and 1700 °C, respectively.

We may speculate that the observed changes of the lattice parameter of titanium nitride are attributed, on the low-temperature section, to formation of a solid-state solution based on TiN (TiN_xO_y or $TiC_xO_yN_2$) [5] and on the high-temperature section to decomposition of the solid-state solutions. This is evidenced by a substantial increase of the relative intensity ($I_T / I_{1100\,^oC}$) of the *X*-ray peak (422) and a decrease of the relative half-width of the lines ($W_T / W_{1100\,^oC}$) for all compositions of the composite material on the latter section.

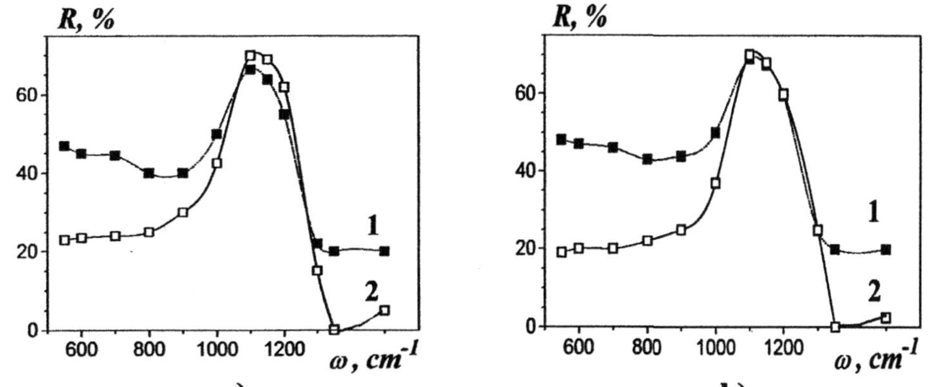

a) b)

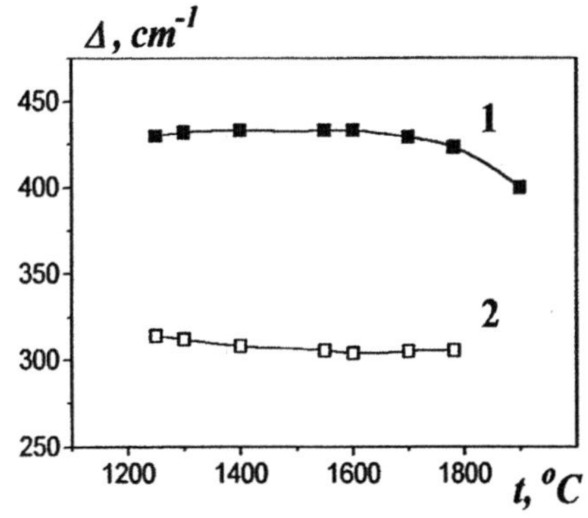

c)

Figure 3. IR reflection spectra of the samples with 60 wt.% cBN (curves 1) and 95 wt.% cBN (curves 2) sintered at T_S =1250 (Fig.3a) and 1850 ^{0}C (Fig.3b); halfwidth of the IR reflection band in spectra of the samples with 95 wt.% cBN (curve 1) and 60 wt.% cBN (curve 2) versus the sintering temperature (curve 3c)

Figures 3a and 3b show the reflection IR-spectra of the samples with minimum (curve 1) and maximum (curves 2) cBN concentrations sintered at T_S =1250 (Fig.3a) and 1850 ^{0}C (Fig.3b). Figure 3c exhibits the sintering temperature dependence of the band halfwidth in the IR reflection spectra the samples with the maximum (curve 1) and minimum (curve 2) cBN concentrations.

These spectra are typical for cBN and represent a wide structureless band in the frequency range 900 ÷ 1350 cm^{-1}. The band intensity and its frequency maximum are almost independent of a sintering temperature. However, with an increase of the filler amount in the samples an intensity maximum of the spectra displaces relatively a frequency toward the longwave region (max by 25 cm^{-1}) thus testifying either to formation of the solid-state solution based on A_3B_5(BN + AlN) or to a decrease of the oscillator strength , the parameter of the dispertion model for cBN [6]. In this case, a background of the spectrum increases strongly and, as a consequence, a value of the limiting reflection at the longwave and shortwave boundaries of the spectrum increases simultaneously by, approximately, 20%. It is known [6] that in the one-oscillator model for cubic boron nitride the equality $(w_l/w_t)^2 = \delta_S / \delta_\infty$ holds where w_t, w_l are the frequencies of transverse and longitudinal optical phonons, δ_S and δ_∞ are the state and high-frequency dielectric constants. Using the parameters of the dispersion model calculated in the present work to a first approximation, we have found that the above equality does not hold for the investigated composite samples:

$(\delta_S / \delta_\infty)_{comp.} > (\delta_S / \delta_\infty)_{cBN}$. In view of this fact, the investigated IR-spectra of the samples with a high concentration of the filler should be considered as a superposition of the IR-spectra of cBN, AlN [7] and TiN [8].

In the samples with high concentration of the filler a halfwidth of the IR-reflection bands is weakly dependent on a sintering temperature, while in the samples with a low filler concentration it markedly decreases on the high-temperature sintering section. Such a change in the halfwidth of lines at their different intensities and unchanged position of the maximum is indicative of an increase of the oscillator strength or of the contribution of fully adequate dipoles (an increase of their amount) to substance polarization under IR-interaction conditions [6]. The mentioned behavior of the parameter of the dispersion model points to a possibility of perfecting the cBN structure in the high sintering temperature range by changing a stoichiometric composition of a material.

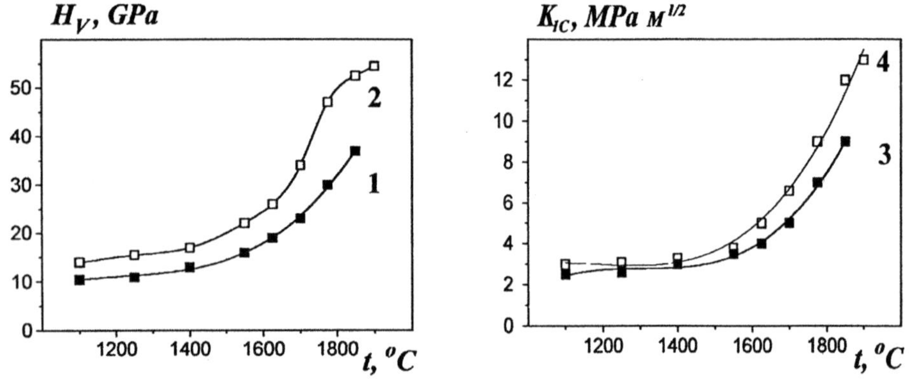

Fig.4 Sintering- temperature dependence of the microhardness (curves 1,2) and the crack viscosity coefficient for the samples with 60 (curves 1,3) and 95 wt.% c BN (curves 2,4)

Figure 4 shows the dependencies of microhardness (curves 1,2) and a crack viscosity coefficient (curves 3,4) of the composites on a sintering temperature and a composition of the samples. As seen, with an increase of T_S and the cBN concentration in the samples, the above characteristics increase and attain the highest values at the maximum T_S and the minimum filler content. For instance, at T_S =1850 ^{0}C for the 60 wt.% cBN composition H_V =37,5 GPa, K_{IC} = 9 MPa·m$^{1/2}$, while for the 90 wt.% cBN composition those are 55.0 GPa and 12 MPa·m$^{1/2}$, respectively. The samples with different compositions have intermediate H_V and K_{IC} values.

It is known [9,10] that under pressure sintering conditions with an increase of T_S ($T_S \geq$1400 K) link formation between grains in cBN due to plastic strain of the material, annealing of nonequilibrium defects and postdeformation recovery by immobilization of dislocations by vacancy complexes occur which leads to narrowing of *X*-ray lines and improvement of mechanical characteristics of polycrystals.

This mechanism seems to be predominant in sintering of the investigated composite since with an increase of T_S the improvement of cBN structure perfection in the absence of phase interaction as well as an increase of a degree of direct contact of separate cBN grains as well as an increase of H_V and K_{IC} have been detected. As judged from the microstructure analyses, a concentration of the contacting grains increases with a decrease of a filler content.

We can suppose that a filler can be a regulator of a stress level in cBN grains, from the one hand, and a nitrogen source, from the other hand. Therefore its relatively low concentration in the samples will cause an increase of the local inhomogeneous stresses between cBN grains and, consequently, favour the process of nitrogen atom diffusion into a volume of a cBN grain by, for instance, the vacancy mechanism.

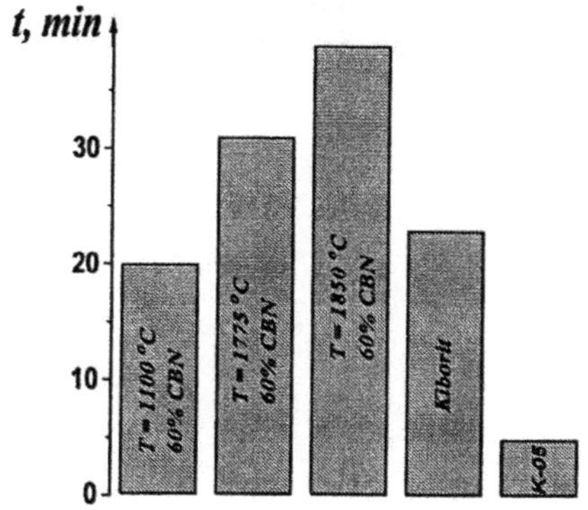

Figure 5. Diagram of wear resistance of different composites based on cubic boron nitride for cutting of hardened steels (HRC = 50)

In order to optimize cutting properties of the obtained composite material with respect to the composition of the initial charge and sintering conditions, the material has been tested as blades for cutting of hardened steels.

Turning was accomplished with the velocity V=180-200 m/min, at the cutting depth t = 0,2 mm with the feeding rate s = 0,05 mm/rev. to steels grade Sh ×15 (FSU) hardened to HRC =50.

The results obtained and comparative tests of the known tool materials based on cubic boron nitride allow a conclusion to be made that the composite material developed for tool purposes outperforms the "Belbor" and "K-05", "Kiborit" composites produced in the FSU with respect to the cutting properties and confirm the best foreign analogs (Amborit, Sumiboron, etc.) (see Figure 5).

References

1. Trotsyuk, A.V. (1984) An influence of aluminium nitride dopants on sintering of cubic boron nitride powders, in "Preparation of Superhard Materials Investigation of Their Properties and Application", Proc.of the Institute of Superhard Materials, Ukr.Academy of Sciences, Kiev, .345-359
2. Gromyko , S.N., Ostrovskaya, N.F., Pilyankevich, A.N., and Frantsevich,. I.V. (1986) Some features of the structure formation of a boron nitride matrix in the composite-12, Fiz.Tekh.Vys.Davlen. 22, 59-64
3. Shulzenko, A.A., Bozhko, S.A., Sokolov, A.N. et.al. (1993) Sintering of cubic boron nitride powders in producting tool materials , in " Cubic Boron Nitride:Synthesis, Sintering and Its Properties", Naykova Dumka, Kiev, .95-146
4. Novikov, N.V., Dub, S.N., and Bulychov (1988) Methods of crack resistance microtests, Zavod.Lab. 54, 7, 60-67
5. Gogotsi, Ju.G., Grigor`ev, O.N., and Tikush, V.L. (1988) Structure condition and

mechanical properties of the material in the system sialone-titanium nitride after radiative heating,, Poroshkov.Metallurg. 5 , 60-66

6. Ukhanov, Ju.I. (1977) Optical Properties of Semiconductors, Nauka, Moskva
7. Pasternak, J. and Hegda, B. (1969) Infrared absorption of AlN in the one-phonon region and finite effects, Phys.Stat.Silidi 35, .941
8. Vlasov, A.G. (1972) Infrared Spectra of Inorganic Glasses and Crystals, Nauka, Leningrad
9. Shishonok, N.A., Shipilo, V.B., and Anichenko, N.G. (1990) in "High. Pressure Effect on Properties of Materials" , Institute of Semiconductor Materials, Ukrain. Academy of Sciences, .36-42
10. Novikov, N.V., Bondarenko, V.P., Kocherzhinssky, Ju.A., et .al .(1985) Investigation of plastic deformation of cubic boron nitride, Sverkhtverd. Materials 2, 17-20

COMPARISON BETWEEN DC AND RF MAGNETRON SPUTTERED ALUMINUM NITRIDE FILMS

C. MOROSANU, V. DUMITRU, ELENA CIMPOIASU
Institute of Physics and Technology of Materials,
P.O.Box Mg7, Bucharest, Romania
CRISTINA NENU
Zecasin S.A., Bucharest, Romania

C-axis oriented and disordered aluminum nitride layers have been deposited by RF, respectively DC magnetron reactive sputtering mode. The same deposition conditions were used for gas flow and pressure. The differences between the two growing modes are explained by the influence of ion bombardment upon the growing layers, which is stronger in the RF mode in comparison with the DC one, at the same deposition rate.

1. Introduction

Aluminum nitride films have been deposited by various authors on amorphous and oriented substrates using the DC or RF deposition modes. Amorphous to highly oriented layers have been reported, depending on the deposition conditions. The quality of the layers is very important for applications like SAW devices [1].

In the present work aluminum nitride films were deposited using the magnetron sputtering reactive method. A comparison between the RF and DC sputtered layers is done from the point of view of growth mode, c-axis oriented or disordered. In order to understand this, plasma probe measurements were performed.

2. Experimental details

The experimental work was done in a YBN-75P-I russian deposition system, equipped with an oil diffusion pumping system. The magnetron cathode (home-made) with a plasma ring of about 55 mm, was used in the RF (1.78 MHz) mode, as well as in the DC one. The target was an aluminum (99.99%) 10 cm diameter disk. First, the sputtering chamber was evacuated below $3*10^{-3}$ mTorr and then spectral argon (99.999%) and nitrogen (99.995%) gases were admitted through needle valves, the gas flow being measured by rotameters. The total pressure was obtained by adding the partial pressures of argon and nitrogen measured with an ionic gauge (PMI-2),

M. A. Prelas et al. (eds.), Diamond Based Composites, 127–132.

calibrated for nitrogen. The argon partial pressure was obtained as a ratio of the indicated pressure and 1.4, the calibration factor for argon [2]. Three total pressures were used, 1.4, 2.0 and 3.0 mTorr, for total gas flows of 35 respectively 50, 50 sccm. For all the depositions a 30% ratio of nitrogen to total flow was used. A presputtering more than 30 min. was done in the same conditions like the deposition. A day before deposition, the substrates were outgased in vacuum by heating them at about 200 °C with an infrared lamp, and they were etched in an argon plasma for 3 min. before deposition to improve the adherence of the AlN layers. The substrates, glass or Si(100), were not heated during the deposition and the substrate temperature was only dependent on the self-heating of plasma.

In the following the samples will be labeled AAx.xBB, where AA stands for the substrate, glass (GL) or Silicon (Si(100)) and BB for the deposition mode (RF or DC). x.x is a number representing the deposition pressure in mTorr. The deposition conditions for all the samples are displayed in Table 1. A regime of constant electrical forward power P, which was not measured, was used for the RF deposited layers.

Optical transmission measurements were performed by means of a Specord M40 spectrophotometer. We have estimated the thickness of the layers deposited on glass from the interference fringes in the visible range. For the layers deposited on silicon we have considered the same value of the thickness as for the layers on glass deposited in similar conditions.

The structure of the films was investigated by X-ray diffraction. A DRON-2 diffractometer provided with cooper target tube was used.

The plasma in the proximity of the substrate was investigated using a rectangular mobile plasma probe.

Tabel 1. Deposition parameters and characteristics of some AlN thin films

Sample	*Thickness (nm)*	*Deposition rate (nm/min.)*	*Electrical power (W)*	*(002) peak normalized intensity*	*FWHM*
GL1.4RF	910	15.1	P	4	0.4
GL2.0RF	905	15.0	P	100	0.27
GL3.0RF	1080	10.5	P	11	0.34
GL2.0DC	790	12.1	79	12	-
GL3.0DC	900	8.75	70	-	-
Si1.4RF	905	15.0	P	2	0.48
Si2.0RF	970	16.1	P	48	0.35

3. Results

3.1 STRUCTURAL RESULTS

We have measured the X-ray diffraction diagrams in similar conditions for all the aluminum nitride samples, in order to make comparison between the intensities of the diffraction peaks of the various samples. Two qualities of layers have been obtained, c-axis oriented normal to substrate surface and non-oriented layers. For the oriented ones, the (002) peak was carefully measured and a value of the full width of half the maximum intensity (FWMH) has been estimated. In order to compare the (002) peak intensities for films of different thicknesses between 800-1000 nm, a normalized intensity was used for 1000 nm layer thickness, considering that the X-ray diffraction intensity is proportional with the layer thickness.

3.1.1 RF deposited layers on glass

The X-ray diffraction pattern for the RF aluminum nitride films on glass revealed only the (002) hexagonal plane, which indicates that these layers have a strong c-axis oriented phase. For the GL2.0RF layer, the X-ray diffraction diagram measured in the center of the substrate, is displayed in Fig.1. A low value of 0.27° for the full width at half maximum (FWHM) was obtained. From Fig.2 one can see that at 2.0 mTorr the normalized value of the (002) peak intensity is maximum, while the value of the FWHM is minimum. This behavior at 2.0 mTorr might be explained by a maximum of crystalline phase percent in the layer and also by a strong orientation of the crystallites with c-axis normal to substrate. A rocking curve measurement and a transversal SEM image is necessary in order to clarify this.

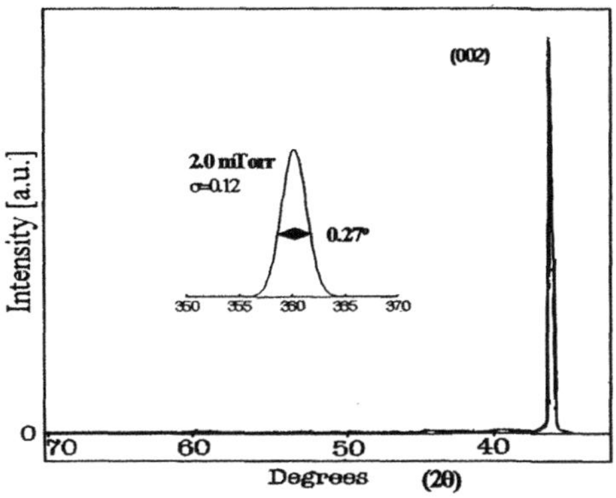

Degrees (2θ)

Fig.1 X-ray diffraction pattern for GL2.0RF in the substrate center

We have also checked the uniformity of the growth mode of the AlN layers on glass surface, by doing diffraction measurements at 0, 10 mm and 20 mm distance from the substrate center. The layers remain c-axis oriented on the entire substrate surface, but the (002) peak intensity is strongly decreasing from the center to the border (Fig3).

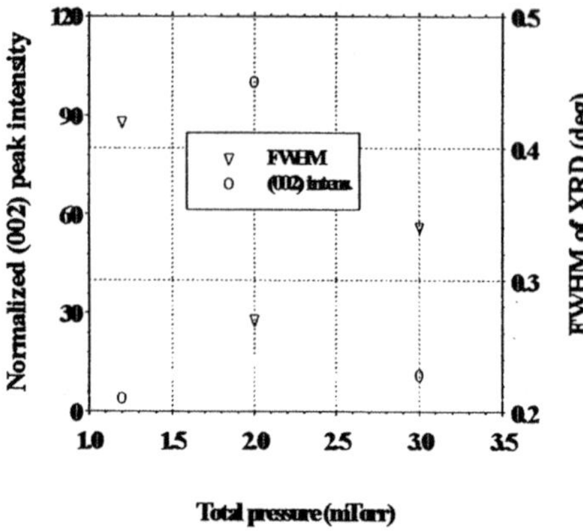

Fig.2 Dependence of (002) peak intensity and FWHM vs. deposition pressure

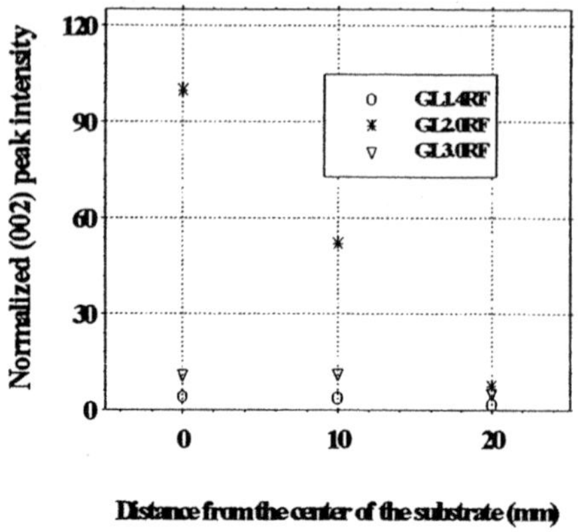

Fig.3 Variation of (002) peak intensity vs. distance to substrate center

3.1.2 DC deposited layers on glass

The DC layers deposited on glass, GL2.0DC (Fig.4) and GL3.0DC (Fig.5), have revealed a disordered structure in the center of the substrates. For the GL3.0DC, X-ray diffraction measurements in the range 33°-38° , were also done at distances (d) of 10, 20, 30 mm from the substrate center (Fig.5). From the center to the border, the angle between the c-axis and the substrate normal is increasing, at a 30 mm distance from the center the (002) peak intensity being practically zero.

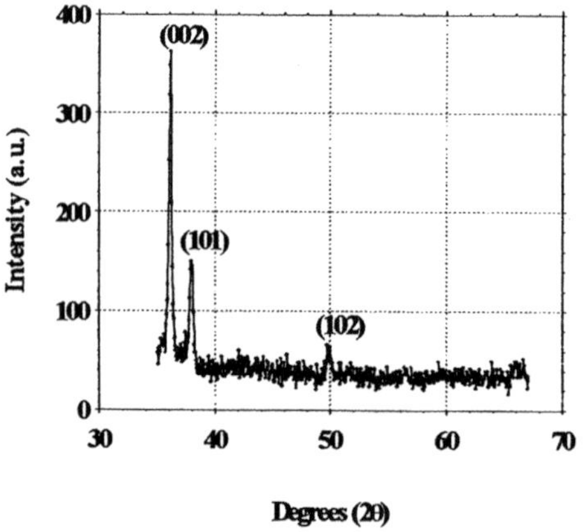

Fig.4 X-ray diffraction spectra for GL2.0DC sample

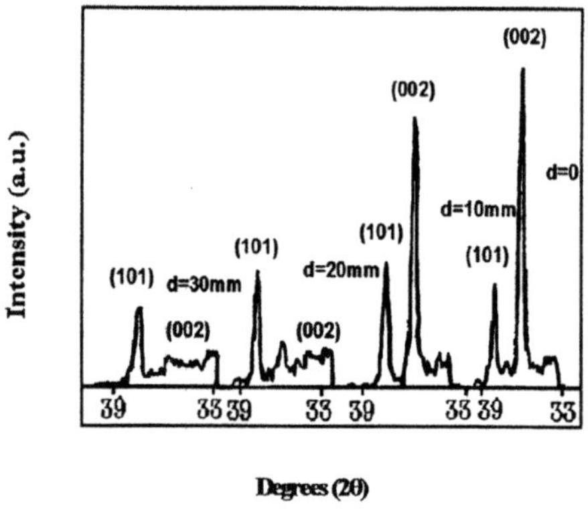

Fig.5 X-ray diffraction patterns in various substrate locations for GL3.0DC sample

3.1.3 RF deposited layers on Si(100)

C-axis oriented aluminum nitride layers have been deposited on Si(100) using the RF deposition mode. From Fig.6, one can see the same tendency of decreasing the crystallization of the layers by increasing the deposition pressure for glass and silicon.

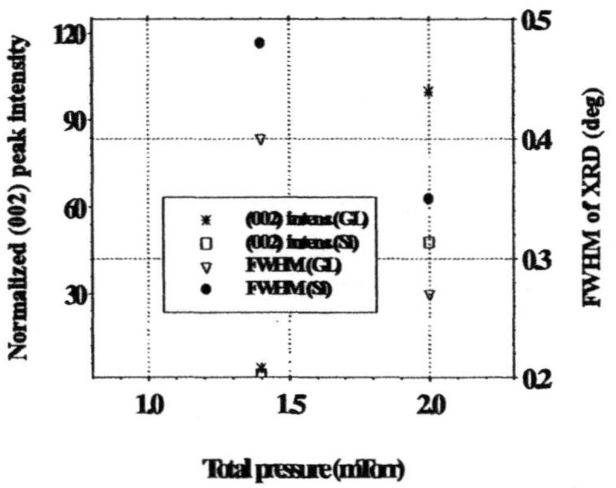

Fig.6 Comparison between AlN layers on glass and Si(100)

3.2 PLASMA PROBE MEASUREMENTS

It is known that the structure of the AlN layers depends on the bombardment with ions during deposition [3]. In order to investigate it, we replaced the substrate with a rectangular mobile plasma probe of a 0.8 cm² area, which could be moved along a vertical line passing through the substrate center. We have measured the current drawn by the probe at a negative bias of 40V respect to ground, which is an indication of the positive ion density in front of the substrate. We have sputtered the aluminum target only in argon atmosphere using similar sputtering conditions of pressure and electrical power with that used for depositing AlN layers.

3.2.1 Dependence of plasma current vs. deposition rate

Fig.7 displays the dependence of the probe current density measured in the center of he substrate vs. aluminum deposition rate, at a 3.0 mTorr argon pressure. One can notice that for the same deposition rate the value of the probe current is three

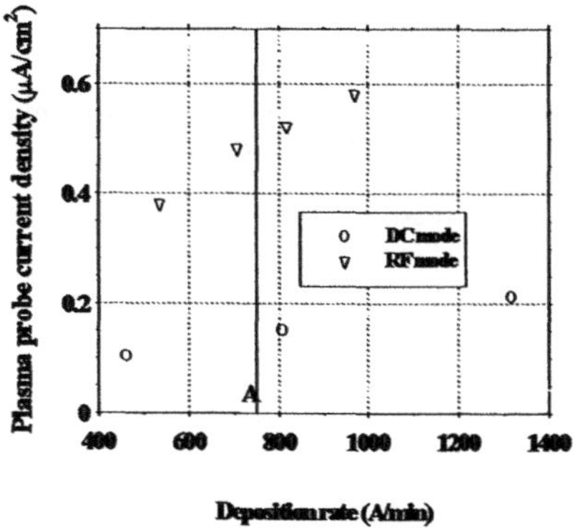

Fig.7 Dependence of plasma current *vs.* deposition rate

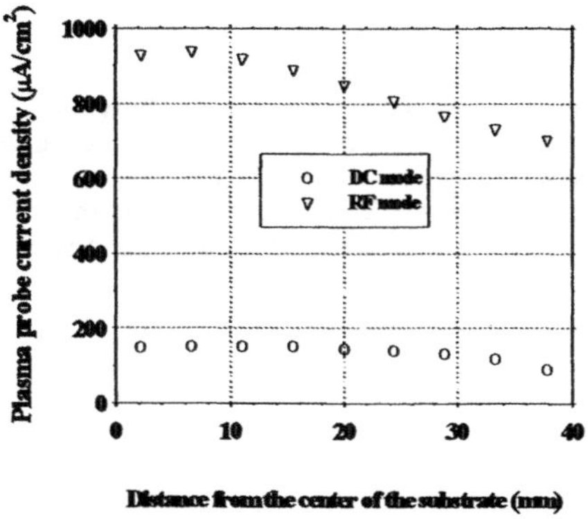

Fig.8 Dependence of plasma current density vs. distance from the center of the substrate

to four times higher in the RF mode compared to the DC one. In the point A, corresponding to a value of electrical power on the target of 80 W in the DC mode, the ratio is about 3.5.

3.2.2 Dependence of plasma current vs. distance from the substrate center

Fig.8 shows the values of plasma current density measured at various points on the substrate surface, for RF and DC sputtering regimes of different deposition rates. It was obtained a continuous decreasing of the probe current from the center to the border of the substrate, similarly for the two deposition modes.

4. Conclusions

Layers of various qualities from c-axis oriented to disordered, have been prepared by reactive sputtering without substrate heating. We investigated the dependence of some AlN thin-film properties on sputtering mode, total pressure and substrate quality. Results were as follows.

(1) The layers sputtered in the RF mode have a better crystalline orientation in comparison with that deposited in the DC one.

(2) There was obtained a pronounced maximum of (002) peak intensity and a minimum value of the FWHM at a pressure of 2.0 mTorr in the RF mode.

(3) The quality of the layers RF deposited on glass and Si(100) is similar.

(4) We have noticed a degradation of the crystalline orientation from the center to the substrate border for the RF and DC samples deposited on glass.

(5) Differences between the DC and RF growing modes are explained by differences between the ionic bombardment upon the substrates during deposition.

Acknowledgments

The authors wish to thank to Dr. C. Popescu for helpful discussions.

References

1. Okano, H., Takhashi, Y., Tanaka, T., Shibata K. and Nakano, S. (1992) Preparation of c-Axis Oriented AlN Thin Films by Low-Temperature Reactive Sputtering, *Jpn. J. Appl. Phys.* 31, 3446-3451.

2. *Edwards Vacuum Products* 1984, pg. 83.

3. Tanaka, N., Okano, H., Usuki, T. and Shibata, K. (1994) Preparation of aluminum nitride epitaxial films by electron cyclotron resonance dual-ion-beam sputtering, *Jpn. J. Appl. Phys.* 33, 5249-5254.

ALUMINUM NITRIDE COMPOSITE FILMS

B.V. SPITSYN[1)], A.F. BELYANIN[2)], L.L. BOUILOV[1)], A.N. BLAUT-BLACHEV[1)], V.P.STOYAN[3)]

1) Institute of Physical Chemistry RAS, 31 Leninsky Prospekt, 117915 Moscow, Russia
2) "Technomash", 4 Ivana Franko Str., 121355 Moscow, Russia
3) Chemical Department, Moscow State University, Vorobjovy Gory, 119899 Moscow, Russia

Abstract

The paper is concerned with thermodynamical, structural and kinetic factors responsible for the formation of aluminum nitride composite films. As such we consider AlN films deposited or grown on the substrates with nonidentical chemical composition and structure made of single crystal (Al2O3, 6H-SiC, Si), polycrystalline (polycrystalline diamond films, silicon nitride ceramics) and amorphous (glassy carbon, quartz glass) materials. The peculiar features of AlN film formation and growth depend on the deposition technique used (RF magnetron sputtering and CVD), as well as on the gas phase composition and temperature of the process. It was shown that the purity of CVD film grown by thermal decomposition of NH_4AlCl_4 and NH_4AlBr_4 vapors in Ar flow is strongly affected by the chemical nature of the substrate and the reactor wall materials. The stoichiometric composition may vary from $AlN_{0\cdot92}$ for the chloride to $AlN_{1.03}$ for the bromide process. Heteroepitaxial AlN films on the sapphire and 6H-SiC substrates were grown only by CVD technique.

1.Introduction

Two - and multilayer film can integrate the functional capabilities of individual film materials and also exhibit extra functional properties of the composing material which lack in the individual film components. Besides, such film, which we call the composite film, is an object relatively easy to obtain and to investigate experimentally, as compared to the ordinary composite materials. From other side the composite film, when used as examplifying model, can give an opportunity to study the early and/or relatively simple processes which determine synthesis and behaviour of composite

M. A. Prelas et al. (eds.), Diamond Based Composites, 133–149.

materials. Among such composite films, the ones based on diamond and on other wide band gap semiconductor materials are of major interest [1].

2. The aim of the study

The aim of the study was the possibility of combining AlN films grown by different techniques with the use of materials of various chemical composition and of different crystalline and band structure.

Taking into account the relative deficiency of the works which deal with AlN composite films, we found it appropriate to make some preliminary thermodynamical calculations (See sect.4) which enabled us to determine the potential reactivity of AlN in relation to the used substrate and reactor wall materials. To determine the most appropriate and actual conditions for thermodynamical equlibrium calculations, not only the kind of substrate and reactor wall materials were taken into account, but also the typical conditions of AlN film growth, which will be considered in the next Section.

3. Experimental

For obtaining AlN films two techniques were used.

3.1. RF magnetron sputtering

RF magnetron sputtering of high purity (99.99%) aluminum target was performed in the medium of mixed nitrogen and argon.

The magnetron sputtering permits to grow highly ordered, close to single crystalline AlN films at substrates made of crystalline and amorphous materials at low (100-300^{0}C) temperature.

AlN films by RF magnetron sputtering of Al in Ar - N_2 gas mixture have been deposited. We used specially designed laboratory and also commercial plants (UVN - 62P-3, UVN - 014, KATOD - 1M and LND - Plazma), with specially constructed inside deposition cameras of planar and cilindrical magnetron sputtering systems adapted for AlN film growing in the plants RF generators (13.56 MHz) and output pawer 1 to 4 kW was used.

Typical parameters of magnetron sputtering system listed in the Table 1.

Table 1. Typical parameters of magnetron sputtering system

Parameter	Value
Preliminary vacuum	$7 \cdot 10^{-4}$Pa
Gas composition	Ar + N_2 (40:60)

Gas pressure	0.6 Pa
Target	Al (99.99%)
Substrate temperature	300 ^{0}C
Target-substrate distance	90 mm
RF power	0.6 -1.5 kW
RF voltage	300 V
Bias on substrate holder	+25 to +40 V
Growth rate	2.5 mm/h

3.2. CVD system

In CVD technique for AlN film deposition the source of aluminum and nitrogen were ammonium tetrahaloaluminates (NH_4AlHal_4, where Hal - Cl or Br). Initial tetrahaloaluminates were obtained by interaction at 310^{0}C equimolar amounts of the corresponding anhydrous aluminum halide and ammonium halide. After subsequent fractional distillation small amounts of pure product were keeped in sealed glass ampoules.

The chemical analysis of initial substances confirms its theoretical chemical formulas with accuracy of 0.05 wt% [2].

NH_4AlCl_4 and NH_4AlBr_4 was evaporated in the flow of pure Ar (99.993% of basic substance at about 300^{0}C in the reactor setup. Usual content of initial substance in Ar flow at 2.7 liter per minute was about and less 1 vol %. Ammonia tetrahalohaluminate vapors decompose in reaction zone at about 1000^{0}C according to the reaction :

$$NH_4AlHal_4 \text{ -------> } AlN_s + 4\,HHal$$

The substrates for AlN film regrows were flat parallel plates with the size of the order of 1*10*10 mm. AlN film was deposited on one of the external surfaces of the substrate which was typically specular - smooth. The thickness of the film was measured by two - beam interferometer and/or by the weighting at microbalance.

4. Thermodynamical calculations

As follows from the previous Section, the growth of AlN composite films occurs in the conditions (especially in CVD technique) when one should or contrary should not (depending on the results of thermodynamical calculations) take into account a set chemical interactions of the solid AlN. The latter include the interaction and with other (besides AlN) solid phases which constitute the material of substrates, holders and reactor walls.

The above interactions are interesting in two aspects. Firstly, if the "solid - solid" type interaction between AlN and solid material of the substrate is possible, this will promote the formation of adhesive bonding between them, which is, in turn, the necessary condition for the formation of the based on AlN film composite.

Secondly, one cannot neglect a series of interactions which are possible at the "gas medium - solid phase" interface, including other solid phases or materials present in the working zone of the reactor. Additionaly to interactions, presented at Fig.1 thermodynamically allowed interactions:

$$SiO_2 + 2/3AlCl_3 = SiCl_2 + 1/3Al_2O_3 + 1/2O_2 \quad (9)$$
$$AlCl_3 = AlCl + Cl_2 \quad (10)$$

can affect the primary process of AlN film formation by doping of AlN by silicon and oxygen, and also hinder, through oxidation or etching, the formation of perfect adhesive bond at the interface between the film and the substrate. It is obvious that the process of this type will be particularily essential on early stages of formation of discontinuous island AlN films, when surface bulk ratio is particularily large and the role of uncovered by AlN surfaces of the substrates and reactor walls is significant.

There are numerous chemical interactions which are interesting from the above point of view. Calculations according to CHET program were made for the temperature range 600 - 1600 K (Fig.1 and 2). The calculations were made with no account of possible mutual influence of different interactions and of formation of solid phases with variable composition, as like $Al_xN_yO_z$.

Using the known expression $Kp = \exp(-DG_T^0/RT)$, which relates the term ΔG_T^0 with equilibrium constant (Kp), we calculated Kp values for the reactions (2) and (3) and the equilibrium composition of gas phase, and after this the thermodynamical yield of aluminum nitride $<Y>=(N/N_0)\cdot 100\%$ (where N is the number of AlN moles formed at equilibrium conditions, and N_0 is the number of moles of initial $AlHal_3$) as a function of temperature and of initial NH_4AlHal_4 partial pressure (Fig.3).

From the graph presented in Fig.2 it follows that reactions of AlN with H_2O and O_2, within the wide range of temperatures, are characterized by essentially larger negative values of Gibbs energy as compared to those for the AlN formation reactions . This can lead to practically complete interaction of AlN with H_2O and O_2 at equilibrium conditions. From Fig.2-3, it is seen also that interaction of AlN with graphite is characterized by a large positive value of $-DG_T^0$ in the wide temperature range. From this it follows in particular that used graphite holders can not to suffer of interaction with AlN.

As it was noted above, at the temperature of solid AlN synthesis the thermodynamical possibility of interaction between aluminum halide and the material of quartz reactor Fig.1-2. For this reaction a negative change of Gibbs energy is characteristic at the vaporization temperature of initial compound and a small positive one - at the temperature of the AlN synthesis.

Thus, the formation of AlN through the reactions (2) and (3) can be performed with high thermodynamical efficiency in the wide temperature (Fig.3) and partial pressures of NH_4AlHal_4 ranges (Fig.4). Our data are in good agreement with those presented in the work [3]. The authors of the latter made thermodynamical calculations using the total Gibbs energy minimization technique for the $AlBr_3$ - NH_3 - H_2 system, on the assumption of ideality of gas phase.

5. Study of the process kinetics

The results of the studies on AlN chemical crystallization rates for fused quartz substrates at different temperatures are presented in Fig.4. It can be seen that the film growth rate changes nonmonotoneously with temperature and is nearly independent of the halogen nature. In carrying out these runs the following regularity was observed: at temperatures of deposition below T_{max}, the growth rate of AlN in the case of chloride system is higher. At temperatures exceeding T_{max} it is higher for bromide system.

The specific temperature dependence observed for the deposition rates of AlN layers (Fig.5) may result from the following causes. Since it is impossible, for the most cases, to measure the film thickness by optical interference technique, we have estimated it by weight method. In doing this it is necessary to take into account that aluminum nitride and the substrate material (SiO_2, sapphire) undergo etching by the reaction products (HHal mainly), which will lower the weight gain calculated value as compaired to the real thickness of the grown films. Therefore only in the temperature range up to T_{max}, an exponential rise of the deposition rate is observed.

Above the maximum temperature growth rate decreases, since the etching process affects significantly on the substrate and AlN film. Besides, as the temperature rises the role of homogeneouus formation of AlN increases, thus lowering the potentially attainable rate of heterogeneous growth of the film. It is important to note that homogeneously formed particles of AlN may be captured by growing AlN film, reducing its quality. By these reasons it is preferable to grow good quality composite film, including AlN, at the temperatures lower than that of the maximal growth rate for AlN film.

The variation of Gibbs energy for the interaction between AlN and HHal is positive (see Fig.1). However, for HCl $-DG_T^0$ will be somewhat lower as compared with that for HBr , i.e. in HCl etching will be thermodynamically more probable. In fact, at temperatures exceeding T_{max} the measured rate lower for the chloride system. Below T_{max} the formation of AlN in the chloride system occurs with higher rate than in the bromide system. Apparently, at these temperatures the growth rate for low - temperature chloride tends to be higher, due to the lower energy of Al - Cl bond, than that for the Al-Br.

Using the temperature dependences of AlN deposition rates in lnV - ln(1/T) coordinates shown in Fig.6, the energy of activation E_{exp} was estimated as 90+4 kJ/mole for the chloride and 92+4 kJ/mol for the bromide systems. The values of activation energy obtained for AlN growth from NH_4AlCl_4 agree with the earlier ones 100 kJ/mole [4] and 70 kJ/mole [5].

In addition to above consideration, we calculated the number of collisions (Z) of gas particles with the substrate surface, according to [6], which amounted to $7,6*10^{19}cm^{-2}s^{-1}$ (at NH_4AlCl_4 flow rate of 0,01 mole/h, argon flow of 2.7 l/min and the temperature of 1000^0C). On the assumption of 100% efficiency of each collision the maximal possible growth rate of 5,9*104 min/h have been calculated. The growth rate observed experimentally under these conditions was about 22um/h. However, not all impacts are effective since only the particles with energies of E>Ea react. Therefore the activation energy, calculated independently from equation V=Zexp(Ea/RT), amounted to 85 kJ/mole. This value is in good agreement with experimentally determined Ea values.

6.AlN film characterization

6.1.Stoichiometry

Despite of measured by chemical analysis aluminum to nitrogen ratio is close to stoichoimetric, the specific deviations for bromide and chloride systems have been observed. Aluminum nitride free standing polycrystalline films was prepared by two versions of CVD - chloride and bromide process at temperature of about 1273 K. For the analysis standard Kjeldal procedure have been used. Average contents of the films in chloride and bromide process was determined are equal to $AlN_{1.03}$ and $AlN_{0.92}$, respectively, with uncertainly equal to $\pm$ 0.5% [2].

6.2. X-ray diffraction

X-ray study of specimens of AlN thin films with about 5 μm thickness deposited by CVD on the sapphire substrates and quartz glass was performed. In optimal growth conditions highly textured (0001) normal to substrate films on quartz glass and single crystalline (heteroepitaxial) films on Al_2O_3 and 6H-SiC was obtained. Crystallite sizes was from 14 to 110 nm in dependence of crystallization conditions was changed.The larger figures are the characteristic usually for higher crystallization temperature.

6.3. Reflection high energy electron diffraction

Reflection high energy electron diffraction (RHEED) demonstrate that single crystalline (heteroepitaxial) films grown at 1223 - 1273 K on single crystal Al_2O_3 and 6H-SiC was obtained.

6.4. Optical properties

By IR - spectroscopy the deposition temperature influence on structure and chemical contents of the films was demonstrated. Maximal structural perfectness of the films at 1273 K was established. Substrate nature (diamond, sapphire, silicon, silicon carbide, glassy carbon, quartz) seriously influenced morphology and quality of the films.

Band gap was determined by UV-spectra is equal to 6.1 eV (according to literature data $E_{optical}$ is 6.0 to 6.2eV) .

6.5 Heterojunction sharpness

Fracture through the boundary of the textured AlN film obtained by CVD technique on single crystalline Si subsrates was investigated by scanning electron microscope (SEM). The observation in SEM demonstrate with resolution about 0.1 mm that substrate - film bondary do not contain voides and third phase innclusions.

6.6.AlN film morphology

Thich AlN films (about 20 and more mm) have not preferential crystallite orientation if grown at Si - waters. But several mm thick AlN films, grown on α-Al_2O_3 with the same CVD conditions.

7. Electronic devices based on AlN films

AlN is a piezoelectric which is held to be one of the most prospective material for creating of surface- acoustic - wave (SAW) devices for electronics. It is characterized by: low insertion loss, high coupling (0.8%), large SAW velocity (~6.103 m/s). When α - Al_2O_3/AlN layer sound duct is used for fabricating interdigital transducers (IDT) by ordinary contact printing, the signal - processing frequency of ~1GHz is achievable. Both AlN films grown by epitaxial and by sputtering techniques [5,7,8] can serve as piezoelectric layer. Sometimes piezoelectric characteristics of RF sputter - deposited AlN films exceed the analogous parameters of epitaxial films [9].

By using metal substrates together with dielectric crystalline AlN films, the heat - spreading multi - layer substrates are fabricated with characteristics superior to those for (highly toxic) BeO based ceramics. It give to AlN films and layers good praspective for e.g. high power resistor elaboration.

AlN is used as oxidation resistive and protective coating for unpackaged circuits. For instance, polycrystalline AlN films are employed to protect thin - film heating elements and conductors of thermoprinting devices from oxidation and mechanical damage. The life time of an impact thermoprinting matrix protected by AlN

- based coating is by a factor of ~10^3 larger then that for Al_2O_3 or SiO_2 coated ones [10]. Besides, AlN is a potential protective coating for the measuring optical devices.

It is promising to use AlN in microwave devices as a buffer layer (0,4 - 0,6 μm thick) on the fused quartz substrates. The deposition of AlN buffer layer enables to preclude the detachment of conductors without inserting losses the stripping signal over the 8 mm range.

8.Diamond/AlN layer structures.

Diamond and AlN are the materials which posess somewhat similar properties; e.g. high hardness and thermal conductivity. Both materials are the wide band gap semiconductors which can work at severe circumstances.

In the Table 2 exposed similarities and dissimilarities in properties and pequliarities of diamond and aluminum nitride, what may predetermine the opportunities of its application in two - and multifilm composite structures.

Table 2. Diamond and AlN similrities and dissimilarities

SIMILARITIES			DISSIMILARITIES		
BEHAVIOR OR PROPERTY (UNITS)	DIAMOND	AlN	BEHAVIOR OR PROPERTY (UNITS)	DIAMOND	AlN
Band gap (eV)	5.5	6.2	Type of crystal	elemental	compound
Microhardness ($kg\cdot mm^{-2}$)	$1\cdot10^4$	$2\cdot10^3$	Bonding	covalent	covalent plus ionic
Thermal conductivity at 300 K ($W\cdot m^{-1}K^{-1}$)	2100	280-320	Atoms belong to period of Periodic system	same	different
Debye tem- repature (K)	1860	1150	Possible electronic state of atoms	sp^3, sp^2, sp^1	sp^3
Difficulty in the shallow level doping	n-type	p- and n-type	Shortest interatomic distance (Å)	1.54	1.89
Breakdown voltage	$2\cdot10^7$	$8\cdot10^6$	Structure type	sphalerite	wurtzite

($V \cdot cm^{-1}$)					
Oxidation resistance (K)	<873	<1073	Centre of symmetry	yes	no
Thermal stability in neutral atmosphere (K)	2000*)	2500**)	Piezoelectricity	no	yes
Radiation stability	high	high	Stability at 1 atm	metastable	stable
Opportunity of NEA***)	yes	yes	Native oxide	no	yes
			Interband transition	nondirect	direct

*) graphitization

**) dissociation

***) NEA - negative electron affinity

The closeness and peculiar features of diamond and AlN enable to use them in electronics in the form of layer structures. Zone of promising area for diamond and AlN layered structures are the fabrication of micorowave acoustoelectronic devices and the heterojunction semiconductor units. Some estimations show that the most fitting application field for polycrystalline diamond layers in electronics is SAW acoustoelectronics [11]. The advantage of diamond/AlN layer structures over other sound ducts in fabrication of microwave SAW devices is related to the rise of the sygnal - processing frequency by 1,5 - 3 times, size and shape of IDT being the same.

We have fabricated filters and delay lines based on W(Si)/diamond/AlN/IDT layer structures. AlN films were grown on polished (surface roughness R<0.04 mm) and unpolished after (regrowth surface) of 20 to 50 mm in thickness layers of on polycrystalline diamond deposited on W and Si substrates of size 8x10x1 mm. The diamond layers used for creating SAW devices consisted of crystalline (with typical content of 90 - 100%) and amorphous phases, with the crystalline phase formed by <111> and <110> axially textured grains of size up to 200 μm. In fabricating the substrate/diamond/piezoelectric structure, it is necessary to grow piezoelectrical layer of over polycrystalline diamond. The only possibility to grow a regularly structured film an a polycrystalline substrate is to use the deposition techniques by magnetron - sputtering.

Using the ordinary contact printing to form IDT on Si/diamond/AlN/IDT structure, we have fabricated SAW filters with the operating frequency of 814.9 MHz (k^2 = 0,09; insertion loss in a mismatch regime was 46 dB). With the use of projection optical lithography SAW filters were constructed for the operating frequencies higher than 1,5 GHz [12]. From the measurements of acoustic synchronism frequency, the velocity of the Rayleigh mode for the diamond/AlN sound duct was estimated as 8820 m/s.

Diamond/AlN layer structures have considerable promise for creating p - n heterojunctions in production of electron devices resistant to heat and radiation. In such structures AlN may provide n - type conductivity which is hardly attainable by doping the diamond. We have shown the possibility of manufacturing wide band gap UV source (photovoltage > 1V), based on the boron - doped diamond layers grown by arc discharge technique on Zn - doped AlN films deposited on Si substrates [13].

9. Conclusion

The study performed demonstrated the apartunities of deposition of coherent and nearly stoichiometric AlN films at substrates of single crystal, polycrystal and amorphous nature. Single crystalline (heteroepitaxial) AlN films several mm thickness only at single crystalline substrates (e.g. a-Al_2O_3, 6H-SiC) have been grown.

However just based on polycrystalline and nighly textured AlN film are of great interest for exanepee composite structures for application in acoustoelectronics. Hawe by hewed with elaboration better heteroepitaxial processes and with solution of AlN doping problem the aneas AlN composite film applications should be drastically exteuded.

10. Acknowledgement

The authors are appreciated very much V.B.Kalinin for X-ray diffraction, S.V.Bantsekov for estimating electrical conductivity, A.E.Gorodetsky for carrying out RHEED studies. For providing of SiC substrates we are grateful to our colleagues from Prof. Yu.A. Vodakov's Laboratory.

The work was performed with a partial support of Sandier National Laboratories under the contract AN-8800.

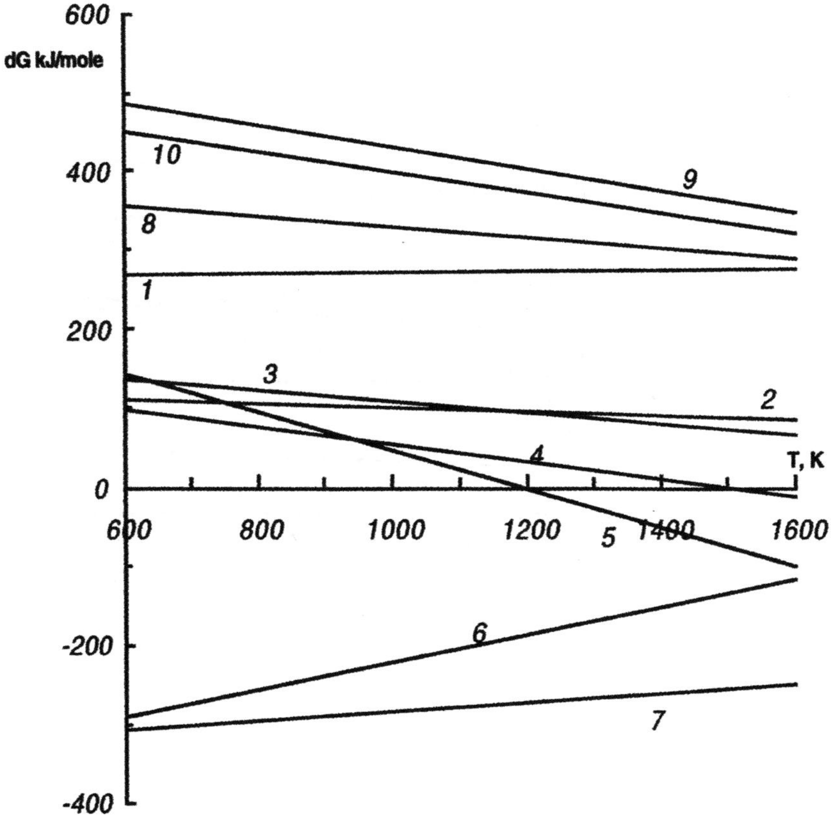

Fig.1. ΔG^{o}_{δ} for reactions:

$$AlN_{(s)} + 3/2H_{2(g)} = Al_{(s,l)} + NH_{3(g)} \quad (1)$$

$$AlN_{(s)} + 3/4CH_{3(g)} = 1/4Al_4C_{3\,(s)} + 3/4NH_{3\,(g)} + 1/8N_{2(g)} \quad (2)$$

$$AlN_{(s)} + 3/8C_2H_{2(g)} = 1/4Al_4C_{3(s)} + 3/8H_{2(g)} + 1/2N_{2(g)} \quad (3)$$

$$AlN_{(s)} + 3/4CH_{3(g)} = 1/4Al_4C_{3(s)} + 9/8H_{2(g)} + 1/2N_{2(g)} \quad (4)$$

$$AlN_{(s)} + 3/8C_2H_{2(g)} = 1/4Al_4C_{3(s)} + 1/4NH_{3(g)} + 3/8N_{2(g)} \quad (5)$$

$$AlN_{(s)} + 3H = Al_{(s,1)} + NH_{3(g)} \quad (6)$$

$$AlN_{(s)} + 3H = Al_{(s,1)} + 1/2N_{2(g)} + 3/2H_{2(g)} \quad (7)$$

$$SiO_{2(s)} + 4/3AlCl_{3(g)} = 2/3Al_2O_{3(s)} + Cl_{2(g)} + SiCl_{2(g)} \quad (8)$$

$$SiO_{2(s)} + 2/3AlCl_{3(g)} = 1/3Al_2O_{3(s)} + SiCl_{2(g)} + 1/2O_{2(g)} \quad (9)$$

$$AlCl_{3(g)} = AlCl_{(g)} + Cl_{2(g)} \quad (10)$$

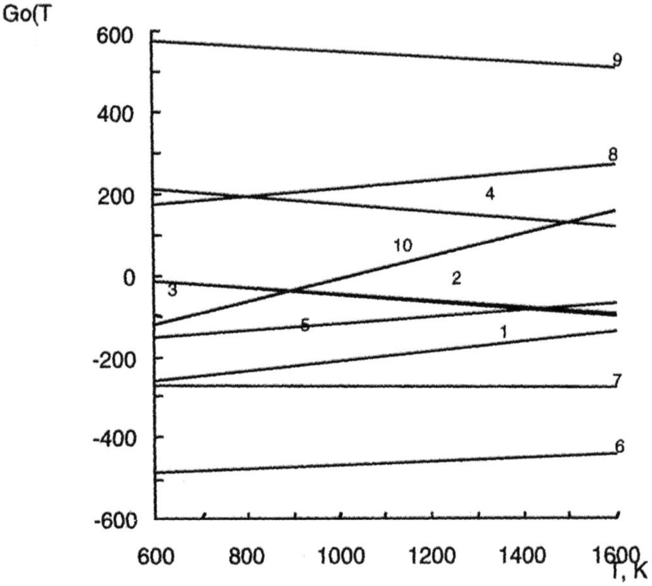

Fig.2. $\Delta G^o{}_\delta$ for reactions 1-10.

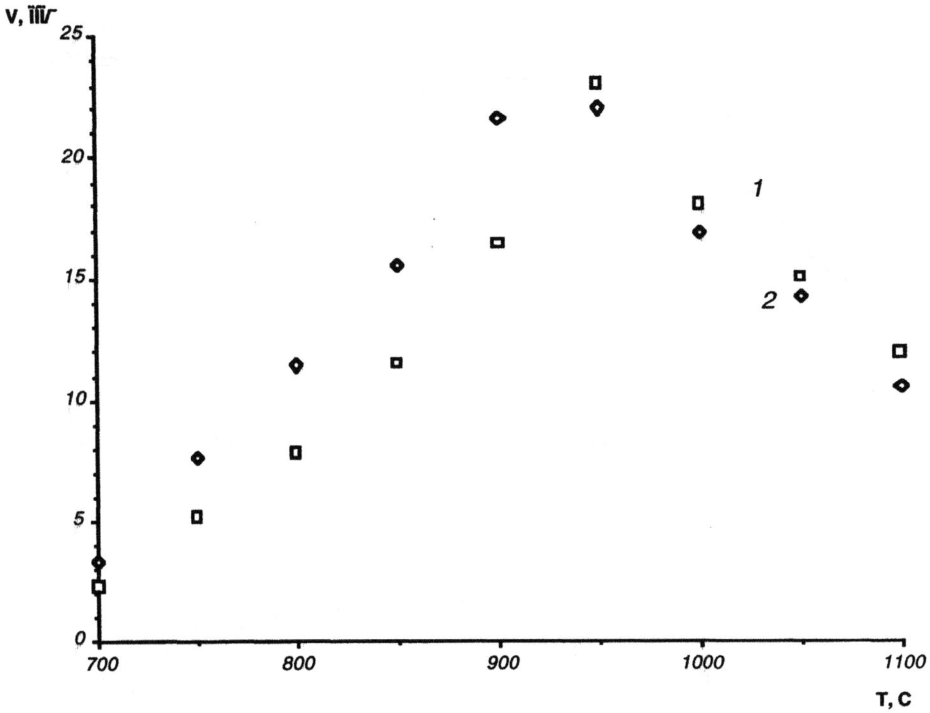

Fig. 3. Yield of AlN as a function of temperature.

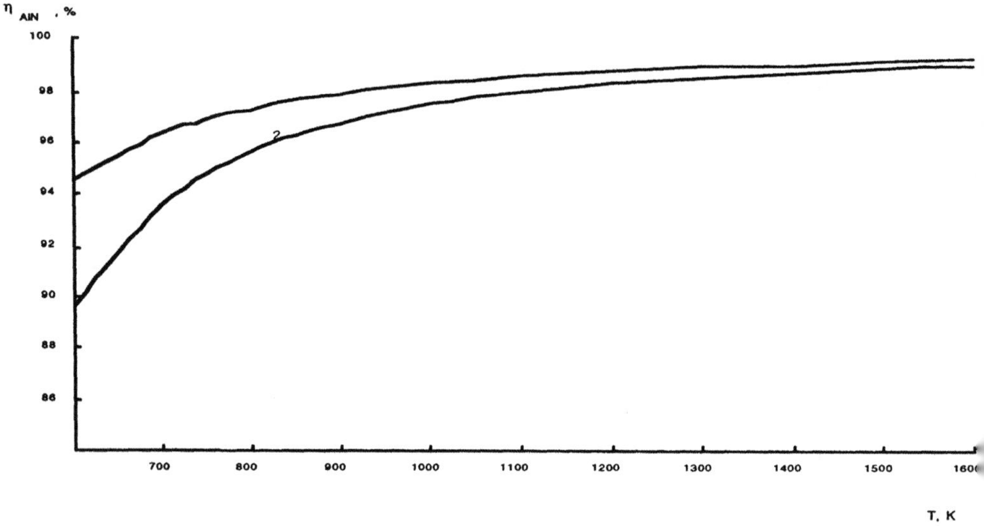

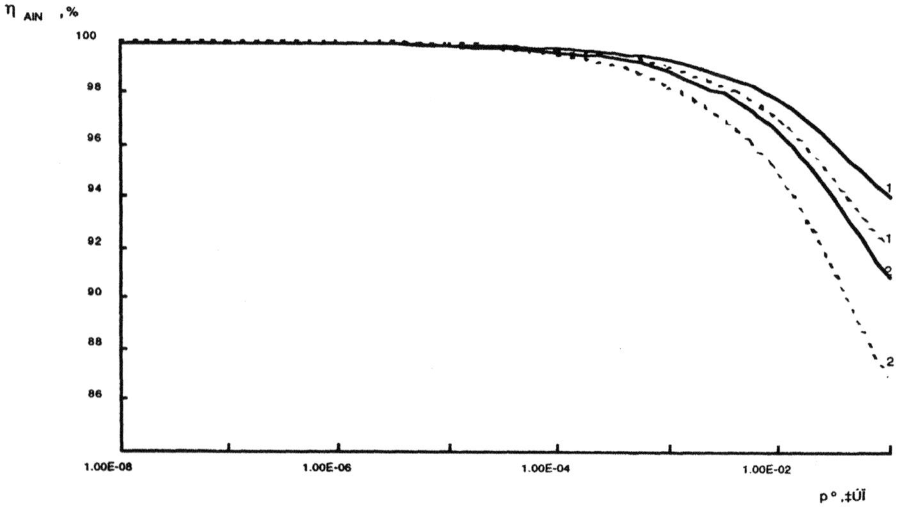

Fig. 4. a) Yield of AlN as a function of temperature b) Yield of AlN as a function of mole fraction of NH_4AlCH_4.

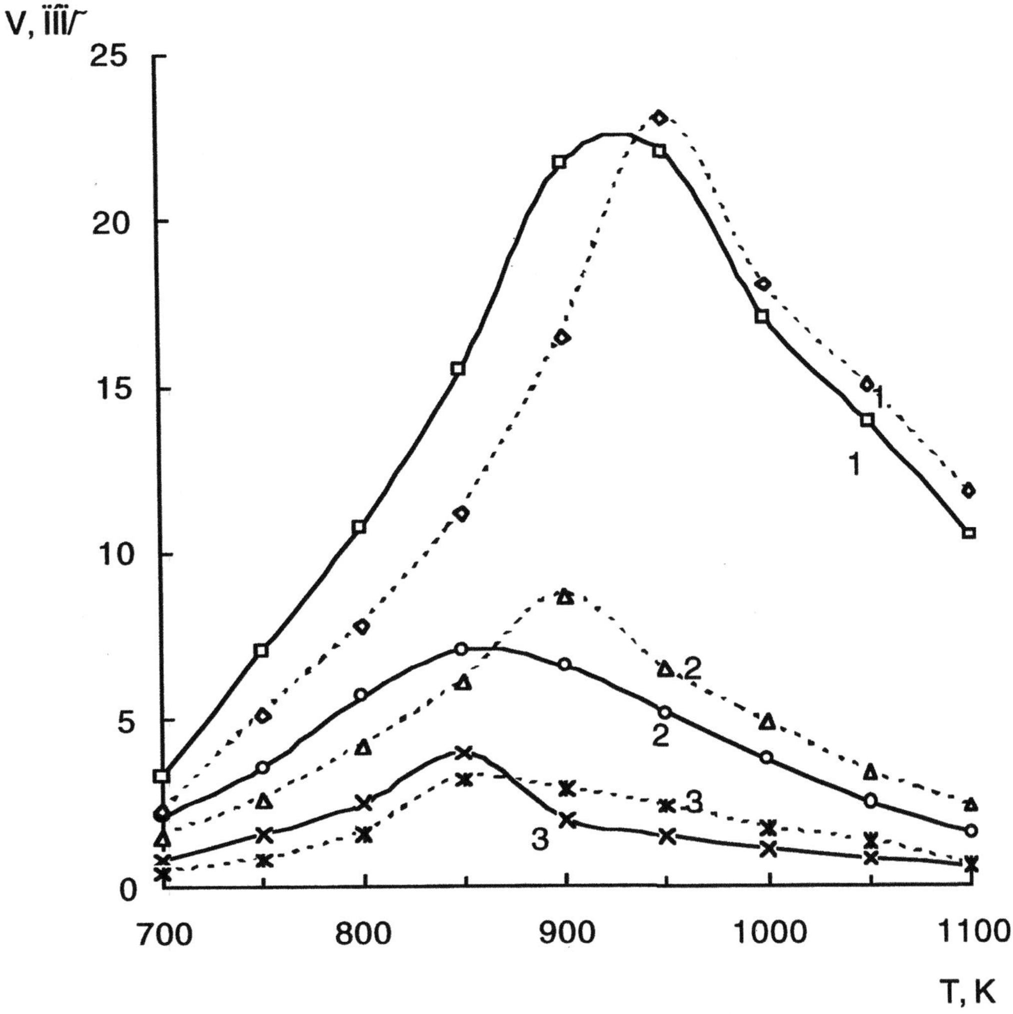

Fig. 5 AlN deposition rate as a function of T.

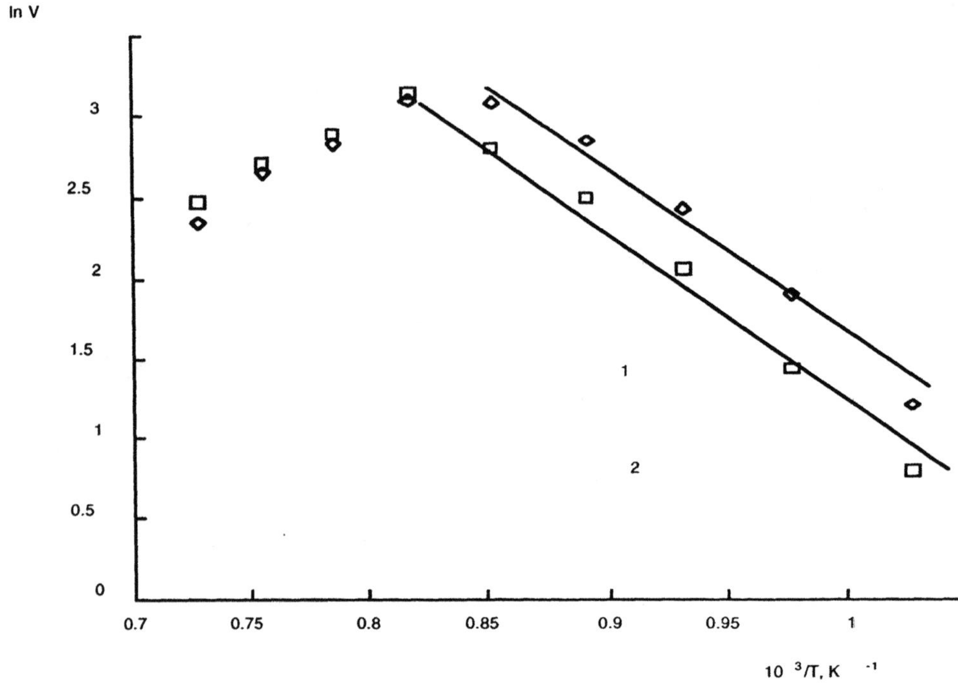

Fig. 6 AlN deposition rate in lnV-ln(1/T).

References

1. Wide Band Gap Electronic Materials. NATO ASI Series. 3. High Technology - Vol.1, 531p.
2. Stoyan V.P., Master Thesis. Moscow State University (1996).
3. T.D.Shermergor, N.N.Streltsova. Piezoelectric films //M.:Radio i Svyaz, 1986, 133pp. (In Russian).
4. Y.Shibata, Y.Kanno, K.Kaya, M.Kanai, T.Kawai. Formation and surface acoustic wave acoustic wave properties of LiNbO3/AlN/Sapphire// Jpn.J.Appl.Phys., 1995, v.34, part 2, N3a, p.L320 - L322.
5. T.S.Shiosaki. Growth and properties of piezo-, pyro-, and ferroelectric thin films// Proc. 6th Conf. "Acoustoelectronics'93", Varna, 1993. S.1, p.41-50.
6. A.F.Belyanin, V.F.Volanski, G.G.Kessernich. Improving of AlN film growth process // Segneto - i piezoelectriki v uskorenii nauchno - technitcheskogo progressa. 1987, M.: MDNTP, pp.51-55.
7. A.F.Belyanin, V.K.Kazarian. Utilization of anodization and RF - magnetron sputtering for fabrication of thin - film thermoprinting matrixes // Pribory i sistemy upravleniya. 199, N6, pp.33-35.
8. T.Nakahara. From business to future dream of new diamond // Advances in New Diamond Science and Technology. MYU, Tokyo, 1994, p.9.
9. A.F.Belyanin, P.V.Pastchenko, L.L.Bouilov, B.V.Spitsyn, N.P.Eliseeva. SAW devices operating at frequencies exceeding 1,5 GHz, based on diamond/AlN layer structure // Fizika i technologiya almaznyh materialov. Materialy VI conf., Moscow, 1996, pp.62-65.
10. I.G.Teremetskaya, V.P.Varnin, V.I.Polyakov, A.V.Khomich, P.I.Perov, N.M.Rossukanyi, A.I.Rukovishnikov, A.F.Belyanin, G.Popovici. Characterization of diamond films grown on AlN/Si substrates and of heterostructures with such films // Applications of Diamond Films and Related Mater.: Third Int. Conf. 1995, USA, p.469-472.

ALLOTROPIC FORMS OF CARBON NITRIDE

A.SOKOŁOWSKA, J.SZMIDT, J.KONWERSKA-HRABOWSKA, A. WERBOWY, A. OLSZYNA, K. ZDUNEK, S.MITURA
Warsaw University of Technology, Koszykowa 75, 00-662 Warsaw, Poland

1. Abstract

On the basis of existing literature and comparison of electric and optic properties of two different materials being the carbon and nitrogen compounds, it was shown that both of them are most probably allotropic forms of hypothetical C_3N_4. Elaborated model of molecular structure of one of these materials is similar to that of chaoite (-C=C-), whereas model of another one shows resemblance to that of diamond. This leads to the conclusion that C_3N_4 is a compound similar to carbon and not to βSi_3N_4, and also, that various C_xN_y materials, which are being mentioned in literature may possibly be allotropic forms of C_3N_4.

2. Introduction

Researches on synthesis of C_3N_4, a compound not occurring in nature, predicted by Liu and Cohen [1] in 1989, analogous to well known βSi_3N_4, have been carried out for 15 years. Alike to βSi_3N_4 [2], C_3N_4 is expected to have single σsp^3 type bonds and crystal structure compatible to cubic system. Therefore, C_3N_4 would be an homologue of diamond, superior to it in respect of mechanical properties due to certain degree of its chemical bonds ionicity, likewise it is in the case of Si_3N_4 and Si.

It should be pointed out however, very characteristic difference between diamond and silicon. The latter exists only in one allotropic form having σsp^3 type bonds, Si_3N_4 whereas occurs in allotropic forms varying only in crystalline structure but not in type of chemical bonding. Carbon on its hand exists in a couple of allotropic forms, depending on type of atomic bonds formed (σsp^1 πp^2, σsp^2 πp^1 and σsp^3 for carbyne, graphite and diamond, respectively). Therefore, it might be expected that also C_3N_4 exists in different allotropic forms, varying in type of atomic bonding, and thus in properties. The goal of this presentation was verification of such hypothesis.

M. A. Prelas et al. (eds.), Diamond Based Composites, 151–159.

3. Carbon nitrides and their properties

There is a rapidly growing number of publications describing attempts of C_3N_4 synthesis. Most of them however, deal with searching for material harder than diamond. That is probably the reason that only few of them bring thorough characterization of chemical bonding of produced material. Table 1 shows an overwiev of so far obtained results in this matter.

TABLE 1. An overview of selected results of carbon nitride synthesis

Ref.	Method	Chemical formula	Multiple bonds	Singular bonds	Hardness	Remarks
[3]	CVD	$(CN)_x$	+	-	soft	polymer-like
[4,5]	Laser abblation in N_2^+ beam	?	+	-	2000 kG/m^2	electron diffr.
[6]	Ion implantation	$C_3N_{0.7}$	+	?	?	
[7,8]	Sputtering	$C_3N_{1.2} \div C_3N_3$	?	?	?	amorphous
[9,10]	I.B.A. PVD	$C_3N_{1.2}$	+	?	6400 kG/m^2	X-Ray diffr.
[11]	I.B.A.D.	C_3N_2	?	?	5560 kG/m^2	X-Ray diffr.
[12]	Carbon dust in N_2 arc	C_3N_4	+	?	brittle	amorphous
[13]	P.P.A. CVD	C_3N_6	-	+	2000 kG/m^2	amorphous

It seems that there should not be attached such importance to the hardness of layers or precipitates being produced, as the real value of this parameter can be easily obscured by properties of simultaneously formed interphase [14], mechanical stress and highly elastic response [15] of fabricated material. Also determination of band gap width (which in itself strongly depends on the way of its evaluation as well as on determination of material structural and molecular perfection) cannot be considered as something unequivocally characterizing material's molecular structure. In addition, it should be also pointed out the fact that produced layers frequently are amorphous and not pure nitride for usually they contain considerable amounts of carbon in a form of various kinds of diamond-like carbon (DLC).

It appears that unequivocal determinant of what kind of chemical bonds have investigated materials are their IR or Raman spectra.

For comparison of different carbon nitrides there were chosen films obtained by means of pulse plasma method under conditions varying mainly in the quantity of plasma energy released in process [13,16]. In general, in this work carbon nitride synthesis was carried out in similar way to DLC synthesis from pulse plasma [17].

In Figure 1 is presented scheme of deposition system, whereas in Table 2 are collected parameters of two used here different carbon nitride layers synthesis processes.

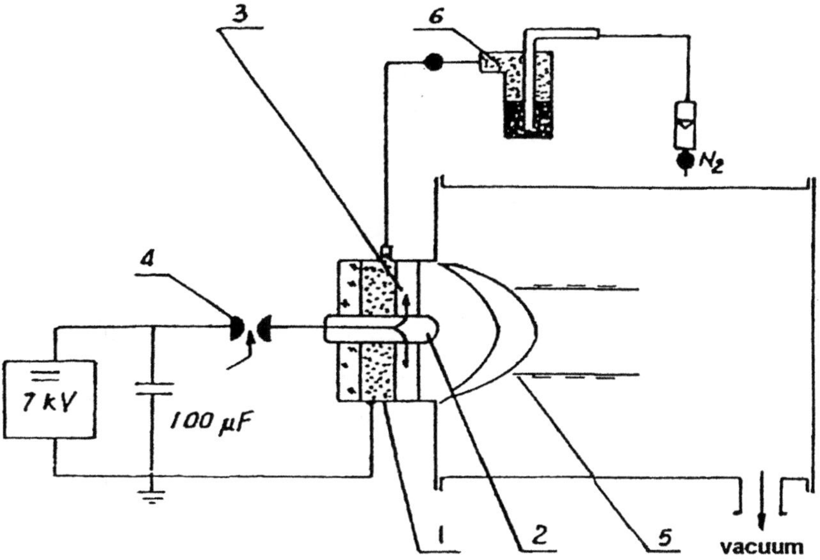

Figure 1. Scheme of deposition system: 1 - external electrode, 2 - internal electrode, 3 - plasma disc, 4 - trigger switch, 5 - substrate, 6 - carbamide.

TABLE 2. Parameters of carbon nitride compounds deposition processes No.1 and No.2

Process	Pressure (Pa)	Gas	Energy released per impulse (J)	Pulse duration (μs)	Internal electrode
Process No.1 (C_5N)	60	N_2	50	100	C
Process No.2 (CN_2)	60	N_2 carbamide	1250	100	Ta

Produced films were investigated in "as deposited" state and then they were isothermal annealed for 3 hours in N_2 atmosphere at the temperature of 600K. Research framework included measurements of: IR absorption spectra in the 400-4000 cm^{-1} wavenumber range (by Perkin Elmer FTIR 2000 spectrometer), UV absorption spectra in the 190-1000 nm range (by Shimatzu UV 2100 spectrophotometer), refractive index values (by Gaertner L116 ellipsometer, λ = 632.8 nm). Carbon to nitrogen atoms ratio was determined by means of RBS method at proton beam energy of 1100 keV. Phase composition of obtained material was determined by means of electron diffraction method (TEM Jeol). Electrical properties were analyzed on the basis of measured electrical characteristics (current-voltage I-V and capacitance-voltage C-V) of MIS capacitors with investigated layers as dielectrics.

All produced films, as deposited, were amorphous and contained small amounts of impurities originating from material of deposition system electrodes. This amorphous structure was also preserved after annealing. Annealing at temperatures higher than 600K and carried out longer than 3 hours, resulted in layers decohesion and their subsequent transformation into powder. Unfortunately, obtained amounts of material were not sufficient to conduct its investigation with use of microcalorimeter and dilatometer.

Investigations of chemical composition of fabricated layers by RBS method allowed determination of their summary chemical formulae as C_5N for material obtained in process No.1 and CN_2 for compound created in process No.2.

Figure 2 and Table 3 present results of investigations of optical properties of produced films.

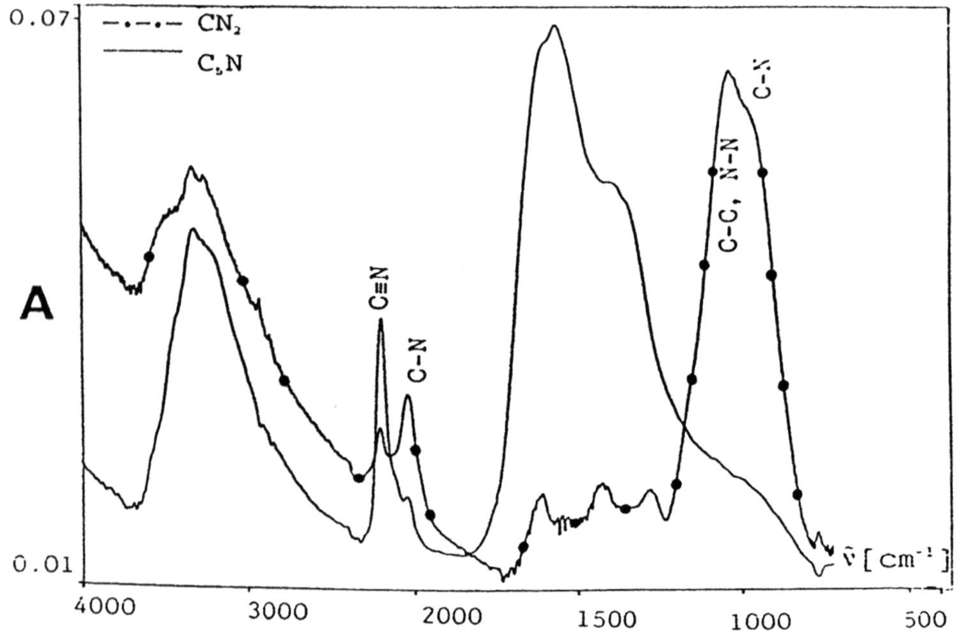

Figure 2. IR absorption spectra of investigated materials.

TABLE 3. Results of investigation of optical properties of materials obtained in processes No.1 and No.2

Determination method	Process 1 (C_5N)	Process 2 (CN_2)
$\alpha^{2/3}$ hν = f(hν)	straight forbidden transitions	straight forbidden transitions
E_g^1 (eV)		
E_g^2 (eV)	2.5	-
E_g^3 (eV)	2.8	2.8
	-	4.25
dN(E)/dE = max		
E_g^5 (eV)	5.0	6.5

The results indicate substantial differences in chemical bonding occurring in both materials. Vibrational spectrum brings the evidence that in C_5N layers C=N and C≡N multiple bonds are predominant, whereas in CN_2 films, basically single C-N bonds occur. In both cases also different band gap widths were obtained, as irrespective of determination method used, band gap values of C_5N material were lower, than of material CN_2 (see Table 3).

There were measured current-voltage (I-V) and capacitance-voltage (C-V) characteristics of metal-insulator-semiconductor (MIS) structures. As substrates *p*-type silicon wafers of resistivity 3÷4 Ωcm were used, onto which were deposited carbon-nitride layers in the course of processes described above. Subsequently on top of these layers circular (ϕ = 0.75 mm) metal (Al) contacts were evaporated.

Prior to electric characteristics measurements, ellipsometric measurements of refractivity of carbon-nitride films were carried out as well as of their thickness, which ranged from 150 to 300 nm. All measurements were performed for "as deposited" as well as for annealed layers. Obtained results are collected in Table 4.

TABLE 4. Results of investigation of electrical properties of obtained materials

Investigated material	*n*	ε_{ri}.	ρ [Ωcm]	I-V curves
C-N (CN_2)	~1.8	3 ÷ 10	$10^6 \div 10^{10}$	asymmetric
C-N (CN_2) annealed	1.6 ÷ 1.7	?	$10^4 \div 10^7$	diode-like
C=N, C≡N (C_5N)	1.8 ÷ 2.2	1.3	$10^3 \div 10^5$	symmetric

Absence of dielectric constant ε_{ri}. values of annealed structures results from the fact that measuring of C-V curves in this case was not possible as characteristics obtained after annealing turned out to be degenerated, probably because of local changes of layers morphology, what could be also seen in metallographic microscope.

Analysis of results presented in Table 4 leads to the conclusion that properties of CN_2 films changed due to annealing, as drops in their refractivity index ε_{ri}. values as well as in resistivities ρ can be clearly seen. Annealing also limited electrical field penetration into layers what resulted in degradation of C-V curves too.

In effect, properties of annealed CN_2 became close to those observed in the case of C_5N films.

Investigation of electrical properties of both CN_2 and C_5N layers show that these materials distinctly vary in mechanisms limiting their bulk as well as junction electric current flows. After "soft" annealing process CN_2 films present intermediate electrical properties.

4. Assumed molecular structure of carbon nitride forms

Numerous simulations were carried out using the CHEM-X software (Chemical Design Ltd., License No.0822), taking into account results of IR spectra measurements performed for investigated compounds. An assumption of the presence of π type bonds only in the C_5N was additionally supported by UV spectra measurements, which clearly showed the differences in the band gap widths between both materials. The CHEM-X computer program enables the two-dimensional space to be easily converted into the three-dimensional, and also allows results to be visualized together with the Van der Waals volumes calculated by it. Therefore, using this software we were able to verify, quickly and easily, various hypothetical molecular and chemical structures of investigated compounds.

Molecular model of the C_5N carbon nitride (process No.1) in the "as deposited" state is shown in Figure 3.

It is a flat, stiff molecule, which, when multiplying itself forms a flat imine chain with cyanide groups joined alternately to it. The ratio of the C to N atoms is here 1:1. In the basic unit there are 6 carbon atoms: 3 of sp^3 hybridization, linked alternately with 3 nitrogen atoms of sp^2 hybridization in the imine type bonds, and 3 of sp hybridization linked with nitrogen by a triple bond in the cyanide groups. Free pairs of nitrogen electrons can stabilize such lattice, which is similar to that of chaoite (a form of carbon having -C=C- bonds) rather than to that of diamond.

Molecular model of the CN_2 carbon nitride (process No.2) in the "as deposited" state is shown in Figure 4.

It is a three-dimensional lattice of only short range ordering, built of carbon and nitrogen atoms linked by single bonds with angles conformable to the structure of sp^3 type orbital. It should be pointed out that this structure does not necessary have to be stoichiometric. Such lattice is similar to that of amorphous diamond.

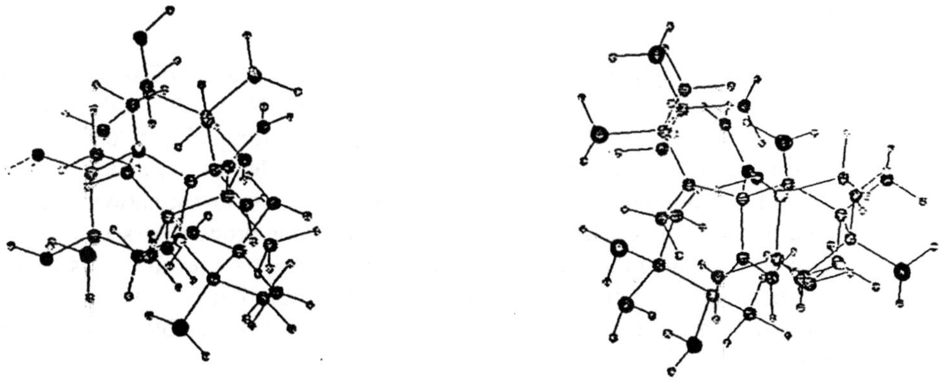

Figure 3. Fragment of the flat polycyanoimine molecule

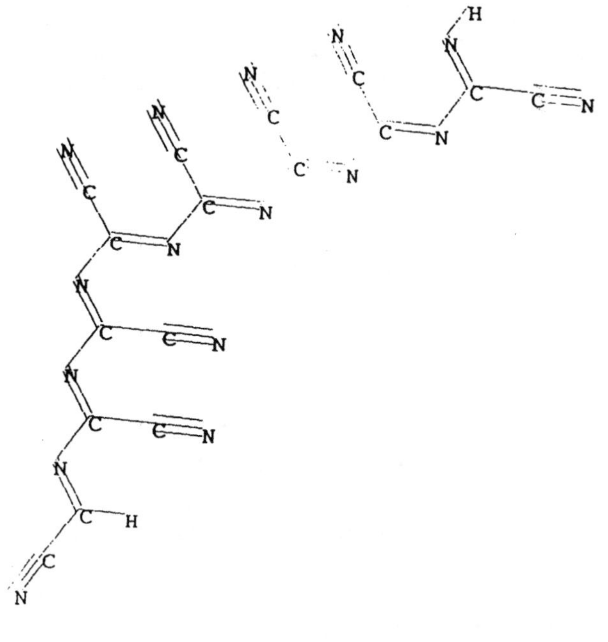

Figure 4. Fragment of the CN_2 molecule seen from two directions.

5. Conclusions

If material produced in process No.2 (i.e. CN_2) is an homologue of such carbon allotropic form as diamond, whereas product of process No.1 (i.e. C_5N) is an homologue of another carbon allotropic form, chaoite, annealing of the former one should result in its transition into compound similar to the C_5N. As it was mentioned earlier, its amount was not sufficient to enable microcalorimetric investigations. There were performed however, investigations of its electric and optic properties (see: Table 4 and Figure 2) which revealed that they change after annealing. As a result CN_2 is being transformed into material having properties close to C_5N. Similarity to transition of metastable carbon forms (like diamond) into thermodynamically stable forms (like graphite) caused by annealing, appears to be here quite clear.

Also characteristic seems to be the fact that CN_2 films (corresponding to diamond) were produced under conditions of intensified energetic activation of synthesis process, as compared to the process leading to production of C_5N. In the case of low pressure synthesis of diamond films this phenomenon can be observed, too.

Thus it seems that a carbon nitride is always the same substance, but existing in different allotropic forms, unlike βSi_3N_4 and similar to carbon.

According to the idea of C_3N_4 as a polymorphous material, carbon nitride layers obtained in the course of various processes may be characterized as composites of both carbon nitride allotropes as well as additionally quite often of some carbon allotropes of undefined composition. These components, differing in mechanical properties and in specific volume values, may lead to obtaining of the exceptionally hard and wear resistant material.

6. Acknowledgments

This work was supported by grant No.7T08A02109 of the Polish State Committee for Scientific Research

7. References

1. Liu, A.Y. and Cohen, M.L. (1989) *Science* **245**, 841.
2. Hardie, D. and Jack, K.H. (1957) *Nature* **180**, 332.
3. Cuomo, J.J., Leary, P.A., Yu, D., Reuter, W., and Frisch, M. (1979) *J. Vac. Sci. Technol.* **16**, 299.
4. Ren, Z.-M., Du, Y.Ch., Ying, Z.F., Qiu, Y.X., Xiong, X.X., Wu, J.-D., and Li, F.M. (1994) *Appl. Phys. Lett.* **65**, 1361.
5. Polo, M.C., Aguiar, R., Warela, M., and Esteve, J. (1995) Paper No.11.091 presented at the 6-th European Conference on Diamond, Diamond-like and Related Materials "Diamond Films'95", 10-15 September 1995, Barcelona, Spain.

6. Hoffman, A., Gouzman, J., and Breuner, R. (1994) *Appl. Phys. Lett.* **64**, 845.
7. Tong, C.J., Siversten, J.M., Judy, J.H., and Chang, C. (1990) *J. Mat. Res.* **5**, 2490.
8. Chen, M.J., Liu, X., Dravid, V.P., Cheng, Y.W., Wong, M.S., and Sproul, W.D. (1993) *J. Vac. Sci. Technol.* **A11**, 521.
9. Fujimoto, F. and Ogata, K. (1993) *Jap. J. Appl. Phys.* **32**, 463.
10. Kolitsch, A., Möller, W., and Ricter, E. (1995) Paper No.11.092 presented at the 6-th European Conference on Diamond, Diamond-like and Related Materials "Diamond Films'95", 10-15 September 1995, Barcelona, Spain.
11. Song, H.W., Su, X.W., Cui, F.Z., and Li, H.D. SMMiB'95 Abstract Book, P-61-B.
12. Matsumoto, S., Chattopadhyay, K.K., Mieno, M., and Ando, T. in press, *J. Chem. Vap. Dep.*
13. Sokołowska, A., Olszyna, A., Konwerska-Hrabowska, J., and Rajchel, B. in press, *J. Chem. Vap. Dep.*
14. Veprek, S., Weidman, J., and Glatz, F. in press, *J. Vac. Sci. Technol.*
15. Li, D., Chung, Y.W., Wong, M.S., and Sproul, W.D. (1993) *J. Appl. Phys.* **74**, 219.
16. Konwerska-Hrabowska, J., Zdunek, K., Sokołowska, A., and Szmidt, J., (1994) *J. Chem. Vap. Dep.* **3**, 156.
17. Sokołowska, A., Olszyna, A., Michalski, A., and Zdunek, K., (1991) *Surface and Coatings Technology* **47**, 144.

PHASE TRANSITION IN C:N FILMS UNDER SHOCK WAVE COMPRESSION

M.B. GUSEVA[a], V.G. BABAEV[a], V.M. BABINA[a], V.V. KHVOSTOV[a] , A.Z. ZHUK[b], A.A. LASH[b], I.A. FEDORININ[a]

[a] *Departmen t* of *Physics, Moscow State University, Moscow 1 19899, RUSSIA*
[b] *Institute for High Temperature RAS Moscow, 12741 7 R , RUSSIA*

1. Introduction

Hypothetical compound **β-C_3N_4** is found to be a good candidate for a new low compressibility material, with bulk modulus comparable to that of diamond [1]. In this compound the local order is such that carbon atoms occupy slightly distorted tetrahedral sites while nitrogen atoms sit in nearly planar triply coordinated sites.

It was suggested that **β-C_3N_4** can be synthesized by application of high pressure and temperatures to amorphous carbon nitride or other combination of carbon and nitrogen. This treatment could induce a phase transition to a solid with at least partial tetrahedral bonding.

The basic idea of this work is the application of linear-chain carbon-nitride with **sp^1** hybridization of C-atoms as a starting material for shock wave experiments. High temperature and pressure necessary to induce phase transition **$C(sp^1)$:N** to **$C(sp^3)$:N** are achieved in shock wave experiments.

2. Background

This approach is based upon the experimental results obtained in our laboratory earlier which shows that **sp^3** hybridization can be directly obtained from **sp^1** hybridization. This hypothesis was proved in 1993 in collaboration with dr. Yu. P. Kudryavtsev and dr. T. D. Varfolomeeva produced diamond crystals from amorphous carbyne under high pressure and temperature without catalysts.

In 1995 the transition of amorphous carbyne into highly dispersive diamond at T=120 °C was performed [2]. This results is shown in fig. 1 where 1 - X-ray diffraction

M. A. Prelas et al. (eds.), Diamond Based Composites, 161–169.

pattern of initial amorphous carbyne and 2 - X-ray diffraction pattern of shock wave compressed material as a result of phase transition.

The second point is specific features of the phase transition mechanism under conditions of short-duration exposure to high pressure and elevated temperatures.

3. Experimental

3.1 FILM SYNTHESIS

Linear chain carbon nitride was prepared as films, deposited by ion-assisted carbon condensation method under conditions of nitrogen-carbon glow discharge.

The vacuum chamber was evacuated with a turbomolecular pump to a pressure of 10^{-7} Torr. After evacuation nitrogen and argon were introduced into the chamber in a definite ratio up to a pressure $p=10^{-1}$ Torr. A high negative bias voltage was applied between the graphite target and the working chamber. An additional negative voltage relative to the plasma was applied to the substrate in a range 0... -200 V.

3.2 SHOCK WAVE EXPERIMENTS

In a shock wave experiments we used a 10-μm thick quasiamorphous carbon nitride films, which were deposited on a cooper disk. The disk was mounted in a stainless-steel container surrounded by a cooper ring (lateral momentum trap).

The ring with the

container were placed on a massive copper substrate. The dimensions of the container, the ring, and the substrate were chosen such that both the largest tensile stress caused by the action of unloading wave and, correspondingly, the fracture zone would be localized in the ring and the substrate but not in the container.

The loading was achieved by the impact of a 10-μm thick aluminum plate accelerated to a speed of 2.5 km/s

by detonation products of a condensed explosive material. The pressure applied to the sample reached 14 GPa. The duration of compression phase was about 2 μs. During compression the film and the surrounding metal reach a state chose to the thermal equilibrium. Therefore, the film temperature in the experiment was determined by the temperature of shock compression for copper and steel and as such was about 400 K [3].

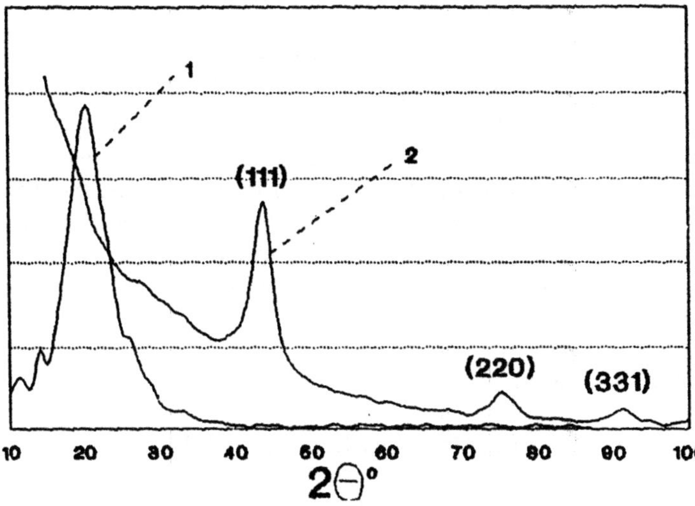

Figure 1. X-ray diffraction pattern : 1 - of starting material amorphous carbyne); 2 - after shock wave compression (highly dispersive diamond).

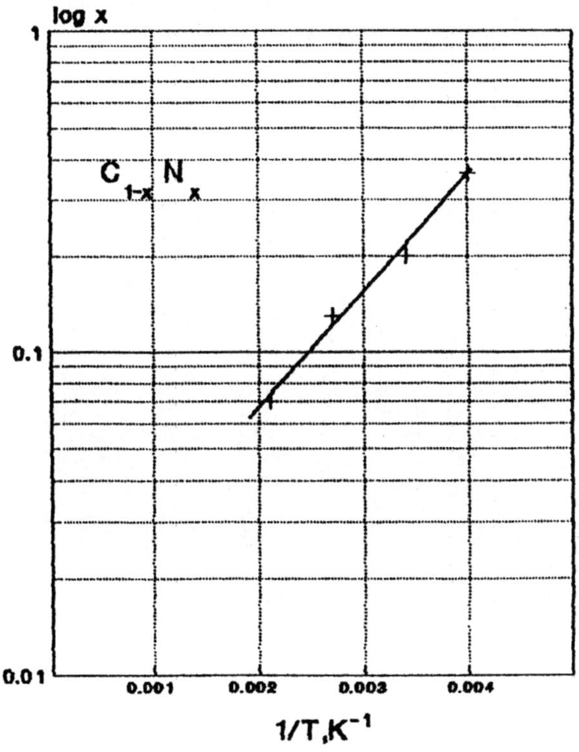

Figure 2. The temperature dependence of nitrogen concentration in **C:N** films.

3.3 METHOD OF CHARACTERIZATION

The chemical composition of the films and nitrogen concentration was determined by Auger electron spectroscopy (AES).

The atomic structure of the films was studied by a TEM technique using an JEM-100 electron microscope in the microdiffraction mode. To prevent radiation damage of the oriented carbon nitride films by the electron beam in microscope, the diffraction patterns were registered at very low electron beam intensity using a special condenser aperture as well as selection of the proper cathode emission current.

The structure of the films was also investigated by Raman spectroscopy with argon laser excitation (λ=484.8 nm).

4. Results and Discussion

4.1 LINEAR CHAIN CARBON NITRIDE FILMS

The experimental results obtained by Anger electron spectroscopy show that the nitrogen concentration in the films depends strongly on the deposition temperature (Fig.2). The highest nitrogen concentrations. equal to 36 at%, was obtained in the films on cooled substrates (down to -20 °C) and the lowest nitrogen concentration was 3 at% (at 200 °C). When the substrate temperature was lower then 20 °C the films were amorphous.

Fig.3 shows the typical electron diffraction pattern from N-doped carbon film at T>20 °C. Six sharp maxima correspond to interplanar spacing d=3.77±0.05 A. This structure is similar to that of Lengmur-Blodget films possessing a good order within a layer.

Figure 3. Electron diffraction pattern of **C:N** film (T>20 °C).

The density of valence state (DOVS) obtained from Auger spectra of carbon nitride films by self-deconvolution of KVV line [4] are presented ins Fig.4. For comparison the calculated energy spectrum of cumulene chains are shown below.

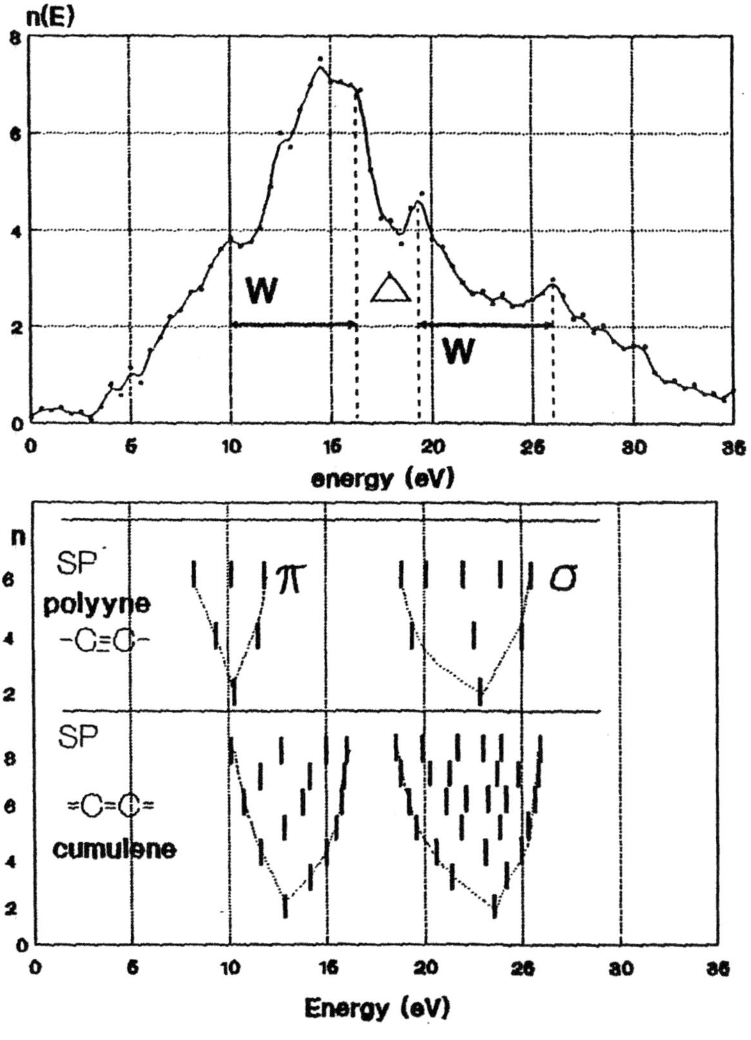

Figure 4. The density of electron states in the valence band of **C:N** film (before shock wave compression).

When the number of carbon atoms in the chain is greater then 8, the edges of π and σ bands become 10 eV. 16.5 eV and 18 eV, 27 eV respectively. The maxima of experimental DOVS curve at 10 and 16 eV correspond to the top and the bottom of π-subband, the maxima at 19 and 26 eV - to the top and the bottom of σ-subband. So,

Auger spectroscopy confirms the linear structure of carbon nitride films.

Raman spectrum of linear chain carbon nitride film is shown on figure 5 (a). As one can see from the figure, a wide band with a maximum at ν= 1525 cm^{-1} is a characteristic of the vibrational frequencies of the carbon skeleton of the chains [5] and two peaks at ν=2060 cm^{-1} and ν=2280 cm^{-1} correspond to stretching vibrations of **(=C=C=)** and **−C ≡ N** respectively.

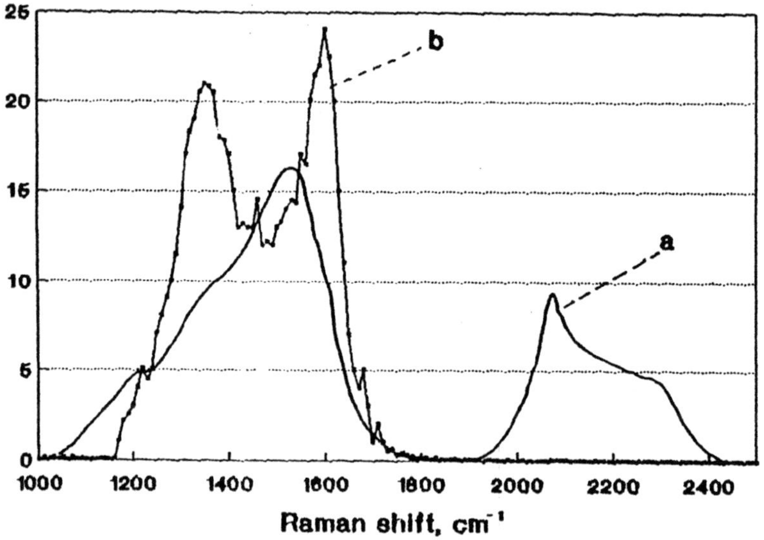

Figure 5. The resonant Raman spectrum of **C:N** films: a) before shock wave compression, b) after shock wave compression.

4.2 SHOCK-WAVE COMPRESSION OF LINEAR CHAIN CARBON NITRIDE FILMS.

Electron diffraction investigation shows, that after shock wave compression the carbon nitride films is quasiamorphous and contain a great deal of crystalline inclusions. The crystal mean size is about 1μm.

Tine typical diffraction pattern from microcrystalline inclusion is shown in fig.6. Measurement of the diffraction pattern yield the d values and corresponding indices given in Table 1. The diffraction pattern of these microcrystal includes the reflections (hk.0), indicating the perpendicular orientation of (00.1) crystal plane to the electron beam. The value for the hexagonal cell parameter "a" was found to be 4.38 A. This lattice parameter is predicted by Liu and Cohen for β-C_3N_4 phase (a=6.429 A) [6]. To determine the location of the carbon atoms in unit cell we have calculated a Patterson function using a conventional procedure [7] and have found the charge density distribution in a (00.1) projection of the crystal unit cell (fig.7).

There are six sharp positive maxima corresponding to C atoms and six smaller positive maxima corresponding to N-atoms placed between them. The positive charge of N atoms is partially screened by its negative electron charge and so they are less pronounced similarly to that one in [6].

TABLE 1 .Comparison of the d spacings determined experimentally from electron diffraction pattern of fig.6 with calculated one.

d_{exp} *A*	**Intensity**	d_{theor} A	**(hk.l)**
3.793	strong	3.798	10.0
1.902	middle	.1.895	20.0
1.263	strong	1.263	30.0
0.942	middle	0.948	40.0
2.19	middle	2.1811	11.0
1.423	middle	1.432	21.0
1.045	strong	1.051	31.0
0.027	middle	0.027	41.0
1.089	middle	1.094	22.0

Figure 6. Electron diffraction pattern of **C:N** film after shock wave compression.

At the center of figure there is another positive maximum corresponding to C atom connected with two neighboring C atoms by two N atoms.

3D crystal structure of this crystal should possess buckled layers where C atoms are connected by N atoms. Such structure can be described as a final stage in a row of **C:N** chain hypothetical crystal lattices one of which is shown in fig.8 in the limit when the number of carbon atoms n become equal.

The probability of such reconstructions is supported by Raman spectrum (fig.5b) of

C:N film recorded after shock wave compression. There are only two peaks iii it v= 1600 cm-1 and v= 1350 cm-^{1} corresponding to vibration frequencies of **sp^{2}** and **sp^{3}** C-C bonds respectively and all frequencies corresponding to **sp^{1}** bonds have disappeared.

Figure 7. Total valance charge density of **C:N** film after shock wave compression in the (0001) plane.

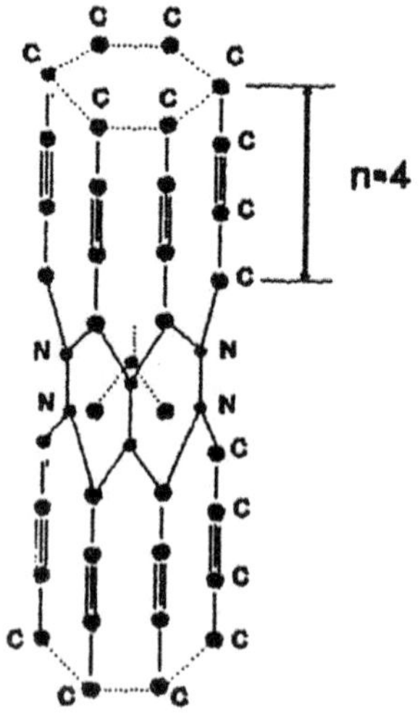

Figure 8. Hypothetical **C:N** crystal structure with **C:N** buckled layers.

5. Conclusions

It was found that **C:N** single crystals can be synthesized by application of high pressure and low temperatures to $\mathbf{sp^1}$ carbon nitride film This treatment induces a phase transition into a solid with $\mathbf{sp^2}$ hybridized N atoms connected with $\mathbf{sp^3}$ hybridized C atoms.

Acknowledgments

This work was supported by RFFR (grant 96-02-03655-a) and Soros Scientific Education Program.

References

1. Liu, A.Y. and Cohen, M.L.(1990) Phys.Rev.B, 41,10727.
2. Guseva, M.B., Babaev, V.G. Valiullova, Z.Kh. et. al.(1995) second International Workshop "Fullerenes and atomic clusters", June, St. Petersburg.
3. Baum, F.A., Orlenko, L.P., Stanyukovich, K.P. et. al.,(1975) Fizika Vzryva (*Physics of Explosion),* Moscow: Nauka.
4. Khvostov, V.V., Guseva, M.B., Babaev, V. G. and Rylova, O. Yu. *(1986) Surf.Sci.169,* L253.
5. Berdugin, V.V., Kudryavstev, Yu. P., Evsucov, S.E., Korshak, *Yu.V.(1989)Dokl.Akad.Nauk SSSR,* v.305, N2, p.362-364.
6. Liu, A. Y. and Cohen.M.L. *(1989) Science* 245,841.
7. Buerger, M.J.(1959) Vector Space and Its Application in Crystal Structure Investigation, Wiley, New York.

COPPER NANOCLUSTERS IN DLC

V.I.IVANOV-OMSKII
A. F. Ioffe Physico - Technical Institute RAS
194021 St. Petersburg, RUSSIA

Abstract. Modification of electrical, optical and structural properties of DLC films through introduction of copper clusters by co-sputtering of graphite and copper targets is considered. Evidence is presented of the existence of various kinds of copper-based clusters. Their structure, size, and size-distribution function are discussed. An outline is presented of those peculiarities of optical and electrical properties of the films which are associated with the presence of the sp2-bonded component.
Keywords: copper/cluster/diamond-like carbon/intercalation/graphite/plasma sputtering/fractals/current percolation/surface plasmon/X-ray/small-angle scattering

1. Introduction

Diamond-like carbon (DLC), which is actually hydrogenated amorphous carbon (a-C:H), offers in itself an example of a composite made up of sp3 - and sp2- bonded carbon nanoclusters. As is well known, this composite structure gives superior strength characteristics to hydrogenated carbon, which to a certain extent justifies the conventional name of this material: diamond-like carbon. In this case the clusters are formed from various allotropes of the same carbon (homoclusters). Incorporation of clusters of different chemical nature (heteroclusters) into DLC may extend the set of its useful properties and give deeper insight into its structure.

We shall survey concisely the recent results of our group's activity in modifying electrical, optical, and structural properties of DLC by loading it with copper nanoclusters. The main feature of this composite is that copper does not interact with carbon chemically, i.e., it does not form stable chemical bonds with carbon. One expects, therefore, that the composite under study is a real heterogeneous system composed of a perfect conductor (copper) and a no less perfect dielectric (DLC). The heterogeneity of the system enables formation of copper-based conductive nanostructures which are in the field of vision of modern interest in problems of copper metalization in ULSI structures [1].

We did not focus our attention on mechanical properties of DLC doped with copper, since appropriate quantitative solution of the problem still remains to be found.

M. A. Prelas et al. (eds.), Diamond Based Composites, 171–189.

However, it is possible to note that, with increasing copper content, the material keeps

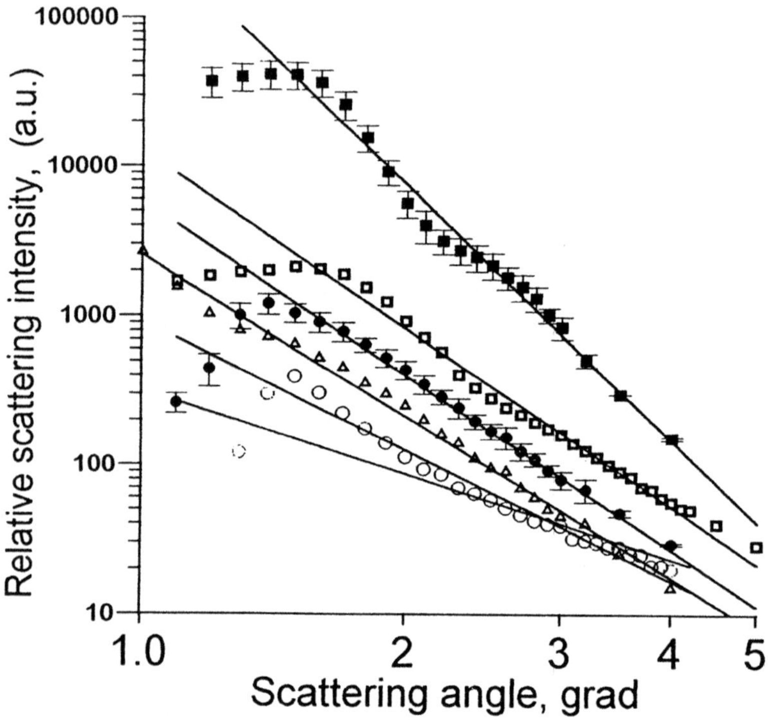

Figure 1. X-ray scattering intensity versus scattering angle Θ for DLC doped with copper: ■ ■ ■ - 4% copper; □ □ □ - 6% copper; ● ● ● - 14% copper; ○ ○ ○ - 14% copper (vacuum annealled sample); Δ Δ Δ - 24% copper.

its high wear resistance up to copper concentrations of approximately 6%, loosing this property rather sharply further on to such an extent that at 24% of Cu a film subjected to two hours vacuum annealing at 2OO°C can be removed from the substrate by fingers.

2. Preparation and structure

The a-C:H films we studied were grown on various substrates by ion (magnetron) cosputtering of graphite and copper targets in argon or argon- hydrogen (80% Ar and 20% H_2) plasma. Single-crystal silicon, fused silica, glass, ceramics, or polished KBr plates were used as substrates depending on specific experimental requirements. All the films had good adhesion. The substrate temperature, gas pressure in the growth cell, and average magnetron power were 500 K, 10 mTorr and 0.4-0.5 kW, respectively. The energy of Ar ions was 350-450 eV. Copper concentration was varied within the range 0-36 at.%. The copper concentration in the growing film was controlled by changing relative areas of graphite and copper targets being sputtered and was occasionally checked by secondary ion mass spectrometry (SIMS). The films thicknesses were within the 0.1-1.5 μm range. Spectral measurements repeated in half

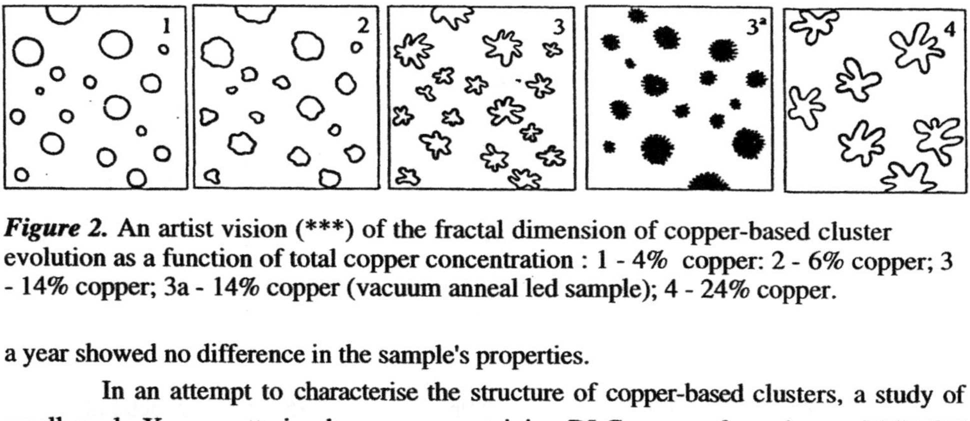

Figure 2. An artist vision (***) of the fractal dimension of copper-based cluster evolution as a function of total copper concentration : 1 - 4% copper: 2 - 6% copper; 3 - 14% copper; 3a - 14% copper (vacuum anneal led sample); 4 - 24% copper.

a year showed no difference in the sample's properties.

In an attempt to characterise the structure of copper-based clusters, a study of small angle X-ray scattering by copper-containing DLC was performed on a RIGACU X-ray diffractometer operating at 1.541 A (Cu-Ka line) [2]. Diffractograms were taken at a potential of 50 kV in reflection geometry. The diffraction patterns were analysed at angles of 1.0 to 4.0 degrees. Copper-based clusters embedded in the DLC matrix were shown to be mainly responsible for X-ray scattering at small angles. Typical angle dependencies of the intensity of X-rays scattered by DLC films with various copper concentrations are reproduced in Figure 1 from Figure 2 of Ref. [2]. As evident from these data, the scattering amplitude I(θ) for all samples measured is a power function of angle θ at not-too-small angles, i.e., it may be written as

$$I(\theta) \sim I_o \theta^{-\alpha} \qquad (1)$$

where I_0 is a proportionality coefficient, and α is the power index whose experimental values are presented in Table 1. Some deviations from the monotone run of experimental points to the right of peaks in Figure 1 are due to the neglect of higher members in eq.(1). The observed dependence of the scattering intensity is typical of scattering at small angles by fractal structures. The power index in eq.(1) may be related to the fractal dimensionality of the structures in question [2]. The corresponding fractal indices are also listed in Table 1. The experimental points in Figure 1 show a maximum at the smallest angles and a similar run of the experimental curves to the right of the maximum. Although the log-log coordinate system smoothes out the maxim, we are nevertheless convinced of their existence except the case of 24% copper. The classical theory of small angle scattering of radiation with wavelength λ by an array of regular spheres [3] predicts a maximum of scattered intensity at the angle $\theta=\lambda/4R$, where R is the sphere radius. In the case of fractais, R may be considered as a correlation radius. Experimental values of R are presented in Table 1. As can be readily seen, the mean correlation radius of scatterers is practically independent of copper content at concentrations less than 24% and thermal treatment has no noticeable effect on the

TABLE 1. The sample parameters.

N	Copper content, %	Index eq.(1), 0	Fractal dimension D=6-a	Scatterer size, A	Comments
1	4	5.4	2.0a	16.	Scatterer with smooth boundary surfaces
2	6	3.9	2.1	1 5.	Scatterer with fractal surfaces
3	14	3.7	2.3	16.	
3 b	14	2.9	2.9c	15	Fractal objects
4	24	3.6	2.4	>24.	Scatterer with fractal surfaces

(a) - $\gamma=7-\alpha=1.6$ (at $\alpha>4$) is defined as index in power law [5].
(b) - Vacuum annealled sample 3.
(c) - Fractal dimension is defined as $D= -\alpha$.

average radius of copper-based clusters. As regards the fractal dimensionality D [2,4], it was found to be essentially less than 3 in most cases, except the annealed sample, increasing from D=2 with increasing copper concentration. This may correspond to a transition from scattering by smooth-boundary surfaces (bald scatterers) at low copper concentration to scattering by fractal-boundary surfaces (hairy or starfish-like scatterers) at higher concentrations. The evolution of the copper-induced cluster shapes with increasing of total copper concentration is illustrated in Figure 2 schematically, following to the fractal dimension scenario scripted in Table 1.

It is interesting to note that the average correlation radius is practically constant except the case of the highest copper content. It seems that the copper cluster size is controlled by typical sizes and shapes of pores in DLC and their distribution function. At very low copper concentrations, the mean distances between pores occupied by copper clusters are large enough to allow appearance of the most equilibrium cluster shape which seems to be spherical due to weak Cu-C chemical bonding. Here, too, there exists a typical size of pores which is defined by the thermodynamics of their formation in DLC and occurs to be about 15 A according to the proposed model. As copper content increases, interaction between pores occupied by copper clusters deforms their shape (see Figtire 2). As a result of the interaction, whose symmetry makes allowance for the disordered DEC structure, the cluster surface acquires a fractal character (see Table I and Figure 2). As also seen, vacuum annealing of a sample with 14% Cu reassembles the cluster structure in a way that can be characterised as coagnlation of cluster branches into smoother-surface clusters. Relying on the absence of a peak at 1.0 degree (the instrumental limit) for the highest copper concentration (24%) one arrives to a conclusion that the internal energy of copper clusters becomes high enough to extend the pore size with the surface topology retained.

Transmission Electron Microscopy (TEM) (100 keV) and Selected Area Electron Diffraction (SAED) were also used to characterise copper clusters [5]. For this purpose, free-standing films approximately 100 nm thick were prepared by dissolving KBr substrates in water. The study was performed with a Philips 400FM transmission electron microscope operating at 100 keV. A bright-field TEM image of copper-born clusters in a sample with 9% of copper is presented in Figure 3. This figure shows clusters of 2.9-nm average diameter distributed in the DLC matrix. The results of SALD show a reasonable match of the observed diffose ring spacing to the corresponding picture for copper. The size variation of copper-born nanoclusters was analysed from TFM micrographs directly and the resulting size-distribution function is shown in Figure 4. Analysis of the distribution evidences its Gaussian-like character in accordance with the fluctuation theory of phase nucleation [5]. The most probable size agrees reasonably with the data from Table 1.

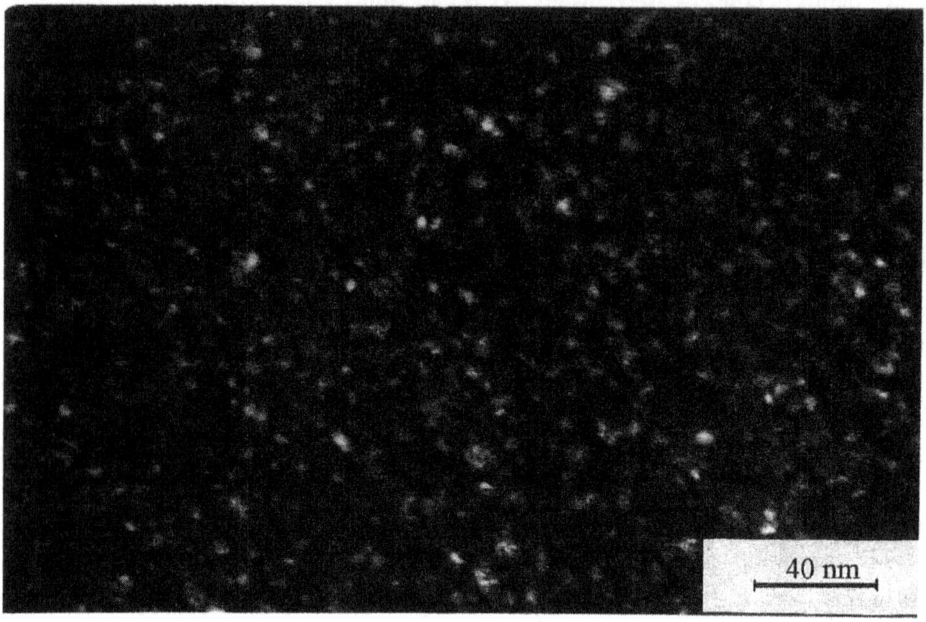

Figure 3. TEM image of copper-born clusters embedded in DLC for sample with 9% of Cu.

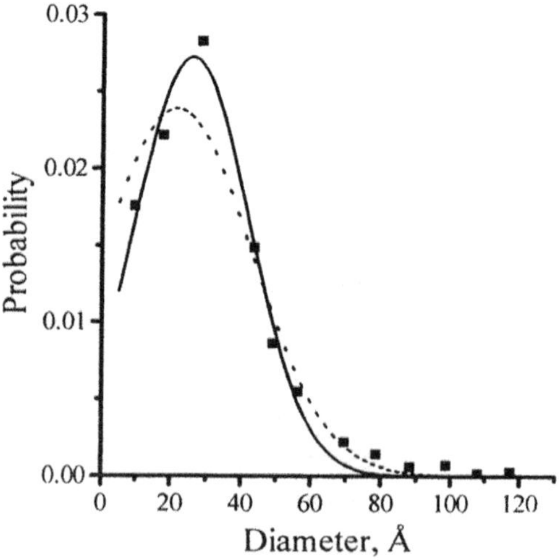

Figure 4. Size distribution function of obtained TEM micrograph (Fig. 3). The full squares stand for experimental points, the full line is Gaussian curve. The dotted curve is SDF obtained from optical data.

Extended X-ray Absorption Fine Structure (EXAFS) and X-ray Absorption Near Edge Structure (XANES) measurements are able to present direct information on the structure of copper clusters also. It is shown [6] that different types of clusters are formed, namely, one with a very small Cu-Cu coordination number and the other with the structure similar to that of the bulk copper which could be either small spheres or flakes. At higher copper concentrations, Cu-Cu interaction increases and eventually bulk-like structure of the clusters is formed as evidenced by EXAFS and XANES spectra for the 14 at.% Cu sample. The fact that the average Cu-Cu coordination number in the cluster is smaller than that in the bulk metal can be understood as follows. In small clusters the relative number of the surface atoms, whose Cu-Cu coordination is half of that for atoms in the vol ume of the cluster, is not negligible and as a result the average coordination number is smaller that 2 (which is the case for bulk copper). It is valid expecially with regard to their fractal surfaces considered above. In the assumption of spherical shape of clusters the average coordination number should generally be larger than 6 while for rather thin flat flake-like formations in which case the average coordination number should be smaller than 6. The obtained value of 6 plus-minus 3 is at the boundary of the two cases which can be interpreted either as due to spheres with small diameter 20 A or two-dimensional flake-like formations with a thickness of several Angstroms and any size in the other two dimensions, or a combination of both. The obtained value of the characteristic size of a few tens of Angstroms agrees with the previous conclusion.

3. Electrical conductivity

The DC electrical conductivity of DLC was studied at room temperatures as a function of total copper concentration [7,8,9]. Since the electrical conductivity of DLC is known to be a strong function of hydrogen content, studies of two sets of hydrogenated (a-C:H) and hydrogen-free (a-C) amorphous carbon samples were performed. The experimental results for both the sets are depicted in Figure 5 and Figure 6. It is easy to verify that pretty sharp and asymmetrical peaks are observed approximately at the same copper concentration, irrespective of the hydrogen content, but the conductivity for both the sets is different. On the whole the amorphous carbon matrix is a fairly perfect insulator; which is especially true for the set of a-C.H samples grown in argon-

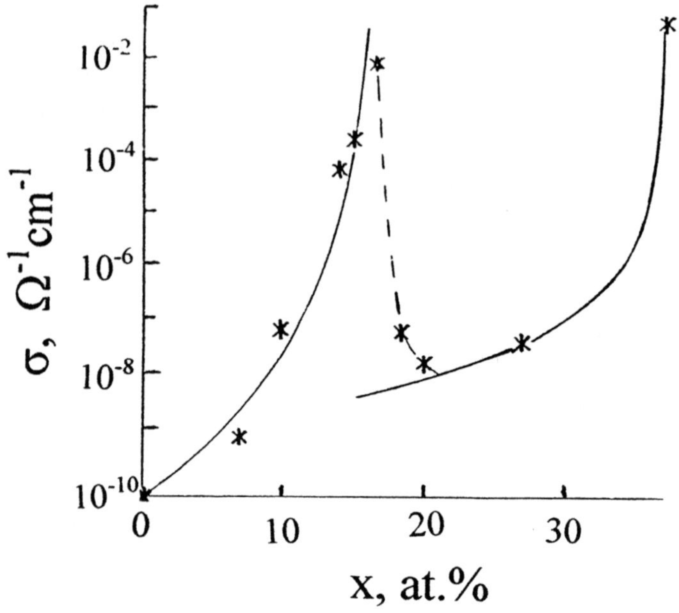

Figure 5. DC electrical conductivity of Cu-doped a-C:H as a function of copper content.

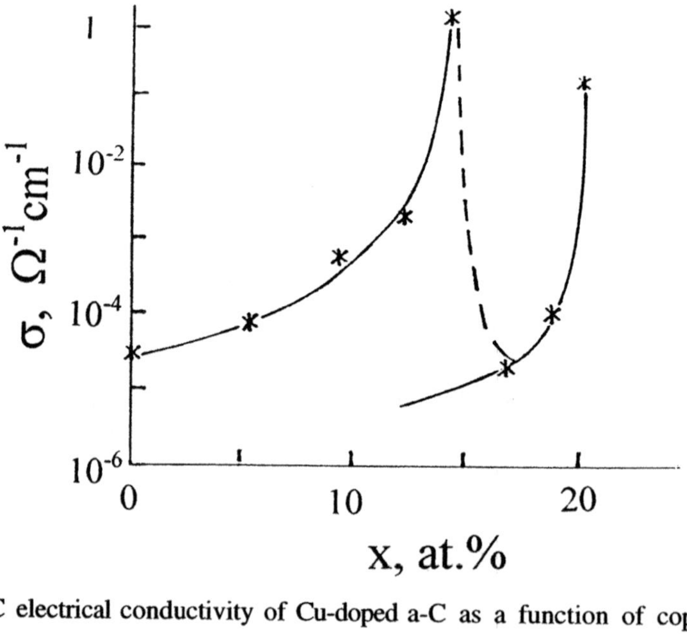

Figure 6. DC electrical conductivity of Cu-doped a-C as a function of copper content.

hydrogen plasma, as it can be seen in Figure 6 at zero copper concentration. This contrasts with the set of a-C samples grown in argon plasma, for which the presence of graphite-like structural elements or graphite-like clusters (GLC) results in a conductivity higher by six orders of magnitude. This is ascribed to percolation of electric current through a random network of conductive GLC embedded in the insulating carbon medium It is also natural to associate the observed dependence of conductivity on copper concentration with current percolation through copper-induced conductive network, as it was done in [8,9]. The rapid growth of the conductivity in the range 0-15% can be well explained by the theory based on the model of electric current percolation through a system of conductive domains randomly distributed over nonzero-conductive medium [10, 11].

Parameters of the percolation system (percolation indices) can be derived from the plots in Figure 5 and Figure 6 by a fitting procedure. The solid curves in Figure 5 and Figure 6 calculated using the equation:

$$\sigma = \text{const} \, (x_c - x)^{-q} \qquad (2)$$

with q=6.4 for a-C:H and q=2.5 for a-C are seen to fit reasonably the experimental results. Here σ is the conductivity, x_c is the percolation threshold, and x is copper concentration. These percolation indices correspond, especially in the case of a-C:H, to a very rapid growth of conductivity with copper concentration, increasing by about an order of magnitude per additional 2% of copper on the left shoulder of the conductivity peak. It is worth noting that not only the q=6.4 value but even q=2.5 are unusually high; typical theoretically predicted values for various percolation systems being less than 2 [101. This means that interaction probably occurs between the graphite-like components of the matrix and copper, e.g., these components are intercalated with copper [1 3], which results in an increase in the conductivity of the percolation network. The values of other critical indices found from the experimental data and reported elsewhere [8] may be ascribed, together with the low values of percolation threshold (x_c = 14-16%), to the formation of percolation nets with a relatively low density or low fractal dimensionalities in the region of 0-15% of Cu (see the preceding section).

Another situation is observed at higher copper contents (above 16 at.%). Dense copper structures (e.g., drops) are formed there (see also sections 2 and 5). This process cannot take place at lower copper concentrations where the probability of copper-copper coupling is small enough and the rapidly stabilised carbon skeleton retards any further coalescence of copper atoms. When copper concentration is about 16%, the probability of copper-copper coupling becomes high enough to make possible the appearance of drop-

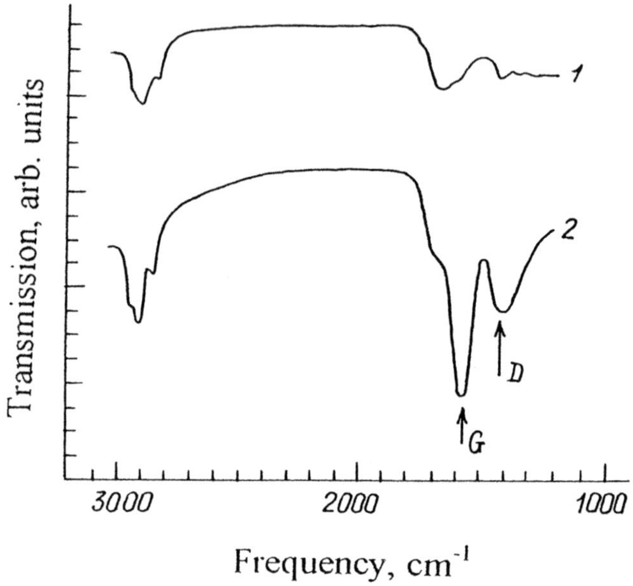

Figure 7. Fragments of IR transmission spectra of copper-free (1) and copper doped DLC.(2).

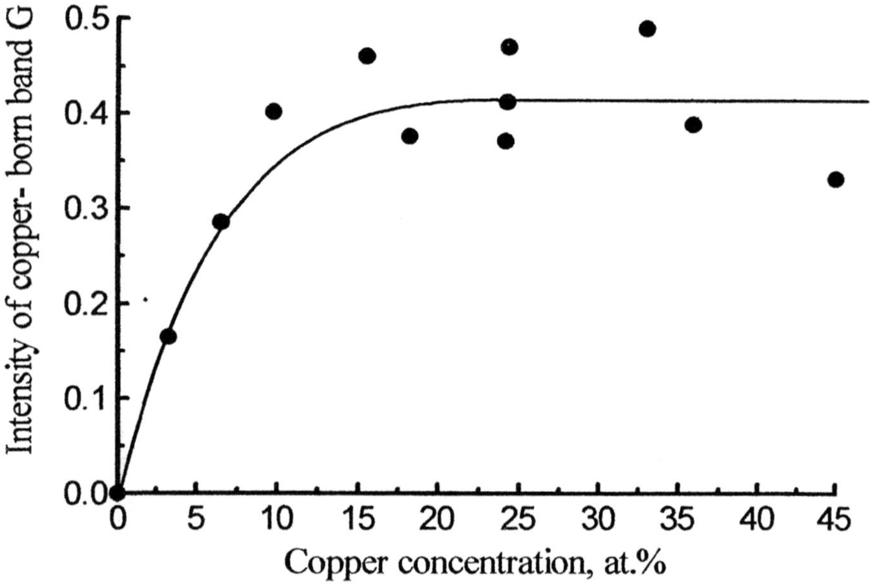

Figure 8. Intensity of Cu-induced band of IR absorption of Cu-doped DLC as a function of copper concentration.

like copper formations. The assembly of copper atoms in drops leads to the appearance of copper-free regions which disrupt the percolation nets, so that the conductivity of the system drops sharply. The material with x > 16% can be also described in terms of the percolation theory, but with another set of percolation indices (q=2.8), which correspond to formation of drop-based percolation nets embedded in a less conductive environment.

4. Vibrational properties

Valuable information on the phenomenology of the copper-carbon interaction can be derived from a study of IR absorption within the vibration frequency range (1000 -4000 cm^{-1}) [12,13]. Although any spectral evidence of the existence of copper-carbon bonds is lacking, in good agreement with the fact that copper is not a carbide-forming metal, copper-carbon interaction nevertheless manifests itself in the activation of vibrational modes of the GLC carbon skeleton, which are otherwise silent in the optical absorption [12]. This is illustrated in Figure 7, where fragments of IR transmission spectra of copper-free and copper-doped DLC films are presented for the frequency region in question. It is apparent that copper excites absorption bands which are practically invisible In copper-free DLC. With increasing copper concentration, the intensity of the copper-induced IR bands first increases and then saturates at about 15 % (see Figure 8). Thus, incorporation of copper into DLC activates in the optical absorption the otherwise silent vibrational modes of the GLC carbon skeleton, labelled G (1580 cm^{-1}) and D (1355 cm^{-1}), which are well-known to he only Raman active in copper-free DLC. This suggests that GLC are intercalated with copper which, intervening between graphite nano-planes, reduces the symmetry of their carbon skeleton vibrations. This is then reflected in an activation of the corresponding vibrational modes in IR optical absorption spectra. Thus, IR absorption measurements permit detection of copper-intercalated GLC forming percolation nets with a relatively low density, as it was suggested in the previous section. Inasmuch as only sp2-bonded GLC occur to be intercalated with copper, this offers an opportunity of analysing the carbon allotrope composition of DLC [12, 13].

In an effort to get additional information on the mechanism of interaction between copper and carbon inside the DLC, the IR absorption of samples annealed at various temperatures in a vacuum of 0.1 mTorr was measured. Figure 9 shows the intensity of copper-born band G as a function of the duration of isothermal annealing. By assuming the intensity of the bands to be proportional to the number of intercal ant copper atoms, one arrives at the conclusion that the temperatures at which the absorption under discussion disappears are about several hundred degrees of Celsius (*). This corresponds to binding-energy values expected for the van-der-Waals bonding which is typical of intercalation

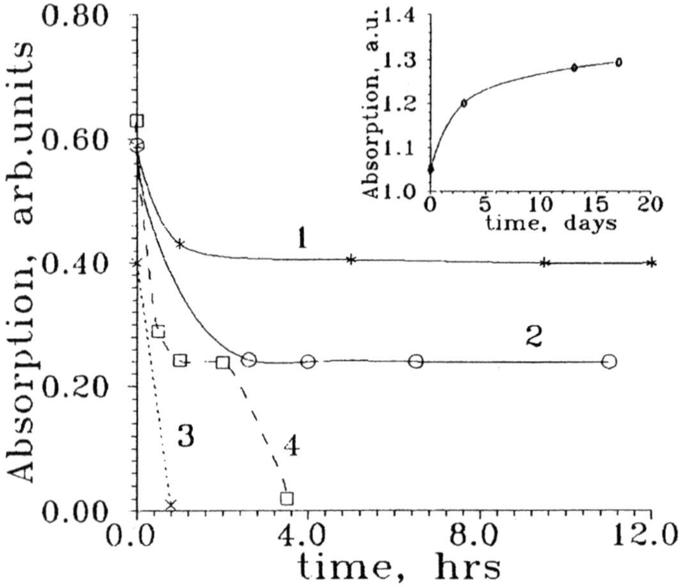

Figure 9. Intensity of copper-born band G as a function of duration of isothermal annealing. The inset shows room-temperature relaxation of G-band absorption.

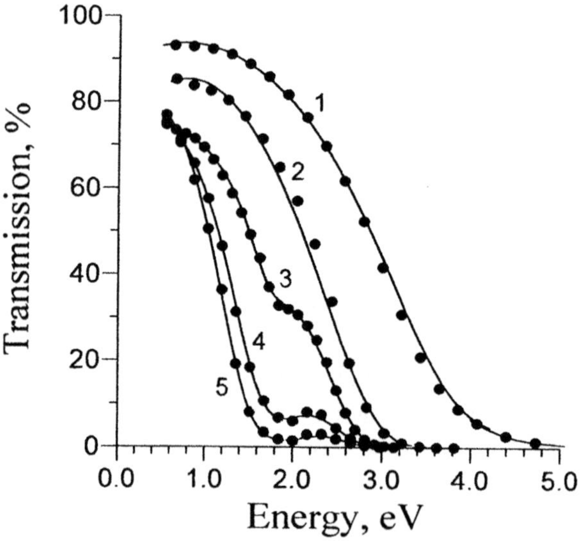

Figure 10. The optical transmission spectra of a-C:H doped with copper with different copper concentration: 1-0%; 2-6%; 3-18%; 4-20%; 5-23%. The points are experimental values. The solid curve is the result of curve fitting.

phenomena. It is also seen that the equilibration time of intercalation in DLC, which is just the time of curve flattening in Figure 9, is rather fast as opposed to a many-hours process of intercalation in bulk graphite. The difference is due to nano-sizes of the GLC in DLC. This circumstance is associated with another striking phenomenon: after intercalation is destroyed by thermal treatment, a well pronounced tendency for the initial level of intercalation to be restored is clearly observed at room temperature. This is indicative of the proximity of the reservoir of copper atoms removed by the thermal treatment to the intercalant sites. In any case, the intercalation of GLC with copper appears to be a fairly reversible process. The room-temperature relaxation of the G-band absorption for a sample with 9% of Cu is shown in the inset in the top right side of Figure 9. (**).The solid line represents exponential decay with a time constant of the order of 70 hours. At higher annealing temperatures and copper concentrations the observed phenomena lose their reversibility. The fast kinetics of CLC intercalation with copper and the reasonable reversibility of the process make the material a promising detector of environmental conditions [14].

5. Electronic excitations

Investigation of electronic excitations in copper-born clusters furnishes information about their structure. Optical transmission in the range 0.6-5 eV was measured in an effort to study electronic excitations in copper-based clusters and the DLC matrix by Ivanov-Omskii et al. [15]. The transmission measurements were made at normal incidence at 300 K with a Hitachi-3410 spectrometer using a rectangular 0.5 cm x 0.5 cm slit. The resulting optical densities for films with various copper concentrations are plotted against photon energy in Figure 10. The fundamental absorption edge in copper-free samples is typical of a-C:H and follows the Tauc rule with E_{opt}=2. 1 eV, which suggests that the a-C:H is in this case somewhat sofier than the more frequently quoted materials [16]. As Figure 10 shows, the absorption edge experiences a significant enhancement with increasing copper concentration. This is accompanied by the appearance of a pronounced peak at higher copper contents. The intensity and the width of the peak depend on copper concentration. An essential peculiarity of the peak is that practically neither its position nor shape depends on temperature in the explored range 77 - 300 K. This is rather a serious argument in favour of relating the peak origin to collective electronic excitations, e.g., surface plasmons in copper particles.

The experimental data in Figure 10 together with the results of ellipsometric measurements were recalculated into dispersion curves of the imaginary and real parts of the dielectric function, presented in Figure 11 and Fignre 12. It is noteworthy that, as seen in Figure 11, copper induces an anomalous dispersion of the real part of dielectric function in the spectral region of the absorption peak (see Figure 12), corresponding to the maximum contribution of spherical copper drops (see below). The data were processed further by performing a best-fit procedure for a model of dielectric medium based on the formalism of Lorentz oscillators expressed by the standard equation:

$$\varepsilon(\omega)= \varepsilon_0 + \Sigma\, \omega_{pj}^2/(\,\omega_j^2- \omega^2+i\, \omega/\tau_j) \qquad (3)$$

where j is an index for an oscillator, ε_o is the frequency independent contribution; ω_{pj},, ω_j, and τ_j are the plasma frequency, the resonance frequency and the relaxation time of electrons of the jth oscillator, respectively. It was assumed that the following four types of Lorentz oscillators would suffice:

(1) a low-frequency oscillator associated with the free-carrier absorption (j=1, ω_1=0);
(2) a mid-frequency oscillator associated with surface plasmon excitations in insulated conductive spherical clusters (j=2);
(3) high-frequency surface plasmon oscillators associated with excitations in non-spherical conductive clusters (j=3); ;

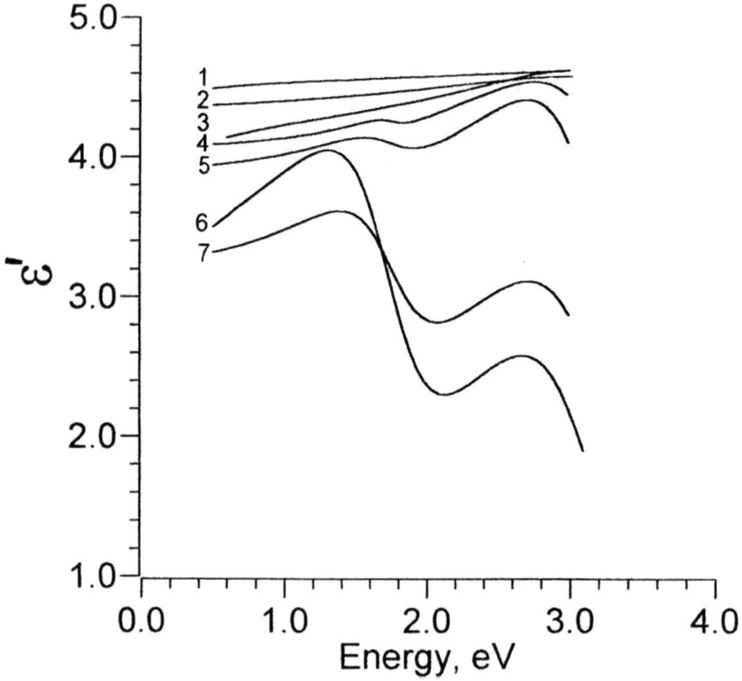

Figure 11. The spectral dependence of the real part of the dielectric function vs copper concentration for CU-doped DLC.1-0%; 2-3%; 3-6%; 4-14%; 5-18%; 6-20%; 7-24%.

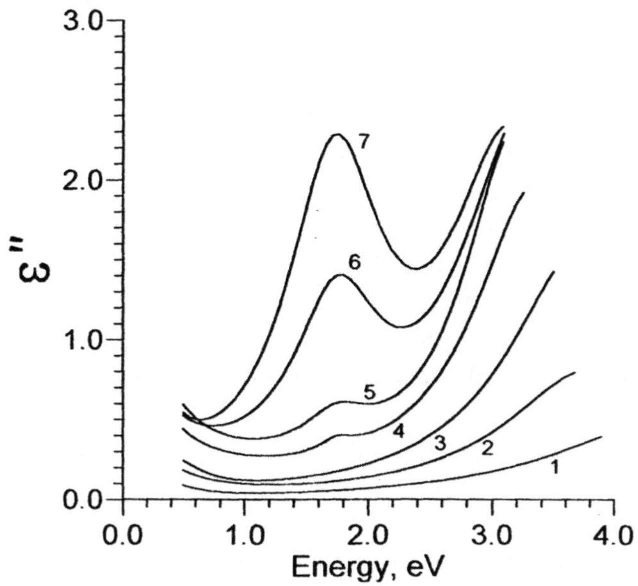

Figure 12. The spectral dependencies of the imaginary part of the dielectric function vs copper concentration for Cu-doped DLC. 1-0%; 2-3%; 3-6%; 4-14%; 5-18%; 6-20%; 7-24%.

TABLE 2. Optical results of the best-fit procedure.

N	Cu, at.%	$1/\tau_{eff3}$ eV	ω_3, eV	ϑ_3	$1/\tau_{eff2}$ eV	ω_2, eV	ϑ_2
1	*3.3*	1.56	3.89	0.028			
2	6	1.20	3.80	0.06			
3	14	0.63	3.41	0.14	0.26	1.76	0.0003
4	18	0.312	3.23	0.18	0.59	1.78	0.0017
5	20	0.29	3.32	0.19	0.76	1.79	0.0080
6	24	0.32	3.20	0.22	0.87	1.78	0.0200

(4) high-frequency oscillators associated with the fundamental ahsorption in the a-C:H matrix (j=4),

The authors of Ref 14 preferred to use the formalism of Lorentz oscillators instead of effective medium models, because the wide spectral range and large concentration interval used in their experiments permitted the contribution of different oscillators to be separated fairly well. They omitted the contribution of low frequency interband transitions in copper, because no evidence of it could be seen in experimental spectra in the 2.2 - 3.1 eV range (see Figure 12). The contribution of the 4th oscillator was accounted for by eq.(3), as measured directly for the copper-free sample. As to the low-frequency oscillator associated with free-carrier absorption, its parameters were estimated for the most heavily doped sample in the frequency region where the contribution of oscillator is of primary importance. The quantitative difference between the contributions of oscillators of the types (2) and (3) is due to differences in their concentrations and form factors (f_j in eq.(4) in [15]). The latter are to account properly for depolarisation effects in conductive clusters of different shapes; they are put f_2=l/3 for spherical and f_3=l/2 for non-spherical clusters. The last value corresponds exactly to an infinite conductive cylinder. This may be also a small fragment of conductive plate or tubular fttllerenes intercalated with copper like that described in [17]. In any case it is worth noting that the estimated value of the resonance frequency of the type (3) oscillator, is fairly close to the surface plasmon frequency for copper (3.5 eV) [18], as seen in Table 2. This means that the type (3) oscillator is very likely to be associated with copper-based clusters of reduced dimensional ity

The final result of the fitting procedure is presented in Figure 10 by solid lines, which are constructed using precisely the same data as in Figure 11 and Figure 12. The copper value of plasma frequency was used during the entire fitting procedure. Therefore, the fact that the absorption can be described through the whole spectral region (for all oscillators !) with a single value of plasma frequency, which is actually the copper plasma frequency, indicates the copper origin of the oscillators of the types (1), (2), and (3). The optimal values of the parameters fitting the experimental data are listed in Table 2, where τ_j is the relaxation time, v_j is the volume fraction, and ω_j is the

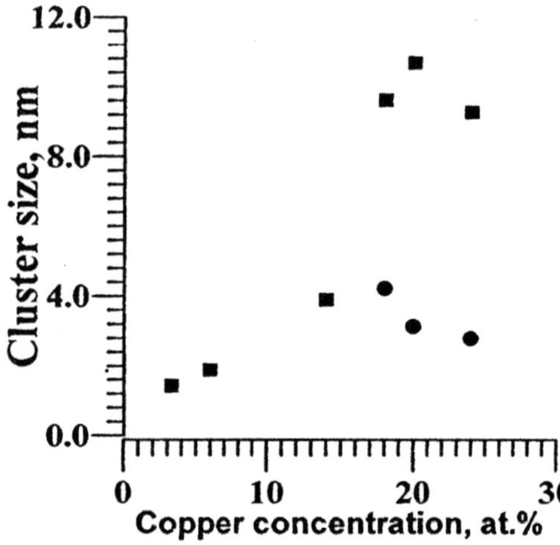

Figure 13. Sizes of copper-born clusters embedded in DLC matrix.

plasma frequency of the jth oscillator. The fairly low values of the volume fraction v_2 suggest that the spherical drop-like clusters are far from being a single kind of copper cluster embedded in the a-C:H matrix. There is much more room for clusters of another shape, as is evident from Table 2.

The relaxation time estimations cited in the Table for the 2-nd and 3-rd oscillators may be used to evaluate the sizes of copper-induced clusters associated with these oscillators by the procedure described in more detail elsewhere [15]. The resulting sizes of copper clusters are shown in Figure 13 as functions of copper concentration. It can be easily seen that the size of non-spherical copper clusters depends on copper concentration, whereas that of copper drops is practically constant. Moreover, the concentration region of a sharper increase in cluster size correlates with the concentration interval where switching from one percolation system to another occurs (Sec. 3). The success of the outlined procedure, achieved despite its multi-parameter nature, originates from the fact that the spectral ranges where the oscillators of the types (1), (2), and (3) are of primary importance are well-resolved, as seen in Figure 12. Therefore, the conditions of negligible overlap of contributions from different oscillators may be found to deterinine the oscillator parameters properly.

Finally, it may be added that a procedure for determining the size-distribution function of copper-based clusters from optical spectra was proposed in [5]. It was shown there that the distribution function determined in this way is in reasonable agreement with that presented in Figure 4.

6. Summary

The DLC medium is favourable for formation of copper-based clusters owing to the lack of stable chemical C-Cu bonding. The non-trivial electrical and optical properties of the DLC-Cu composite originate from the sp2 bonding inherent in the graphite-like component.

There were observed a variety of kinds of copper-based nano-clusters which, however, can be reduced to the following two main types: (i) more or less pure copper, and (ii) sp2-bonded carbon clusters (the graphite-like component) intercalated with copper.

The importance of the intercalation phenomena is well known from both practical and scientific standpoints. The intercalation observed in DLC can offer some advantages in both senses owing to nano-sizes of the sp^2-bonded carbon clusters. Their tiny sizes accelerate the process of intercalation many-fold. Incorporation of copper-based clusters in DLC is capable of extending the set of its useful properties, such as readily controlled electrical conductivity, sensitivity to environment pollutants, possible catalytic activity, etc.

The sizes of copper-based clusters and their size-distribution functions are found experimentally by small angle X-ray scattering, TEM, electric current percolation, and surface plasmon excitation. A reasonable agreement is observed between the obtained results.

Acknowledgements

This survey is just a sum of contributions of my co-authors whose activity has made it possible. The constructive participation of Dr. S.G.Yastrebov deserves to be acknowledged separately.

This work was supported in part by Arizona University.

References

(*) In [I 3], the temperature at which copper induced absorption disappears is indicated erroneously as 280-350 K instead of 280-350 C.
(**) In [13], in the inset of Figure 4 hours are indicated erroneously instead of days.
(***) The coutesy of Dr.V.I.Siklitsky

1. Li, J., Seidel,T.E., and Mayer, J.W.(1994) Copper-based Metallization in ULSI structures, *MRS Bulletin* **Xix (8),** 15-18.
2. Ivanov-Omskii, V.I., Siklitskii V.I, and Baidakova, M.V 1996) Fractal Structure of Copper Clusters Embedded in DLC, Diamond (and other wide band-gap) Composites, *NA TO Advanced Research Workshop* St. Petersburg.
3. Guinier A. and Fournet, G.(1955) *Srnall-Angle **Scattering*** of X-Rays, John Wiley & Sons, Inc., New-York.
4. Bale, H.D. and Schmidt, P.W.(1984) Small-Angle X-Ray-Scattering Investigation of' Submicroscopic Porosity with Fractal Properties, *Phys. Rev. **Leit.*** 53(6), 596-599.
5. Ivanov-Omskii, V.1., Tolmatchev, A.V.,Yastrebov, S.G., Suvorova, A.A., and Sitmkova, A.A. (1996) Optical and TEM Study ofCopper-Born Clusters in DLC, The 3d *Int. Symp. on Diamond Films,* St.Peterburg (to be published)
6. Kolobov, A.V., Oyanagi, H., Yastrebov, S.G., Ivanov-Ornskii, V.1., Tanaka, K.(1996) X-Ray Absorption Study of Copper Clusters Embedded into Hydrogenated Amorphous Carbon, Diamond (and other wide band-gap) Composites, *NATO* ***Advanced Research*** *Workvhop,* St.Petersburg.
7 Ivanov-Omskii, V.I., Abaev, M.l., and Yastrebov, S.G. (1994) Optical properties of amorphous carbon doped with copper, in C.I. Renschler et al.*(eds), Novel Form of Carbon II, MRS Symposium Proceedings Series* 349, MRS Publishers, Boston, USA, 525-528.
8. Ivanov-Omskii,V.I., Lodygin, A.B., and Yastrebov, S.G(1995) Chacteristics of conducting structures in copper-doped diamond-like carbon, *Phys.Solid State* 37 (6), 920-922.
9. Ivanov-Omskii,V.I., Lodygin, A.B. Yastrebov, 5.6., and Clielnokov, V.(1995)

Formation of conductive network in diamond-like carbon by intercalation with copper, in A. Feldman et al.(eds.), *Application of Diamond Films and Related* ***Materials:*** *Third International Conference,* NIST Special Publication 85, Gaithersburg, USA, 779-782.
10. Efros A.L.,and Shklovskii B.I.,(1976) Critical Behawour of Conductivity and Dielectric Constant near the Metal-Non-Metal Transition Threshold *Phys. Stat. Sol. (b),* 76, 475-485.
11. Efros A.L.,and Shklovskii B.I.(1984) ***Electron Properties of Doped Semiconductors,*** Springer-Verlag, New-York.
12. lvanov-Omskii,V.I., and Frolova, G.S.(1995) Activation of Raman frequency in the infrared absorption of a-C:H(Cu), ***Tech.Phys.*** **40(9),** 966-967.
13. Ivanov-Omskii,V. I. and Frolova,G. S.(1995) Detection of graphite-like clusters in DLC by intercalation with copper, in A.Feldman et al.(eds.), *Application of Diamond Films and Related Materials: Third hitei.iiatioiial Conference,* NIST Special Publication 885, Gaithersburg, USA, 779-782.
14. Galdikas, A., Mironas, A., Setkus, A., Dapkis, L., Kazlauskiene, V., Miskinis, J., Prichodko, A., and Ivanov-Omskii, V. I.(1996) Gas sensing properties of CuO-based films, ***Lithuanian Journ. Phys.,*** 35(4), 3 14-320.
15. lvanov-Omskii, V.1., Tolmachev, A.V., and Yastrebov, S.G.(1996) Optical absorption of amorphous carbon doped with copper, *Phil. Mag.* B73, 715-722.
16. Robertson, J., (1986) Amorphous carbon, *Adv. Phys., 35, 317-374*
17. Ivanov-Omskii, V., Sitnikova, A.A., Suvorova, A.A.,Yastrebov, S.C,,and Zvonareva, T.K.(1996) Carbon Cluster Growth under Ion Sputtering of Graphite, *Mol. Mat.* **8,** 99-102
18. Bohren C., and Hull'inan D.(1983) *Absorption and Scattering of Light* hy *Sniall Particles,* John Wiley and Sons, Inc.

-RAY ABSORPTION STUDY OF COPPER CLUSTERS EMBEDDED NTO HYDROGENATED AMORPHOUS CARBON

A.V. KOLOBOV AND K.TAKANAKA
Joint Research Center for Atom Technology - National Institute for Advanced Interdisciplinary Research,
1-1-4 Higashi, Tsukuba, Ibaraki 305, Japan

H. OYANAGI
Electrotechnical laboratory,
1-1-4 Umezono, Tsukuba, Ibaraki 305, Japan

AND

S.G. YASTREBOV, V.I. IVANOV-OMSKII AND V.I. SIKLITSKY
A.F.Ioffe Physical-TechnicalInstitute
Polytechnicheskaya 26, St.Petersburg, 194021, Russia
e-mail: yass@ivom.ioffe.rssi.ru

bstract. Structure of copper nanoclusters embedded into hydrogenated norphous carbon has been studied by means of extended x-ray absorption ne structure and x-ray absorption near-edge spectroscopies. Two different ructures of copper clusters are observed with the average coordination umber for Cu-Cu interaction being 0.5 and 6.

eywords: copper nanoclusters, hydrogenated amorphous carbon, X-ray osorption fine structure

. Introduction

ivestigation of metallic clusters in insulating media is of much interest] which is partly stimulated by possible applications to single electron evices. Recently, successful confinement of copper nanoclusters in a maix of hydrogenated amorphous carbon, a-C:H, was reported [2,3]. The ructure of a-C:H presents a mixture of sp^2 and sp^3 bonded carbon clusrs [4]. Presence of sp^3 bonding ensures such properties of the material

. A. Prelas et al. (eds.), *Diamond Based Composites*, 191–196.

as a high hardness, high optical transparency and strong chemical inertness. The combination of these properties resulted in reference to a-C:H as diamond-like carbon. On the other hand, sp^2 clusters form a conductive network. Detection of sp^2 constituents of a-C:H by doping it with copper (a-C:H⟨Cu⟩) was recently discussed [5]. It was reported that the conductivity of a-C:H⟨Cu⟩ exhibits unusual behavior at concentrations of 14 at.% copper [1] including some evidence of high-Tc superconductivity [6]. Two kinds of clusters were reported for a-C:H⟨Cu⟩ which was confirmed recently by optical studies [7].

In this paper we report the results of Extended X-ray Absorption Fine Structure (EXAFS) and X-ray Absorption Near Edge Structure (XANES) measurements which present direct information on the structure of copper clusters.

2. Experimental

The a-C:H⟨Cu⟩ films were grown on fused quartz and/or crystalline silicon substrates by ion (magnetron) co-sputtering of graphite and copper targets in argon-hydrogen (80% Ar and 20% H_2) plasma. The substrate temperature, gas pressure in the growth cell and average magnetron power were 500K, 10 mTorr, and 0.4 - 0.5 kW, respectively. The energy of Ar^+ ions was 350 - 450 eV. The copper concentration was varied from 0 to 24% which was controlled by changing the relative areas of graphite and copper targets. Film thicknesses were in a range of a few thousand Ångstroms.

The measurements were performed at BL13B station at the Photon Factory [8] using a 27-pole wiggler in a fluorescence mode. An array of 19-element high-purity Ge solid-state detectors was used to detect the fluorescence. EXAFS spectra with good signal-to-noise ratio were obtained for very diluted thin films with a thousand Ångstrom thickness within one hour. The sample was mounted on an aluminum holder in an evacuated cryostat equipped with windows (KAPTON) for incident and fluorescent X-ray beams. A closed-cycle He refrigerator with a cooling power of ~9W at 20 K was used.

Cu K-edge EXAFS and XANES spectra of a-C:H⟨Cu⟩ were measured in the temperature range from 30 K to 300 K. As reference samples we have also measured the spectra for bulk metallic copper and for Cu_2O.

3. Results

Figure 1 (left) shows raw Cu K-EXAFS oscillations for a-C:H⟨Cu⟩ films with 5 and 14 at.% Cu taken at 30 K as a function of photoelectron wavenumber k after subtraction of smooth backgrounds due to the atomic absorption, from the fluorescence yield spectra. The spectrum for bulk

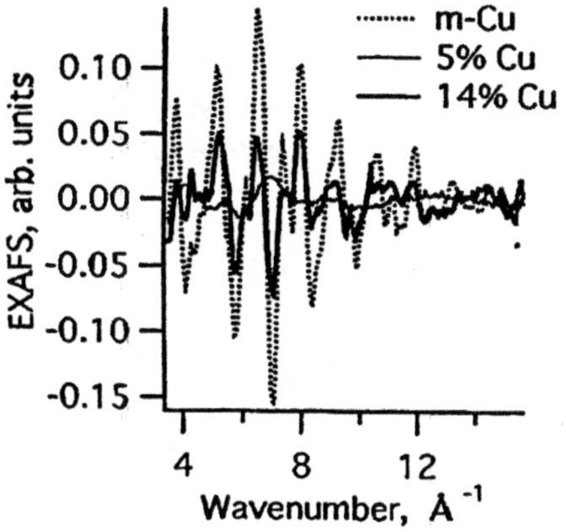

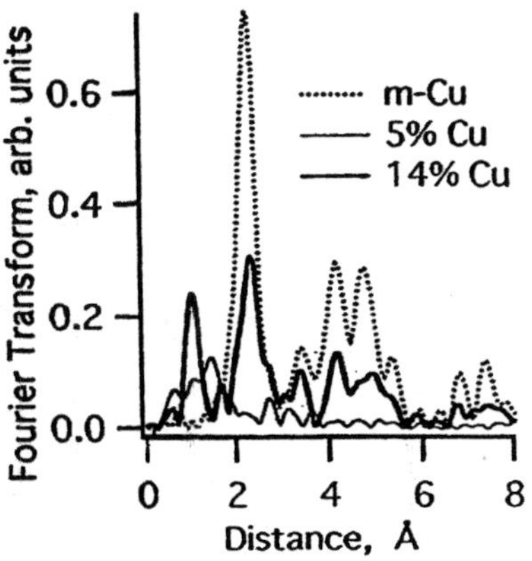

Figure 1. Cu K-EXAFS oscillations (left) and Fourier transformed spectra (right) for a-C:H⟨Cu⟩ films with 5 and 14 at.% Cu. The spectrum for bulk Cu is also shown.

copper is also shown for comparison. The background function given as a combination of the third and forth order polynomials, with tabulated coefficients [9] (Victoreen Function) which smoothly interpolate EXAFS oscillations using a cubic spline method, was normalized to the edge jump and subtracted from the fluorescence spectrum.

The EXAFS oscillations multiplied by k were Fourier-transformed (Figure 1, right) using the region extending from 4.5 to 15 Å^{-1}. One can see that for the sample with 5 at.% Cu there is basically no peak corresponding to Cu-Cu interaction while such peaks clearly appear for the sample with 14 at.% sample. A peak at R$\sim$1 is due to poor background subtraction for a very dilute sample, peaks at larger R correspond very well to peaks for the bulk copper. It should be noted that all samples other than the one with 14 at.% Cu have similar spectra (not shown here). A broad peak is observed in these samples at distances smaller than the Cu-Cu bond length. Curve-fitting analysis using a single-scattering theory [10] and FEFF amplitudes [11] gives the average coordination number of $\sim$0.5 for those samples. For the case of the 14% Cu sample the Cu-Cu correlations are much more strongly pronounced and the average coordination number reaches 6 ± 3.

Figure 2 (left) shows XANES spectra for the measured samples and for the standards (bulk Cu and Cu_2O). It should be noted that the samples were measured in the fluorescence mode while the references were measured in the transmission mode. This difference results in various scales of the y-axis (while the energy axis is obviously the same for the two cases). One can see that the spectra for the samples with concentrations different from 14 at.% Cu resemble strongly the spectrum for Cu_2O implying that copper in the samples exists in the form of copper oxide. The 14 at.% Cu sample is similar to that of the bulk metallic copper which agrees well with the

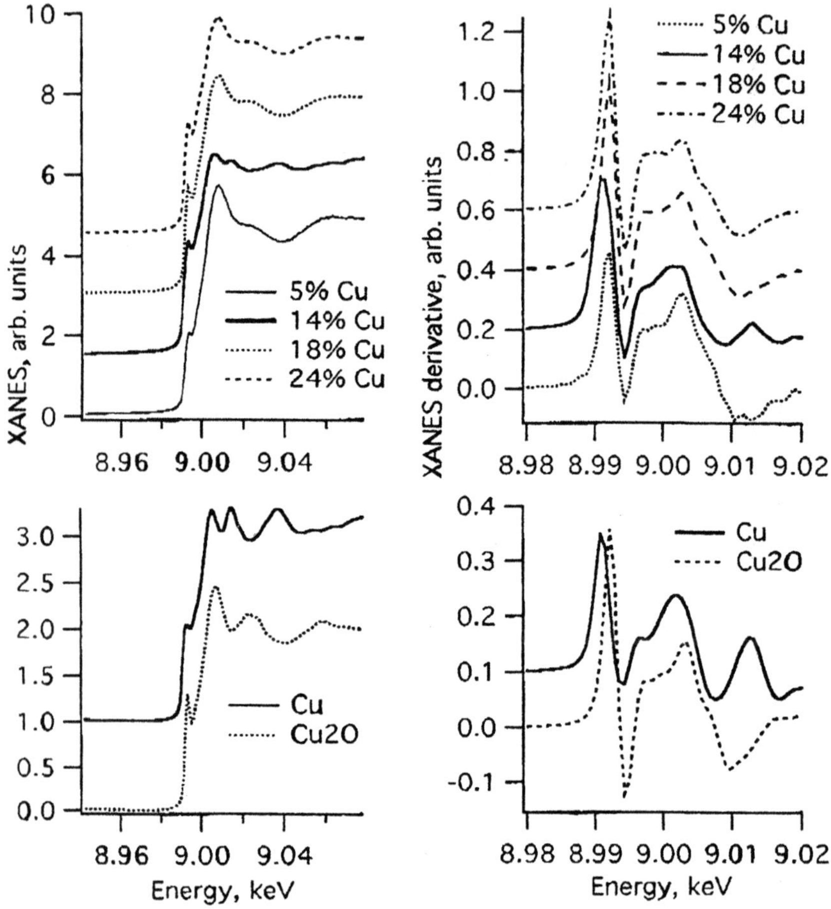

Figure 2. XANES spectra for the measured samples (top left) and references (bottom right) and derivative XANES spectra for the same samples (left)

EXAFS data and suggests that for this concentration the structure of Cu clusters is essentially the same as that of the bulk copper. These correlations can be seen even better from the derivative spectra shown in the same figure (right).

Figure 3 shows on the expanded scale XANES spectra for samples with three different copper concentrations. Isobetic points around 9.0045 and 9.017 keV are clearly seen. One can see a systematic increase in the intensity of peak A and a corresponding decrease in the intensity of peak B with the increase of the copper concentration.

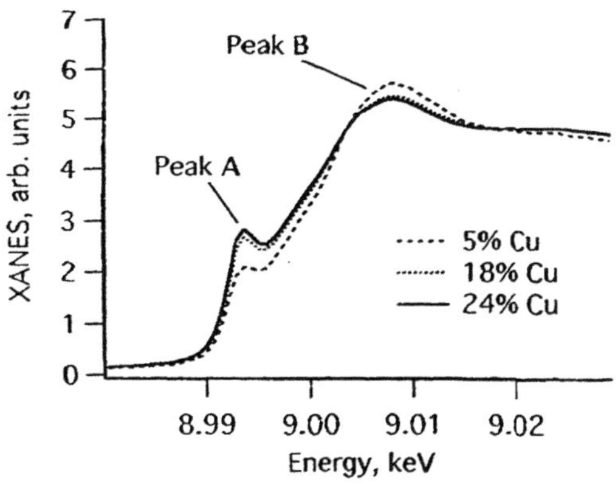

Figure 3. Variation in the peak intensities of XANES spectra with different Cu concentrations

4. Discussion

As suggested earlier from the optical studies, copper in a-C:H⟨Cu⟩ can form two different kinds of inclusions. First, copper can be randomly distributed within the graphite-like constituent of a-C:H⟨Cu⟩ matrix. In such a case the average Cu-Cu interaction is negligible and one should expect a very low average coordination number for this interaction as observed experimentally. The observed coordination number of <1.0 can indicate either a complete absence of Cu-Cu correlations or the formation of very short Cu-wires. It is not quite clear from the present study whether copper in this geometry mainly interacts with carbon or oxygen species of the surrounding matrix. The fact that the XANES spectra resemble that of Cu_2O and non-existence of Cu-C compounds support the latter possibility although the former one cannot be completely excluded.

At higher Cu concentrations, Cu-Cu interaction increases and eventually bulk-like structure of the cluster is formed as evidenced by EXAFS and XANES spectra for the 14 at.% Cu sample. The fact that the average Cu-Cu coordination number in the cluster is smaller than that in the bulk metal can be understood as follows. In small clusters the relative number of the surface atoms, whose Cu-Cu coordination is half of that for atoms in the volume of the cluster, is not negligible and as a result the average coordination number is smaller that 12 (which is the case for bulk copper). In the assumption of spherical shape of clusters the average coordination number should generally be larger than 6 while for rather thin flat flake-like formations in which case the average coordination number should be smaller than 6. The obtained value of 6 ± 3 is at the boundary of the two cases which can be interpreted either as due to spheres with small diameter ($\sim$ 20Å), or two-dimensional flake-like formations with a thickness of several Ångstroms and any size in the other two dimensions, or a combination of both. The obtained value of the characteristic size of a few tens of Ångstroms agrees with the previous conclusion [7].

The redistribution of peak intensities seen in Figure 3 indicates changes in the density of states arising from the increase in copper concentration. Unfortunately, the data are not sufficient for a more detailed conclusion but they show that samples with smaller and larger Cu concentrations are different despite the apparent similarity in the XANES spectra.

5. Conclusion

Using EXAFS and XANES spectroscopies we have obtained direct structural information about Cu clusters embedded into a a-C:H⟨Cu⟩ matrix. It is shown that different types of clusters are formed, namely, one with a very small Cu-Cu coordination number and the other with the structure similar to that of the bulk copper which could be either small spheres or flakes. The size of bulk-metal-like clusters is estimated to be several tens of Ångstroms.

References

1. Hovel, H., Fritz, S., Hilger, A., Kreibig, U., and Volner, M. (1993) Width of cluster plasmon resonances: bulk dielectric functions and chemical interface damping, *Phys.Rev.*, **B48**, 18178-18188
2. Ivanov-Omskii, V.I., Abaev, M.I., and Yastrebov, S.G. (1994) Novel Forms of Carbon II, MRS Meetings Symposium Proceedings Paper T7.61
3. Ivanov-Omskii, V.I., Abaev, M.I., and Yastrebov, S.G. (1994) Optical properties of amorphous carbon doped with copper, in C.I.Renschler et al.(eds), Novel Form of Carbon II, MRS Symposium Proceedings Series, 349, MRS Publishers, Boston, USA, 525-528
4. Robertson, J. (1986) Amorphous carbon, *Adv.Phys.*, **35**, 317-374
5. Ivanov-Omskii, V.I. and Frolova, G.S. (1995) Activation of Raman friquencies in the infrared absorption of hydrogenated amorphous copper graphite, *Tekh.Phys.*, **40**, 966-967
6. Bazieva, N., Yastrebov S.G., Masterov, V.F., and Prihkodko, A.V. (1994) Electrical and magnetic properties of thin amorphous carbon films doped with copper, *Mol.Mater.*, **4**, 143-147
7. Ivanov-Omskii, V.I., Tolmatchev, A.V., and Yastrebov, S.G., (1996) Optical absorption of amorphous carbon doped with copper, *Philos.Mag.*, **B73**,715-722
8. Oyanagi, H., Shyoda, R., Kuwahara, Y., and Haga, K., (1995) Surface sencitive EXAFS in the hard x-ray region with sub-monolayer sensitivity, *J.Synchrotron Radiation*, **2**, 99-105
9. *International Tables for X-ray Crystallography*, (1962), Vol. III, Kynoch Press
10. Stern, E.A., (1974) Theory of extended x-ray absorption fine structure, *Phys.Rev.*, **B10**, 3027-1037
11. Rehr, J.J., Albers, R.C., and Zabinsky, S.I., (1992) High-order multiple-scattering calculations of x-ray-absorption fine structure, *Phys.Rev. Lett.* **69**, 3397-3400.

FRACTAL STRUCTURE OF COPPER CLUSTERS EMBEDDED IN DLC

V.I. IVANOV-OMSKII, V.I. SIKLITSKY AND M.V. BAYDAKOVA
A.F.Ioffe Physical-Technical Institute
Polytechnicheskaya 26, St.Petersburg, 194021, Russia
e-mail: siklitsk@ivom.ioffe.rssi.ru

Abstract. A study of small angle X-ray scattering by the DLC containing copper-based clusters was performed to characterize the cluster structure. Fractal dimensions and sizes of scatterers, which are shown to be the copper-based clusters, are estimated as a function of total copper concentration.
Keywords: copper/cluster/diamond-like carbon/fractal dimension/ X-ray small angle scattering

1. Introduction

The existence of two types of copper-based clusters has been assumed recently as a result of the study of electrical [1] and optical [2] properties of DLC doped with copper. We report on a study of small angle X-ray scattering by the DLC containing copper-based clusters, bearing in mind their structural characterization.

2. Experimental

DLC films doped with copper was produced on silicon substrates by co-sputtering graphite and copper targets in argon (80%) - hydrogen (20%) plasma. The substrate temperature and gas pressure in growth chamber were about 500 K and 10 mTorr respectively. The magnetron power was maintained within 0.4-0.5 kW. Typical film thickness was in the range from 0.1 to 0.6 micron. The copper concentration presented in Table 1 was determined by SIMS.

M. A. Prelas et al. (eds.), Diamond Based Composites, 197–202.

TABLE 1. The sample parameters

N	Copper content,%	Index eq.(1), α	Fractal dimension D=6-α	Scatterer size, Å	Comments
1	4	5.4	2.0^a	16.	Scatterer with smooth boundary surfaces [5]
2	6	3.9	2.1	15.	Scatterer with smooth fractal surfaces [4]
3	14	3.7	2.3	16.	———"———
3^b	14	2.9	2.9^c	15.	Fractal objects [6]
4	24	3.6	2.4	>24.	Scatterer with smooth fractal surfaces [4]

(a) - γ=7-α=1.6 (at $\alpha > 4$) is defined as index in power law [5].
(b) - Vacuum annealled sample 3.
(c) - Fractal dimension is defined as D=-α [6].

X-ray scattering by the samples were studied with a RIGAKU X-ray diffractometer operating at a wavelength 1.541Å(Cu-Kα line) and having high intensity beam with angle divergence $\leq 10'$. Diffractograms were measured at 50 kV in reflection geometry. The diffraction patterns were analyzed at angles falling in the range from 1.0 to 4.0 degrees.

3. Results

It was found by direct comparison of the intensity of radiation scattered by copper-free and copper-doped DLC films that the contribution of the DLC matrix is rather small, i.e., copper-based clusters embedded in the DLC matrix are responsible for the X-ray scattering appearing at small angles. Typical angle dependencies of the intensity of radiation scattered a DLC film containing 6% of copper are shown in Figure 1 for two spectral slit widths. Both the curves demonstrate clearly defined peaks at the smallest angles, and their shapes are similar to the right of the peak. Narrowing the slit clearly improves the peak resolution but has practically no effect on the curve slopes to the right of the peak. The classical theory of small-angle scattering of radiation of wavelength λ by an array of regular spheres [3] predicts the scattered intensity to be at a maximum at the angle $\theta = \lambda/4R$, where R is the sphere radius. The shape and position of the peak in Figure 1 show that the average size of the copper-based cluster assembly is fairly well defined and the mean cluster radius occurs to be equal to 15Å in reasonable agreement with estimations for non-spherical clusters provided by optical measurements [2]. As to the angular dependence of background scattering,

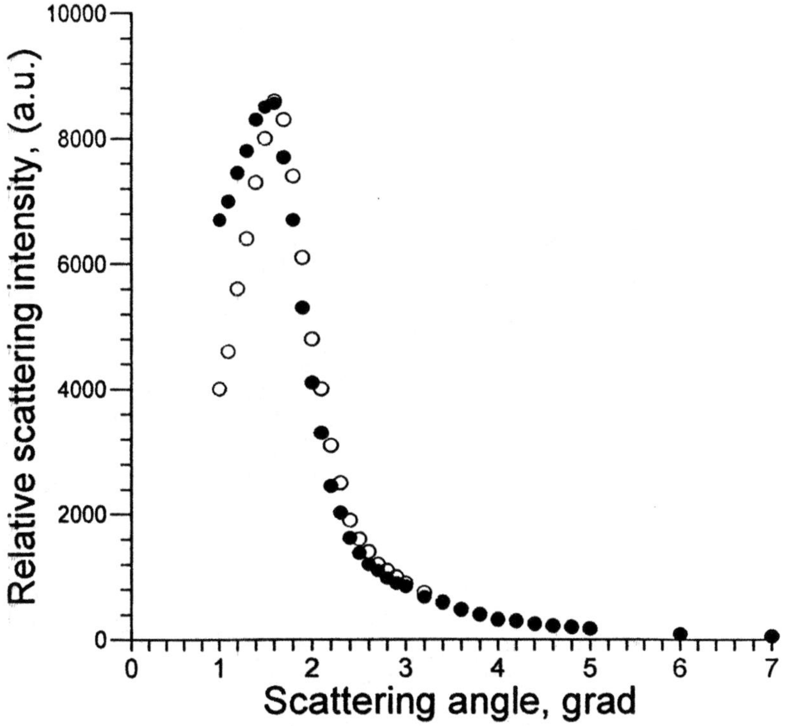

Figure 1. X-ray scattering intensity versus scatering angle θ for DLC doped with 6% copper: • • • - spectral slit 0.32mm; o o o - spectral slit 0.05mm

according to the traditional theory, which assumes that the scatterers are bounded by a smooth surface, the scattered intensity I(q) should be proportional to an integral negative power α of the transferred momentum q:

$$I(q) \sim q^{-\alpha}, \tag{1}$$

where $q = 4\pi\lambda^{-1}sin(\theta/2)$, where λ is the wavelength of scattered radiation. For scatterers bounded by a smooth surface the index α is exactly 4 [3]. This means that log I(q) has to be a linear function of logθ, and, therefore, log-log representation of experimental data was used in Figure 2 to check the applicability of eq.(1). As evident from Figure 2, for all measured samples the scattered amplitude is indeed a power function of angle at not-too-small angles, but the index α is not integer, as is apparent from the Table 1. Although the log-log coordinate system smoothes the peaks (compare Figure 1 and Figure 2), it still possible to observe a weak dependence of the position of the scattered intensity maximum on copper content, except

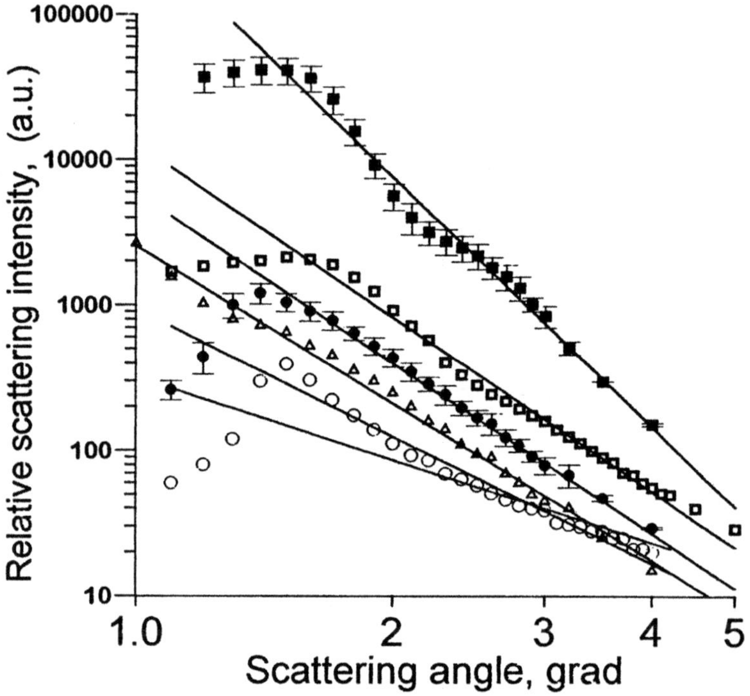

Figure 2. X-ray scattering intensity versus scatering angle θ for DLC doped with copper: ■■■ - 4% copper; □□□ - 6% copper; • • • - 14% copper; o o o - 14% copper (vacuum annealled sample); △△△ - 24% copper.

for sample 4. This means that the mean correlation radius of scatterers is practically concentration independent at copper concentrations less than 24%, as also seen in Table 1. Some deviation from the monotone run of experimental points to the right from the peaks in Figure 2 results from the neglect of higher terms in eq.(1).

Annealing sample 3 in vacuo at 240^0C for 2 hours has no noticeable effect on the peak position (compare samples 3 and 3a in Table 1 and Figure 2). This sample was chosen for annealing because its peak position had been well defined even before annealing, probably owing to unintentional rise of the film growing temperature. In any case thermal treatment has practically no effect on the average radius of copper-based clusters, but improves the peak resolution and changes the curve slope (Figure 2).

4. Discussion

The observed angular dependence of scattered intensity is typical for small-angle scattering by fractal structures [4]. The exponent α in eq.(1) may be related to the fractal dimension of the structures in question. In doing this we shall follow the model approach [4] which ascribes the power-law scattering to independently acting submicroscopic scatterers with a power-law dimension distribution. The uncertainty in the exponent is about ± 0.1 as estimated from data reproducibility. We believe that the fractal structure that scatters X-rays is a material containing scatterers with smooth or fractal boundary surfaces. According to the model, at $3 < \alpha < 4$ the exponent α in eq.(1) is related to the fractal dimensionality D of scatterers by $D = 6 + \alpha$ (scatterers with fractal boundaries)[4]. For the case $\alpha > 4$, the exponent α defines the exponent $\gamma = 7 - \alpha$ describing the distribution P(r) of scatterers with smooth boundary surfaces of radius R [5]:

$$P(R) \sim R^{-\gamma}$$

As it follows from the foregoing and the data displayed in Table 1, the fractal dimensionality D is essentially less than 3 in most cases except in the annealed sample, increasing from D=2 as the copper concentration becomes higher. This may correspond to a transition from scattering by smooth-boundary surfaces (bald scatterers) at low copper concentration to scattering by fractal-boundary surfaces (hairy or starfish-like scatterers) at higher concentrations.

It is notable that the average correlation radius remains practically constant except for the highest copper content. It looks like the copper cluster size is controlled by typical sizes and shapes of pores in DLC and their distribution function. At low copper concentrations mean distances between pores occupied by copper clusters are large enough to allow appearance of the most equilibrium cluster shape which seem to be spherical due to weak Cu-C chemical bonding. There is also a typical size of pores which is determined by the thermodynamics of their formation in DLC and occurs to be about 15Å according to the proposed model.

With increasing copper content, interaction between pores occupied by copper clusters deforms their shape. As a result of the interaction whose symmetry makes allowance for disordered DLC structure, the cluster surface acquires a fractal character (see Table 1; Figure 2 of Ref.[7]). As also seen from Table 1, annealing a sample with 14% of Cu in vacuo reassembles the cluster structure in a way that may be characterized as coagulation of cluster-branches into smoother-surface clusters through relaxation of structural deformations. As follows from the fractal dimensionality D=2.9, this object resembles a snow ball. Relying upon the absence of the peak at 1.0

degree (the instrument limit) at the highest copper concentration (24%), one arrives to a conclusion that the internal energy of copper clusters becomes high enough for the scatterer size to be extended with retention of the surface topology.

5. Acknowledgments

This work was supported in part by the Arizona University grant, the US Department of Defense and the Russian National Interdisciplinary Program "Fullerenes and Atomic Clusters" (grant N94007). The authors are grateful to Dr. N.N.Faleev for providing his experimental data.

References

1. Ivanov-Omskii, V.I., Lodygin, A.B., and Yastrebov, S.G. (1995) Characteristics of conducting structures in copper-doped diamond-like carbon, *Phys.Solid State*, **37(6)**, 920-922
2. Ivanov-Omskii, V.I., Tolmachev, A.V., and Yastrebov, S.G. (1996) Optical absorption of amorphous carbon doped with copper, *Phil.Mag. B*, **73**, 715-722
3. Guinier, A. and Fournet, G. (1955) *Small-Angle Scattering of X-Rays.* John Wiley & Sons, Inc., New-York
4. Bale, H.D. and Schmidt, P.W. (1984) Small-Angle X-Ray-Scattering Investigation of Submicroscopic Porosity with Fractal Properties, *Phys.Rev.Lett.*, **53(6)**, 596-599
5. Glatter, O. (1979) The Interpretation of Real-Space Information from Small-Angle Scattering Experiments, *J.Appl.Cryst.*, **12**, 166-176
6. Kjems, J., Freltoft, T. (1985) *Scaling Phenomena in Disordered Systems.* Plenum Press, New York
7. Ivanov-Omskii, V.I. (1996) Copper Nanoclusters in DLC, Diamond (and other wide band-gap) Composites, *NATO Advanced Research Workshop*, St.Petersburg.

ON THE COPPER -CARBON INTERACTION IN Cu-DOPED DIAMOND-LIKE CARBON

E.A.SMORGONSKAYA and V.I.IVANOV-OMSKII
A.F.Ioffe Physico-Technical Institute, Russian Academy of Sciences
Polytechnicheskaya 26, St.Petersburg 194021, Russia

Abstract

The tight binding approach is applied to study the electronic energy spectrum and bonding in small Cu-containing graphite-like clusters (GLC). The $4s$-$2p_z$ coupling between Cu and C atoms is taken into account as the principal mechanism of the Cu - C interaction in the GLC. Numerical calculations are performed in the nearest neighbor approximation with universal Hassison's matrix elements of the Hamiltonian. Estimations are made for the electronic energy gap, Cu - C-plane binding energy, distribution of the effective charge among the constituent ions, and Cu-induced static dipole moments. The results are used to discuss the electronic and vibrational properties of Cu-doped diamond-like carbon.

1. Introduction

Diamond-like amorphous carbon doped with copper (a-C:H<Cu>) in the process of magnetron-assisted co-sputtering of the components from graphite and copper targets exhibit a number of peculiarities in electrical and optical properties. The behavior of d.c. conductivity and optical absorption spectra of the a-C:H<Cu> films with increasing Cu content were attributed to the formation and rearrangement of Cu-containing cluster systems built in the host a-C:H matrix [1]. In this context, the problem of interaction between the Cu dopant and C atoms in a-C:H on the microscopic level seems to be of considerable interest.

As suggested by the data of IR and Raman spectroscopy [2], the undoped and Cu-doped a-C:H layers prepared by magnetron-assisted sputtering contain both graphite-like and diamond-like microdomains with sp^2- and sp^3-bonding between C atoms, respectively. The presence of graphite-like clusters (GLC) 5 - 20 Å in linear dimensions is typical for diamond-like (hard) carbon [3]. It is reasonable to believe that Cu dopants do not not perturb σ-electrons and interact mainly with π-electrons of the C atoms in 6-membered "aromatic" rings of graphite-like plane sheets of the related GLC.

The principal goal of this work is to gain an insight into the effect of Cu impurity on the electronic π-states of carbons in GLC. By analogy with graphite crystal, it is

M. A. Prelas et al. (eds.), Diamond Based Composites, 203–210.

these states that form the upper edge of bonding (valence) states and the lower edge of antibonding (conduction) states. Really, the bonding and antibonding π-states of the GLC are broadened due to the interaction between the randomly oriented microdomains (homogeneous broadening) and possible difference between them in size and geometry (non-homogeneous broadening). Nevertheless, finite size of the GLC is responsible for the electronic energy gap, making a-C:H into a semiconductor material, as distinct from semimetal crystalline graphite. This is the reason why the problem imposed in this work can be partially solved by treating a rather small plane carbon cluster, or a few adjacent "aromatic" rings consisting of N carbon atoms and interacting with a single Cu dopant.

In this report the results of simple tight binding calculations of the electronic spectrum in the vicinity of the energy gap and effective charge distribution in such GLC are presented. In spite of relative roughness of the approximation, some semi-quantitative and qualitative conclusions about possible Cu-induced changes in the electronic and vibrational properties of a-C:H are drawn and discussed in connection with experimental data.

2. Tight Binding Model

According to the well-known tight binding approach, the wave function of an electronic state in a GLC, $|\psi\rangle$, is represented as a linear combination of atomic orbitals, χ_i (LCAO):

$$|\psi\rangle = \sum_i u_i \, |\chi_I\rangle \, , \qquad (1)$$

where the coefficients u_i are to be found and $i = 1,2,\dots n$ numbers the basis orbitals. The variational procedure commonly used to determine the eigenvalues E_j and eigenfunctions $|\psi_j\rangle$ of the Hamiltonian H in the LCAO basis gives the following system of j equations in u_i:

$$\sum_i \langle\chi_j \,|H|\, \chi_I\rangle \, u_i \, - E_j \, u_j \; = 0 \, , \qquad (2)$$

where $E_j = \langle\psi_j|H|\psi_j\rangle$ with the LCAO basis functions $|\chi_I\rangle$ presumed orthogonal. The electronic energy levels E_j are obtained by diagonalization of the $\langle\chi_j\,|H|\,\chi_I\rangle \equiv H_{ji}$ matrix. For each value E_j the set of coefficients u_i, defining $|\psi_j\rangle$, can be found to estimate the electronic density $|u_i|^2$ on the i-th atom in the j-th state.

3. Numerical Calculations

Numerical calculations were carried out in the nearest neighbor approximation. This means that the LCAO matrix elements H_{ji} were assumed to be non-zero only if j and i are referred to the nearest neighboring atoms. To compute these H_{ji}, we used the well-known universal Harrison formula [4]:

$$V_{ll'm} = \eta_{ll'm} \, \hbar^2 / (\mu \, d)^2 \, . \qquad (3)$$

Here $V_{ll'm}$ is the matrix element H_{ji} taken between the atomic functions with the orbital quantum numbers l and l' and the magnetic number m for two neighboring atoms, d is the interatomic distance, μ is the electronic mass, and $\eta_{ll'm}$ are the universal dimensionless factors tabulated in [4]. For atomic s- and p-states, $\eta_{ll'm}$ takes 4 values correspondent to the $ss\sigma$-, $sp\sigma$-, $pp\sigma$-, and $pp\pi$- interactions.

The calculations were performed for several kinds of Cu-containing GLC that consisted of a few side-shared ideal "aromatic" rings of C atoms in the xy plane and one Cu atom placed on the hexagonal axis z of one of the rings ($z \perp xy$). It was assumed that the interaction between different carbon sheets in the GLC were negligibly small, as in graphite crystal. Besides, it was assumed that the Cu dopant interacted mainly with carbons from only one graphite-like plane. Figure 1a,b schematically shows two examples of such microdomains which involve 10 and 16 carbon atoms, respectively. As mentioned above, we restricted ourselves by one $2p_z$-orbital per C atom and one $4s$-orbital of Cu dopant, assuming that the latter did not affect the strong covalent C - C σ-bonds in the xy plane.

In this case, the (N+1)×(N+1) H_{ji} matrix (N being the number of C atoms in the GLC) involves 4 different parameters: ε_p , ε_s , $V_{pp\pi}$,and $V_{sp\sigma}$. Here the diagonal elements $\varepsilon_p = \langle p_z | H | p_z \rangle = -8.97$ eV and $\varepsilon_s = \langle s | H | s \rangle = -6.92$ eV are the energies of the $2p$- and $4s$-states for C and Cu atoms, respectively. The non-zero off-diagonal elements $V_{pp\pi} = \langle p_{zj} | H | p_{zl} \rangle$ for a pair of neighboring C atoms were calculated by Eq.(3) with $\eta_{pp\pi} = -0.81$ and $d = 1.42$ Å, equal to the C - C distance a in the ideal graphite sheet. The off-diagonal elements $\langle s | H | p_z \rangle$ correspondent to the Cu - C interaction depend on the position z of the Cu atom with respect to the carbon plane:

$$\langle s | H | p_z \rangle = V_{sp\sigma}\, z \,/\, (z^2 + a^2)^{1/2} , \tag{4}$$

where $V_{sp\sigma}$ is calculated by Eq.(3) with $\eta_{sp\sigma} = 1.84$ and $d^2 = z^2 + a^2$.

As easily seen from Eqs.(3) and (4), there exist an optimal position of the Cu atom, z_0, that provides a maximal bonding energy between the Cu atom and the C sheet: $z_0 = a / \sqrt{2} \cong 1$Å. This z_0 value can be accepted as an approximate characteristics of the equilibrium position of a Cu atom in a GLC. The related Cu - C-plane bonding energy, $E_b^{Cu\text{-}C}$, can be assumed roughly equal to the activation energy of removal of the Cu atom from its equilibrium position.

4. Results and Discussion

Figures 2 and 3 show the calculated energy level structure originated from coupled carbon $2p_z$-electrons (the in-plane coupling), not interacting (b) or interacting (c) with the Cu $4s$-electron in the both GLC shown in Fig.1. The related levels of isolated C and Cu atoms are also shown in Figs.2, 3 (a).

First of all it should be noted that the structures exhibit a gap, E_g , between bonding and antibonding states which is due to the "in-plane" C - C coupling in

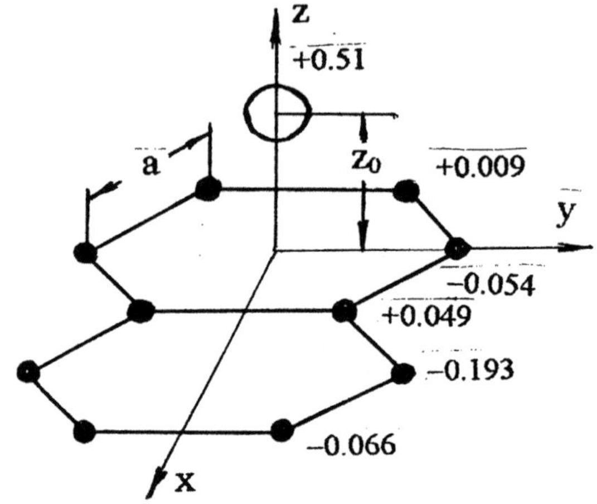

a

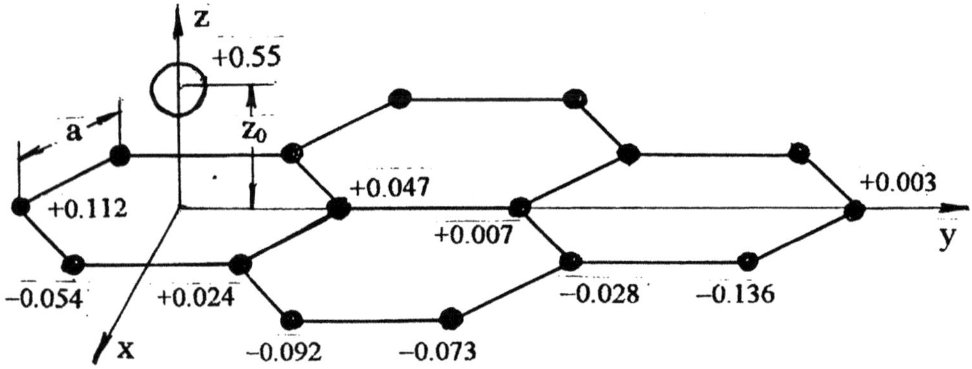

b

Fig.1. Sketches of two small graphite-like clusters of N carbon atoms (solid circles) and one copper atom (empty circle), with $N = 10$ (a) and $N = 16$ (b).
Figures near the atoms indicate the correspondent effective ionic charge $e_i{}^*/e$.

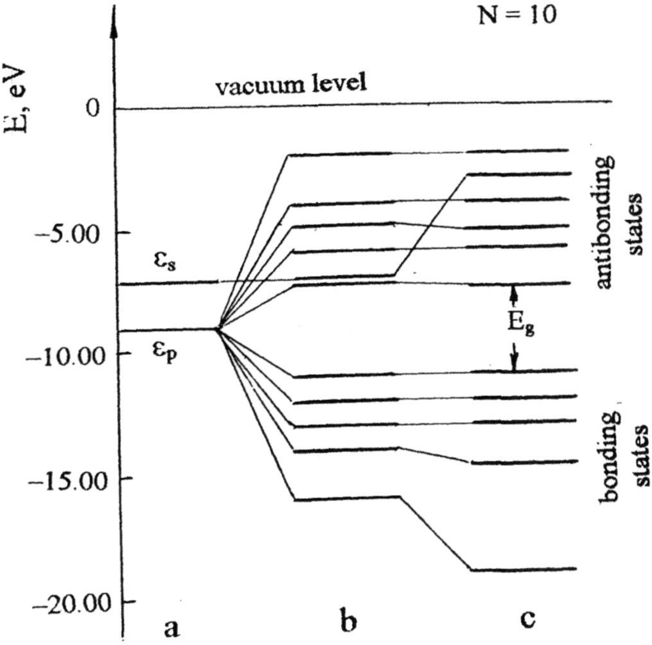

Fig.2. Splitting of the $2p_z$- and $4s$-levels of C and Cu atoms (a) calculated with taking into account the "in-plane" C-C coupling only (b) and the both "in-plane" C-C and "out-of-plane" Cu-C coupling (c).

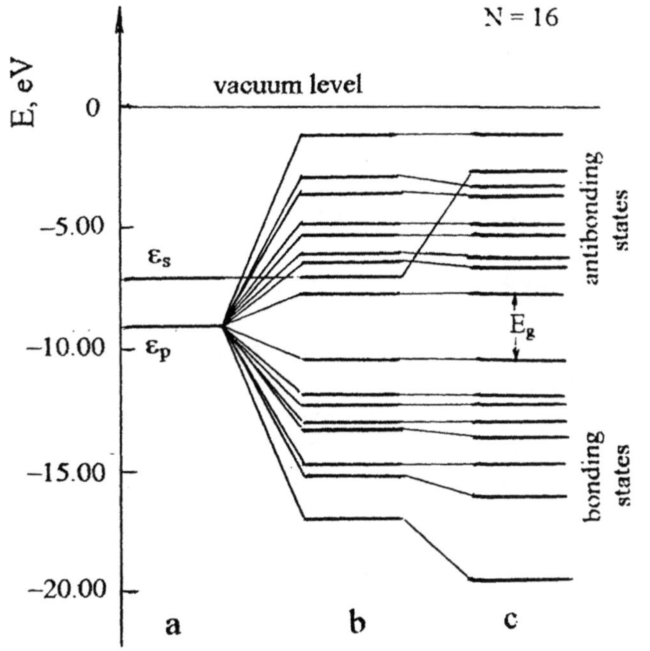

Fig.3. The same, as in Fig.2, for the GLC of Fig.1b (N = 16).

confined GLC. At zero temperature, all of the bonding π -states of undoped GLC are occupied by electrons, and the antibonding π^*-states are empty in the absence of Cu. As expected, the calculated value of E_g turns to be less for the larger GLC (Fig.3). In this case we have $E_g \cong 2.6$ eV, that is in satisfactory agreement with the experimental data for the optical gap in a-C:H [1]. As evident from Figs.2 and 3, the Cu 4*s*-level falls in the energy range of antibonding π^*-states of carbons, so that the extra unpaired electron of the Cu dopant should be localized in the lowest antibonding state, with the localization length confined by the GLC size. In the conditions of sufficiently strong interaction between different GLC, such electrons can make a noticeable contribution to electrical conduction in the Cu-doped a-C:H. A profound increase of the conductivity of a-C:H<Cu>) with increasing Cu content up to ~15 at % was actually observed in the experiment [1]. Thus, copper can serve as an efficient electrically active dopant in semiconductor diamond-like carbon.

The Cu - C coupling yields a lowering of the lower bonding levels in the GLC, in turn resulting in lowering the average energy of the bonding states. At the same time, the lowest occupied antibonding state is hardly shifted by the Cu - C coupling. The absolute value of the Cu - C-plane bonding energy $E_b^{\text{Cu-C}}$ can be estimated as

$$E_b^{\text{Cu-C}} = (N+1)^{-1} \sum_{\text{occupied states}} [2\,(E_b^{\,ud} - E_b^{\,d}) + (E_{nb}^{\,ud} - E_{ab}^{\,d})] \,. \tag{5}$$

Here subscripts *b*, *ab*, and *nb* are referred to occupied bonding, antibonding and non-bonding states, respectively. (The non-bonding state is the 4*s*-state of an isolated Cu atom, i.e. $E_{nb}^{\,ud} = \varepsilon_s$). The superscripts *ud* and *d* denote the levels of undoped and Cu-doped GLC. For the two GLC of Fig.1, $E_b^{\text{Cu-C}} \cong 0.67$ eV ($N = 10$) and 0.43 eV ($N = 16$). Thus, in the larger GLC the Cu dopant atom is weakly bound to the C-sheet, as compared to the smaller GLC. With increasing temperature the Cu atom can be removed from its equilibrium position, with the activation energy of the process of $\sim E_b^{\text{Cu-C}}$.

The effective charges on the Cu and C atoms , e_i^*/e (*e* being the electronic charge), are indicated in Fig.1 a,b near the correspondent atoms in the both GLC under consideration. The estimations were made by the formula

$$e_i^*/e = (N+1)^{-1} \sum_{\text{occupied states}} (2\,|u_{bi}|^2 + |u_{abi}|^2). \tag{6}$$

As indicated above, the Cu dopant acts as a donor impurity with a negative ionization energy and hence it is positively charged even at zero temperature. In the both GLC the effective charge of the Cu ion is slightly higher than + 0.5*e*. Clearly the plane carbon sheet as a whole has a negative charge of the same absolute value, − 0.5 *e*. Consequently, a static dipole moment in the *z* direction is naturally produced by the Cu dopant. Moreover, due to the asymmetry in the position of the dopant with respect to the C plane, a dipole moment parallel to the plane appears as well. The Cartesian projections P_x , P_y , and P_z of the total Cu-induced dipole moment $\boldsymbol{P}$ in the GLC are presented in Table below. Note that the negative electronic charge is typically shifted to the “edge” carbon atoms of the GLC.

N	P_x	P_y	P_z
10	$-0.95\ ea$	0	$0.36\ ea$
16	0	$-1.53\ ea$	$0.38\ ea$

It should be emphasized that the presence of Cu dopants in the both GLC configurations lowers the symmetry of the clusters, simultaneously transforming them from two- to three-dimensional systems with non-zero static dipole momemt. These changes might be expected to manifest themselves in vibrational properties of the Cu-doped GLC and a-C:H<Cu>. In particular, additional low-frequency bands will appear in the IR spectra due to dipole-active local vibrations of the Cu ion with respect to carbons in the *xy* plane. In addition, because of the "out-of-plane" Cu dopant, pairs of neighboring C atoms in the "aromatic" rings turn out to be non-equivalent in charge not only due to the confinement of the GLC (Fig.4). In this case, the C - C atomic vibrations may change the "in-plane" and "out-of-plane" net dipole moments and thus become active in IR absorption.

It is well-known that the Raman-active E_{2g} mode of "in-plane" C - C vibrations of ideal "aromatic" rings in a single crystalline graphite plane is forbidden in IR absorption. More precisely, the related IR band is weakly allowed, since the Raman-active E_{2g} mode is splitted into the Raman-active E_{2g} and IR-active E_{1u} modes, when considering two adjacent ideal planes shifted with respect to each other [5]. In some sense, the role of Cu dopants in the GLC is similar to that of the adjacent C planes in graphite: in both cases the symmetry of a single plane is broken. However, in the case of our confined Cu-doped GLC, the effect of symmetry breaking is much more pronounced, since many of C atoms located even in the same plane become non-equivalent.

The above considerations clarify the mechanism of strong activation of the Raman 1570 cm^{-1} band (the so-called G band) in the IR absorption spectrum observed experimentally on doping a-C:H with Cu [2]. At the same time, the position of the Raman G band remained unchanged with doping. This means that the geometry of rings and hence the frequency of the "in-plane" vibrations are hardly affected by the Cu dopant. It is reasonable to argue that the effect of Cu on the "in-plane" charge distribution and net dipole moments will be stronger, if the Cu dopant is located above a perimeter ring rather than above an internal one in a GLC. This is confirmed by Cu-induced activation of the so-called D Raman 1350 cm^{-1} band in the IR spectrum of Cu-doped a-C:H [2]. In fact, the D band is usually considered as a result of disorder and, in particular, size confinement of the GLC in a-C:H.

The origin of the IR absorption bands G and D in Cu-doped a-C:H is somewhat similar to the related effect of nitrogen doping of a-C:H [6]. However, as argued in [6], the N atoms substituted the "edge" C atoms of microdomains. In our model the Cu dopants intercalate graphite-like plane sheets, inducing dipole-active C - C vibrations even in ideal hexagonal rings of a single sheet.

5. Summary

In conclusion, the simple tight binding model applied to small GLC shows that doping with Cu essentially influences the electronic and vibrational properties of diamond-like carbon films. The doping yields partial filling of the antibonding states of semiconductor GLC, making the clusters "more metallic" in conductivity. The behavior of such single GLC in electric field will resemble to some extent that of a degenerate n-type semiconductor, with the Fermi level pinned in the conduction band. Remember, however, that the antibonding states of a single GLC are localized due to size confinement, so that the hypothetical d.c. charge transport inside the GLC will require thermal activation. In the system of interacting GLC, doping with Cu will result in a sharp increase of conductivity, as observed experimentally [1]. At Cu contents higher than ~15 at % the interaction between the Cu impurity atoms should be taken into account.

The Cu dopants most likely intercalate the sp^2-coordinated carbon plane sheets in graphite-like microdomains, resulting in redistribution of the electronic charge and inducing static dipole moments parallel and orthogonal to the carbon plane fragments. For this reason the Raman-active "in-plane" C - C vibrations in "aromatic" rings become active in IR absorption. The effect of considerable enhancement of well-known Raman G and D bands was actually observed in IR spectra of a-C:H<Cu> [2]. The appearance of low-frequency bands arising from "out-of-plane" dipole-active vibrations can be also predicted. Heating of the samples will promote removal of the Cu dopant from its equilibrium position. The dopants may escape from the GLC and, for example, be trapped by a microvoid. As a consequence, it should be expected that the Cu-activated IR absorption can be suppressed by heating and recovered once the dopants diffuse back.

References

1. Ivanov-Omskii, V.I., Tolmatchev, A.V., and Yastrebov, S.G. (1996) Optical absorption of amorphous carbon doped with copper, Phil. Mag. **73**, 715 - 722.
2. Ivanov-Omskii, V.I., and Frolova, G.S. (1995) Copper-induced activation of Raman bands in the IR absorption spectra of amorphous carbon, Zh. Tech. Phys. **65**, 186 - 189.
3. Robertson, J. (1986) Amorphous carbon, Adv. Phys. **35**, 317 - 390.
4. Harrison, W.A. (1980) Electronic Structure and the Properties of Solids, ed. W.H.Freeman and Company, San Francisco.
5. Nemanich, R.J., Lucovsky, G., and Solin, S.A. (1977) Infrared active optical vibrations of graphite, Sol. State. Com. **23**, 117 - 120.
6. Kaufman, J.H., and Metin, S. (1989) Symmetry breaking in nitrogen-doped amorphous carbon: Infrared observation of the Raman-active G and D bands, Phys. Rev. B **39**, 13053 - 13060.

VERIFICATION OF NANOCRYSTALLINE DIAMOND FILMS' QUALITY

MAŁGORZATA LANGER[1)], STANISŁAW MITURA[1)], JAN SZMIDT[2)], ALEKSANDRA SOKOŁOWSKA[3)]
1)The Institute of Materials Science & Engineering, Technical University of Łódź, Poland
2)The Institute of Microelectronics and Optoelectronics, Warsaw University of Technology, Poland
3)The Faculty of Material Science, Warsaw University of Technology, Poland

Abstract

The films discussed in this paper are deposited by r.f. dense plasma CVD. Amorphous carbon, DLC, and nanocrystalline diamond layers are very interesting candidates for active and passive electronic and optoelectronic applications. It is well known that diamond films possess semiconductor properties. (Though, as by now it is possible to obtain doping in special conditions). Because of their large energy gap, chemical inertness and temperature stability they are very perspective. The principal condition to apply them is not only to gain high quality but also to assure the stability of the obtained properties.
To verify the structure quality it is sufficient to make quite easy measurements: ellipsometric ones (thickness and refractive index) and electric ones (esp. breakdown voltage vs. temperature).

The possibility to control the process parameters to obtain stable and reproducible films with demanded properties has been stated. The influence of the plasmachemical reactor chamber geometry and the electric field distribution on the uniformity of the films have been examined.

1. Introduction

Since the first experiment of S. Aisenberg and R. Chabot [1] a certain amount of methods to obtain diamond and DLC has been developed. All these methods have been described many times elsewhere [2 - 6]. In this work the low pressure nanocrystalline diamond synthesis from hydrocarbon plasma activated in the r.f. electric field, with eventually additional source of electrons has been considered. The films discussed in this paper, amorphous and nanocrystalline ones, are very interesting candidates for active and passive electronic and optoelectronic applications. It is well known that

M. A. Prelas et al. (eds.), Diamond Based Composites, 211–217.

diamond films possess semiconductor properties. Because of their large energy gap, chemical inertness and high values of the temperature conductivity they seem to be very perspective. Taking into account the similarity of some DLC properties, many authors indicated [7÷10] the possibility of some significant applications.
The essential condition to apply them is not only to gain the demanded properties but also to assure the stability of them. The synthesis of diamond under reduced pressure takes place in the hydrocarbon plasma. In the region of plasma, in the gaseous phase, as well as on the substrates, whole the mess of micro- and macroscopic physical and chemical phenomenas take place, that depend on many outside parameters' values and result in the films of determined properties, phase structures, etc. The phenomenological description is insufficient for the process of nonconventional synthesis and there is still a lack of a numerous one.

2. Method

The method applied for the films deposition has been modified by Mitura and is described elsewhere (for example in [11]). The post for the film depositing (Fig. 1) consists of the chamber, that is fed by the carboneous gas (methane - CH_4) and in which low pressure is kept under the control. Owing to the special shape of the designed electrodes, the plasma is created between them with r.f. electrical energy, that is supplied with the frequency of 13.56 MHz. The r.f. generator is connected with the electrode via a capacitor. Thanks to that the r.f. electrode obtains a constant negative self-bias potential V_e. There are placed substrates on this electrode, that the films are deposited on.

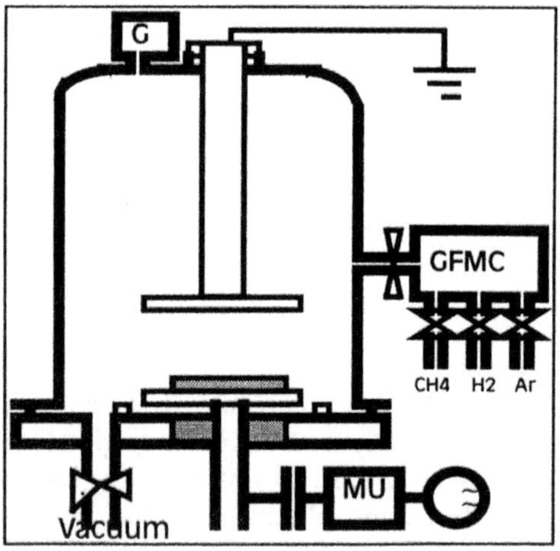

Fig. 1. The scheme of the device for the films' depositing. This device is mounted in the Institute of Material Engineering in The Technical University of Lodz

3. Self-Bias Potential and Modeling the Process

Many authors [12÷15] confirm the significance of the self-bias voltage value as the important technological parameter during the depositing process. It seems to be essential that in spite of the fact that it is not an input parameter, but an indirect one - it can be measured easily and in the same time it shows stable and simple functional dependencies in regard to the obtained films. The figure 2 introduces the function: depositing rate / self - bias voltage vs. discharge power r.f. generator [15]. It appears to be a constant - so the depositing rate is in the linear dependence on the self - bias voltage. To compare the functions, there is also shown a direct dependence of depositing rate on discharge power in the same figure - and it is a nonlinear one.

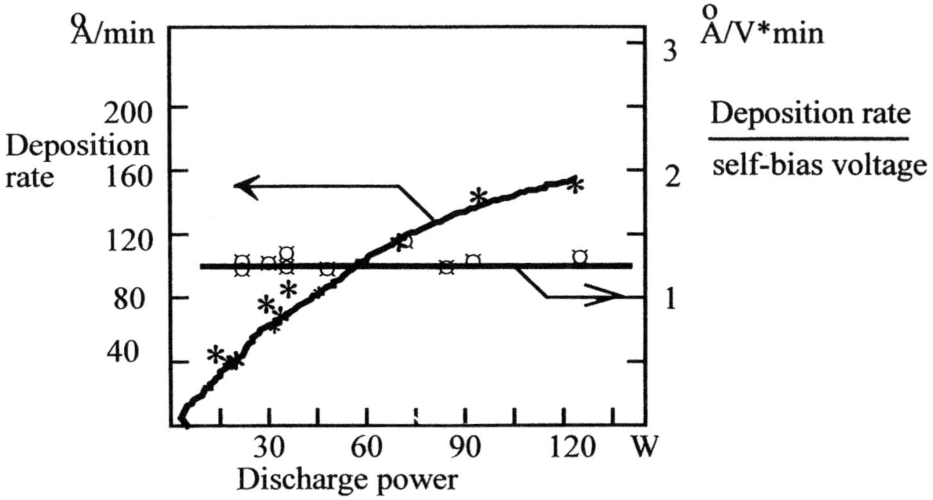

Fig. 2. Evolution of the ratio deposition rate/self-bias voltage (¤); and the deposition rate (*) as a function of the power of a 13.56 MHz discharge [15]]

The figure 3 introduces changes of forbidden gap width [11] in carboneous films that are obtained on the r.f. electrode as the function of the negative self - bias voltage. This function is also a linear one and the functional value shows the film structure properties clearly. To exam these dependencies we have decided [16] to calculate the exact distribution of the electrostatic field in the reaction chamber, to examinate how the chamber shape and wall grounding influence it and what the density of the current is that achieves the r.f. electrode. The last value reflects the number of carboneous ions that arrive at the substrate surface. It has been shown that the r.f. electrical field does not activate other particles than electrons. That is why one may assume that the electrostatic field which value results from the self-bias voltage, reaction chamber shape and its walls' potentials is the only one that influences the positive electrical charges that are inside it according to the classical electrotechnical theory. The charge density (positive

ions that are plasma components) have been calculated in the base of technical parameters that are utilized in the CVD method.
The analysis has been led for two models of the reaction chamber. In both the cases they were models with the central axial symmetry; the first one has been simplified to be a long cylinder with parallel electrodes that are placed in the central part of it, and the other one takes into account the real structure of the generator, with isolating parts and different environment -the regions with plasma and the ones with gas.

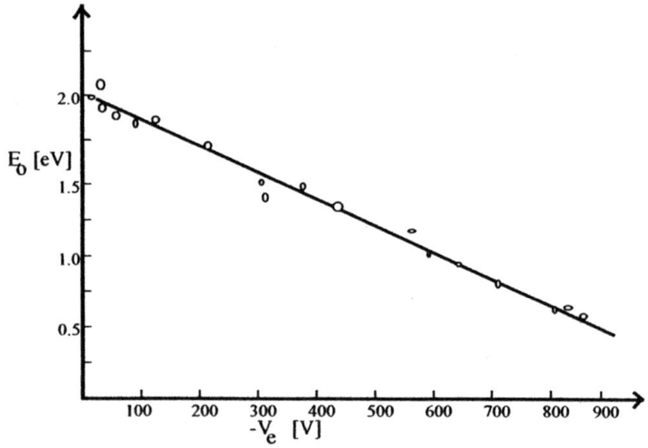

Fig. 3. Changes of forbidden gap width in carboneous fillms that are obtained on the r.f. electrode as the function of the negative self-bias voltage [11]

We have obtained the electrostatic field distribution for our models. The Fig. 5 shows the fragment of the results connected with the edge of the r.f. electrode. The widening of the higher values' regions near the r.f. electrode edge is visible clearly. One must note, that the shape of the field is determined fully - the field is surrounded by the belt with '0 V' potential and any change of the model's dimensions (for example a higher chamber or a wider one) does not influence values of the field.

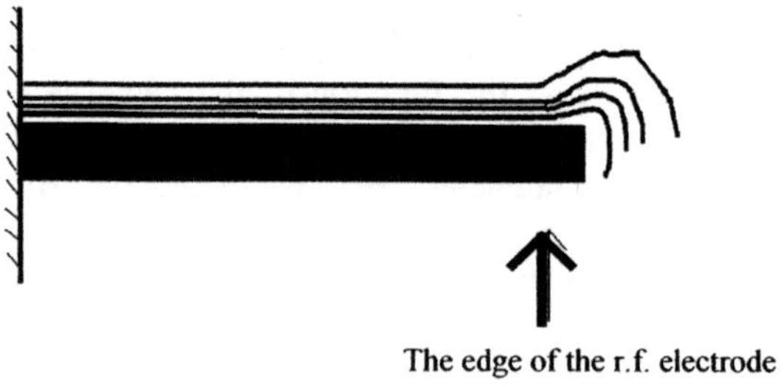

Fig. 5. The shape of the electric field near the edge of the electrode [16]

4. Measurements

To verify the structure quality it is sufficient to make quite easy measurements: for example the ellipsometric ones (thickness and refractive index). The refractive index value is very sensitive for any changes of the chemical composition, especially crystalline structure.

Table 1.

Chamber dimensions		
- diameter	500 mm	
- height	550 mm	
Electrode diameter	200 mm	
Max. Power input to the generator	2 kW	
Radio frequency	13.56MHz	
Gas	Methane (CH_4)	
Velocity of gas flow	15 sccm	
Etching	5 min / 800 V	
Pressure	50 Pa	
Time of deposition	5 min	
DC negative self-bias voltage of the r.f. electrode	I. 60V	II. 200V
Type of carbon film	I. DLC	II. NCD

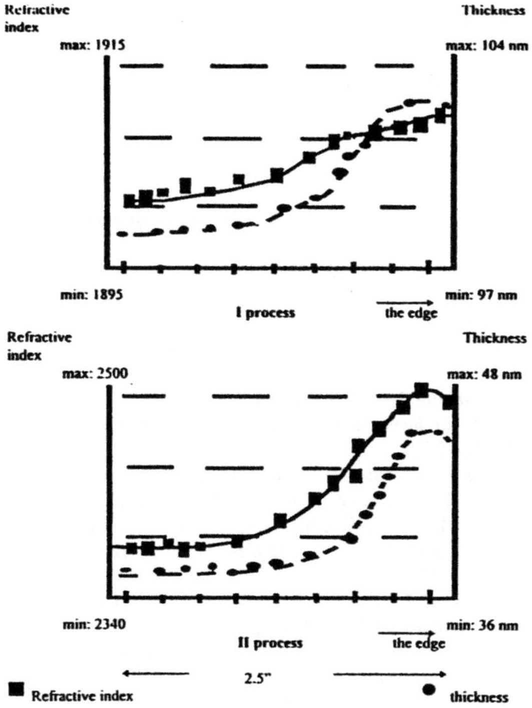

Fig. 6. Refractive index and thickness of the layer vs. distance from the r.f. electrode edge for I - 60 V; II - 200 V (see Table 1)

Fig. 6 shows these average values vs. the distance from the edge (i.e. vs. changing calculated results of the electric field). The parameters of the processes are introduced in the Table 1. The authors introduce two processes leading to obtaining two types of carbon films [11], i.e.: DLC (diamond like carbon), and NCD (nanocrystalline diamond).

5. Results and Conclusions

Our tests show that when it is possible to obtain the homogenous electrical field on whole the surface of the sample, the films will have a homogenous structure and thickness. With the aid of the electrical field model and the computer simulation one may hope to shorten significantly the expensive period of tests how to match the geometry and parameters of the process.

The changes of the thickness are closely correlated to the shape of the electric field in the plasma reactor chamber. The simultaneous changes of the refractive index show the structure changes in the obtained carbon layer. They are correlated with the electric field values.

The value of the electric field decides about the processes that take place in the gas phase and they influence directly on the kind of the crystalline that is created [17] through plasma-chemical reactions with the participation of electrons.

The authors think that the uniformity of the electric field shape gives the possibility to obtain the demanded pure phase of carbon what is the essential condition for any applications in electronics, optoelectronics and many others.

References

1. Aisenberg S., Chabot R. 1971.'Ion-beam deposition of thin films of diamondlike carbon', *J. Appl. Phys.*, 42: 2953-2959
2. Bubenzer, A., Dischler, B., Nyaiesh, A.: 1982.*Thin Solid Films*, No 91, , p. 81
3. Holland, L., Ojha, S.,M.: 1977. *Thin Solid Films*, No 40, , p. L31, No 48, 1978, p. L21
4. Nabot, J.,Ph., André, B., Pad'dassi, S.: 1990. *Surf. Coat. Technol.*, 43/44, , p. 71 - 79
5. Seth, J., Padiyath, R., Baton, S.,V.: 1991. in *Applications of Diamond Films and Related Materials*, , p. 851 - 856
6. Vilnat, M.: 1994. *L'Usine Nouvelle*, No 2439, Janvier, p. 44-47
7. Szmidt J., (1994), "Diamond-like layers as passivation coatings for power bipolar transistors", *Diamond and Related Materials*, 3 s.849-852.
8. Szmidt J., Beck R.B., Mitura S., Sokołowska A., (1994), Application of Diamond-like Layers as Gate Dielectric in Metal/Insulator/Semiconductor Transistor, *Diamond and Related Materials*, 3 s.853-857.
9. Louda P., Lisik Z., Mitura E., Mitura S., Niedzielski P., Szmidt J., Jakubowski A.: (1994). "Heat transfer through DLC and PCD layers"; Ref. (4-3) - *Second International Symposium on Diamond Films*, ISDF-2, 3-5 May, Minsk, Bielarus, s. 45-46.
10. Langer M., Szmidt J., Lisik Z., Mitura S., Niedzielski P.: (1995). Warstwy diamentopodobne w mikroelektronice"; *Conference MATEL'95*; Poland, Dobieszków.
11. Mitura, S.: (1992). 'Znaczenie elektronow w procesie niskocisnieniowej syntezy diamentu', *Professor Thesis, in: Zeszyty Naukowe Politechniki Lodzkiej*, Nr 666, Rozprawy Naukowe, z. 182
12. Bubenzer, A., Dischler, B., Brandt, G., Koidl, P.: (1983). *J. Appl. Phys.* 54, , p. 4590-5
13. Mitura, S.: (1987). *Journal of Crystal Growth*, 80, p. 417 - 424
14. Ohja, S.,M., Norström, H., McCulluch, D.: (1979). *Thin Solid Films*, 60, p. 213 - 25

15. Pastol, A., Catherine, Y.: (1990), *J. Phys. D. Appl. Phys.* 23, p. 799 - 805
16. Langer M.: (1994). Essai de modélisation de phénomenes se produisant lors de la synthčse de films carbone adamantin (DLC) dans un plasma R.F. hydrocarbone dense sous faible pression, Ed. ECAM, Lyon, May.
17. Clapa M., Langer M., Mitura S.: (1991). 'The TEM Analysis of The Carbon Films Obtained By hf r.f. CVD' *in Proc. Of 'Diamond Films '91', Nice, France*

Acknowledgements: We acknowledge the Polish State Committee for Scientific Research (KBN) for support under grants No. 8T 11B 058 08 and No. 8S 501 043 05

THE SURFACE STRUCTURE OF CARBON FILMS DEPOSITED BY DIFFERENT PLASMACHEMICAL METHODS

S. MITURA, E. MITURA, P. NIEDZIELSKI, M. DŁUŻNIEWSKI, E. STARYGA, S. DER-SAHAGUIAN*
Technical University of Łódź, Stefanowskiego 1, 90-924 Łódź, Poland,
**ECAM - Lyon, France*

J. ŻAK
Silesian University of Technology,
Krzywoustego 6, 44-100 Gliwice , Poland

A. SOKOŁOWSKA, J. SZMIDT
Warsaw University of Technology,
Koszykowa 75, 00-662 Warsaw, Poland

A. STANISHEVSKY
Plasmoteg Eng. Centre of the Belarus Academy of Science,
Zhodinskaya 1/3, Minsk, 220141, Belarus

1. Introduction

Diamond-like carbon films on suitable substrates have attractive properties for electronic, optoelectronic, mechanical and biomedical applications. Carbon in a form of diamond is a promising material as a protective coating for high temperature superconducting devices. The excellent chemical inertness of diamond-like film make them a promising material for medical implants, cardiovascular surgery and for coating of certain components of artificial heart valves.

The applications of carbon films impose some special requirements on their quality, purity, phase content and the state of surface. Of particular importance is the smoothness of the surface and good adhesion of the coatings to the substrate.

Defects such as voids, hydrogen, boundary dislocations can reduce the above-mentioned properties of the films. It is well known that the macroscopically quality of films has been improved since more detailed knowledge on a microscopic level became available.

This work is aimed to examine the profile of the sample surface, their structure, grain boundaries.

The investigated carbon films were synthesized by various plasma-ion methods, differing significantly from each other in ion energies reaching the substrate during the deposition process. The carbon coatings were obtained by the following methods: Vacuum Pulse Arc Deposition (VPAD), Ion Beam (IB), Radio Frequency Plasma Chemical Vapour Deposition (RF PCVD), Reactive Pulse Plasma Deposition (RPP).

M. A. Prelas et al. (eds.), Diamond Based Composites, 219–228.

Thin films were prepared on different substrates (silicon wafers, steel AISI 316L titanium, polyurethane, porcelain).

The morphology of the surface of diamond-like coatings obtained at different stages of the deposition processes has been studied using an Atomic Force Microscope (AFM). Additional analysis of physical properties of the above mentioned carbon coatings has been provided.

2. Experimental details

In the paper are presented the results of AFM studies concerning to the manufacture and characterization of carbon films [1] obtained by the methods of:

- Radio Frequency Plasma Chemical Vapour Deposition (RF PCVD),
- Radio Frequency Dense Plasma Chemical Vapour Deposition (RF DPCVD),
- Vacuum Pulse Arc Deposition (VPAD),
- Ion Beam
- Hot Filament Chemical Vapour Deposition (HF CVD).

2.1. RF PCVD METHOD

Various methods of carbon layer production have been developed on the basis of hydrocarbon decomposition in a plasma glow discharge excited by an r.f. electric field. These methods are based on proposals by Holland and Ojha [2]. RF PCVD method has been extensively used to produce amorphous, nanocrystalline or polycrystalline carbon films in the laboratory in Lodz Technical University since 1978. The structure of carbon films depends on the applied parameters. The use of Radio Frequency allows deposition of carbon on any type of solid, even non-conductors. A typical schematic diagram of the RF PCVD apparatus is shown in Figure 1.

Bubenzer *et al.* [3] have shown that the RF deposition process is reproducibly and accurately controlled by two independent parameters: the negative self-bias potential of the r.f. powered electrode and the gas pressure. They applied the method to deposit amorphous carbon coatings containing hydrogen (a-C:H).

Vora and Moravec [4] modified Holland and Ojha's reactor by additional negative bias of the RF electrode. The carbon layers they obtained consisted of an amorphous warp with in-built diamond crystals. Mitura and coworkers [5,6] confirmed the presence of diamond microcrystals embedded in an amorphous film.

2.1.1. Diamond-like carbon (DLC) synthesis

Figure 1 shows typical experimental conditions to produce amorphous hydrogenated carbon (a-C:H) or typical DLC films. It produces films with good adhesion on a wide variety of substrates by a two-stage (etching and deposition) process.

Ionized fragments of hydrocarbon molecules are produced during hydrocarbon decomposition in a glow discharge and are accelerated in the constant electric field created between the positively glow charged plasma and the negatively self-biased RF powered electrode. The RF electrode is negatively biased owing to the higher mobility of electrons relative to that of positive ions in the RF electric field (the Debye effect). The energy of the RF electric field is supplied through a matching box to an electrode which is the substrate holder. The amount of energy transmitted owing to the collisions of electrons with the gas molecules is characterized by the negative self-bias voltage of the RF powered electrode.

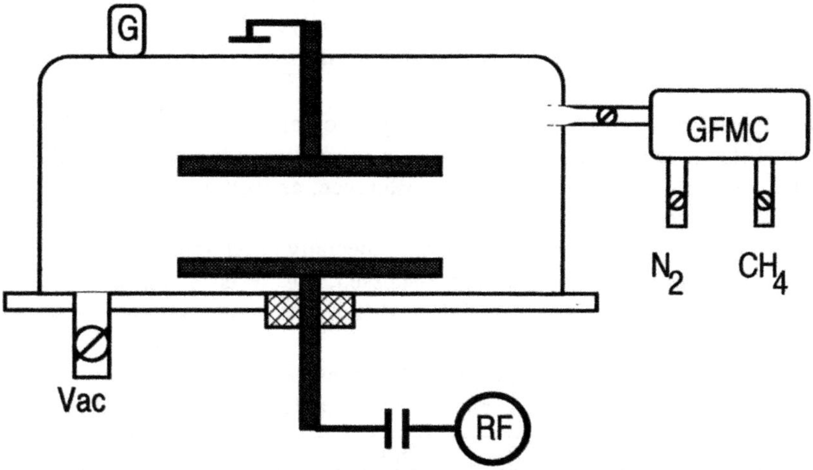

Figure 1: Apparatus for RF methane plasma CVD method: RF, 27.12 MHz generator; GMFC, gas feeder with microcomputer control; G, vacuum gauge; Vac, Vacuum unit.

Has and Mitura [6] have shown that under certain conditions substrates placed in RF hydrocarbon plasmas are subject to etching while under other conditions a polymer film is deposited onto the substrates. In the range of parameters investigated, *i.e.* negative self-bias voltage, electrode-electrode separation and gas pressure, the two main regions have been determined.

For films obtained in two stages, first, the substrates are etched in the range of parameters above the straight line $(V,p)_{crit}$. This phenomenon removes atoms by pulverization and cleans the surface. Then on the substrates cleaned by etching, films are deposited in the range of parameters below the straight line $(V,p)_{crit}$
The processes occurring at the substrate surface during etching are used to produce films of relatively good adhesion with different substrate [6,7].

2.1.2. Nanocrystalline diamond (NCD) synthesis
Investigations on the optimization of production of superhard carbon layers resulted in the development of a new technology, called by the authors dense RF methane plasma method to synthesize a-C films with nanoscale grain size (nanocrystalline diamond films). The idea of this method is to introduce a relatively high gas pressure (50 - 400 Pa) in the RF plasma technique [8].

A schematic diagram of the apparatus is shown in Figure 1. The main parameters for growth of superfine crystalline diamond film (NCD) onto AISI-316L, stainless steel used in surgery, are given in Table 1.

TABLE 1: Main parameters and growth conditions for the RF dense methane plasma CVD process.

Substrate	AISI-316L
Gas	Methane with nitrogen
Pressure	50 - 400 Pa
R.F. generator	27.12 MHz, 2 kW
Negative self-bias voltage	400 - 1200V
Electrode to electrode distance	200 mm

Differences between properties of the layers obtained by the two methods should be stressed. Ellipsometric measurements of the refractive index of typical DLC (n = 1.9) and NCD (n = 2.4) show a significant difference, as well as the results of tribological investigations presented in Table 2 [8].

From Table 2, one can see NCD layers especially good adhesion to steel. Additional analysis of properties of the coatings discussed, *e.g.* tests of corrosion resistance [9], breakdown tests [10], etc., confirmed these differences.

TABLE 2. Results of tribological investigations of AISI 316L steel implants coated with typical DLC and ultrafine grained diamond (NCD)

Test	DLC	NCD
Microhardness	3000 VHN	8400 ± 1000 VHN
Adhesion - Scratch test (critical load)	2 N	50 ± 5 N
Adhesion energy of the layer-surface interface	0.09 Jm^{-2}	4 Jm^{-2}

The carbon films are amorphous. One should underline that these films are chemically homogeneous. Specificity of the layers consisting first in very good adhesion to steel is explained by the authors on the basis of investigations carried out by Auger electron spectroscopy (AES) [9, 13]. The surface layer several thousand angstroms thick consists of carbon. The carbon film passes smoothly into a thick layer (~ micron) consisting of metal carbides of metals which are included in steel [13]. Such a structure of the coatings ensures a higher mechanical strength. Therefore, the coatings are also a good protection for metal against corrosion.

2.2. VACUUM PULSE ARC DEPOSITION (VPAD) METHOD

This method is based on the evaporation of coating material under pulsed arc discharge between eroded cathode and anode of a plasma source [11]. Figure 2 shows experimental conditions to produce amorphous hydrogenated carbon (a-C:H) or carbon films.

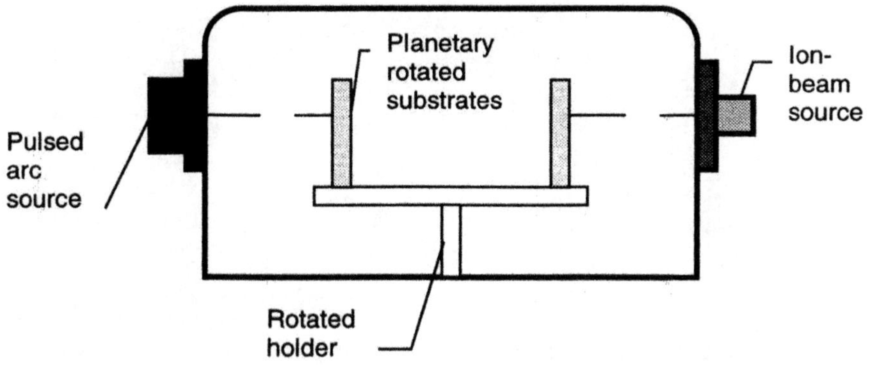

Figure. 2. The apparatus for amorphous carbon film synthesis by Vacuum pulse arc deposition (VPAD) and ion beam (IB) methods

Carbon films by VPAD method were prepared at the integral substrate temperature less than 80°C and the average deposition rate 5÷20 nm/min. Substrates were rotated with planetary motion. Films thickness was up to 1 μm. No bias was applied during experiments.

2.3. ION BEAM (IB) METHOD

Films by ion-beam deposition were prepared at the substrate temperature less than 100°C and deposition rate 20÷30 nm/min. Starting gas was benzene. Carbon films prepared by methods b) and c) have good adhesion and are quite amorphous (near-ordering area less than 1 nm) with embedded crystalline carbon particles [12]

3. AFM investigations

Diamond film growth by CVD processes involves the nucleation and growth of the nuclei to from films which exhibits particular structures. All mechanisms proposed so far recognize the important link between growth processes and surface morphology, however, relatively little has been done to relate observed morphology and growth features to growth mechanisms.

In this study, Atomic Force Microscopy (AFM) is used to examine the morphology of the surface of diamond-like coatings obtained at different types of deposition processes. AFM seems to be one of the most suitable technique for investigating the nucleation and growth of films on substrates, even though atomic resolution is almost impossible to achieve. This technique offers the ability to identify nuclei which are too small to be recognised with their characteristic cube-octahedral shape of diamond crystallites by Scanning Electron Microscopy (SEM) (see Figures 3-8).

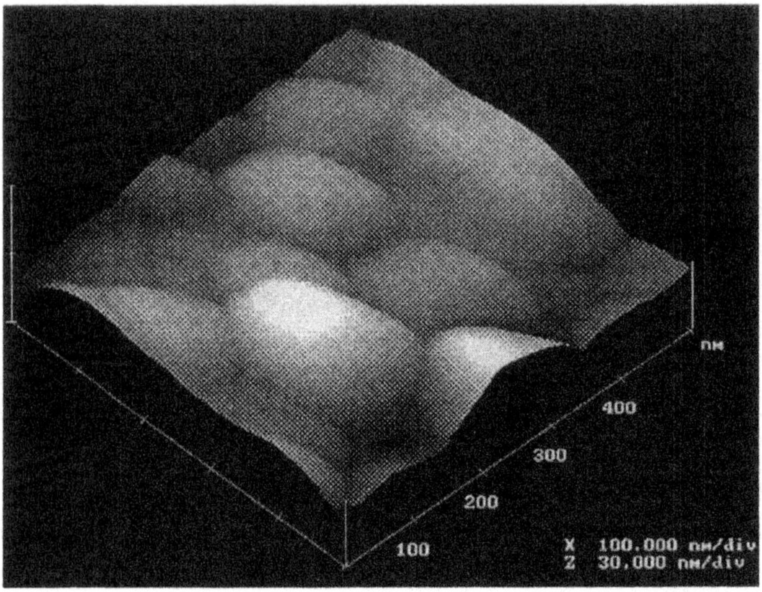

Figure 3. AFM image of carbon film deposited by RF PCVD method.

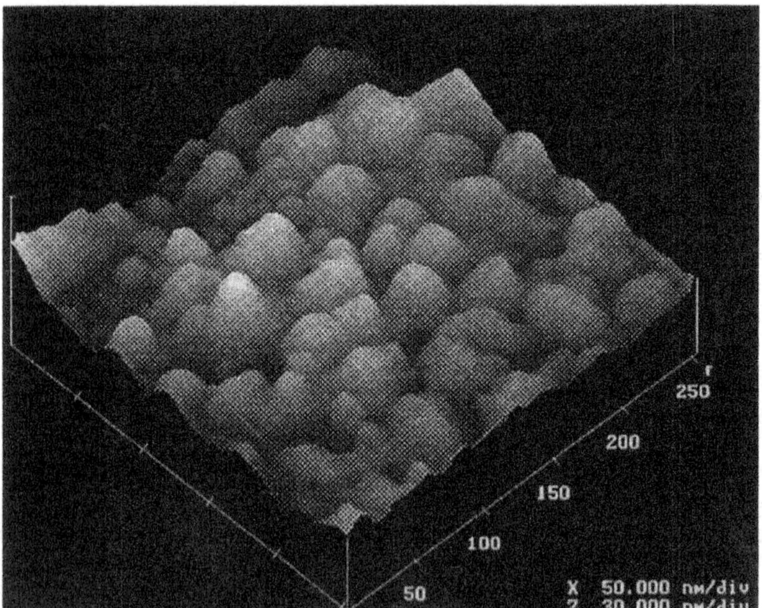

Figure 4. AFM image of carbon film deposited by RF dense PCVD method

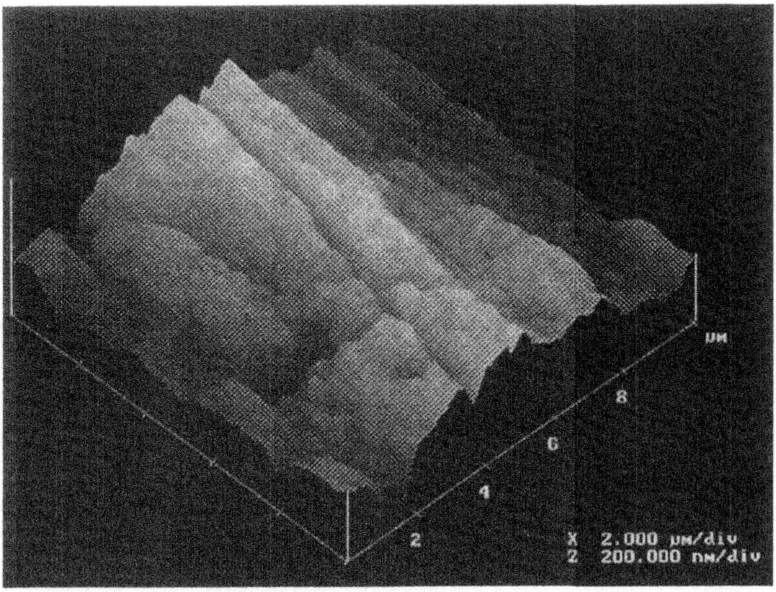

Figure 5. AFM image of carbon film deposited by Vacuum pulse arc deposition (VPAD) method

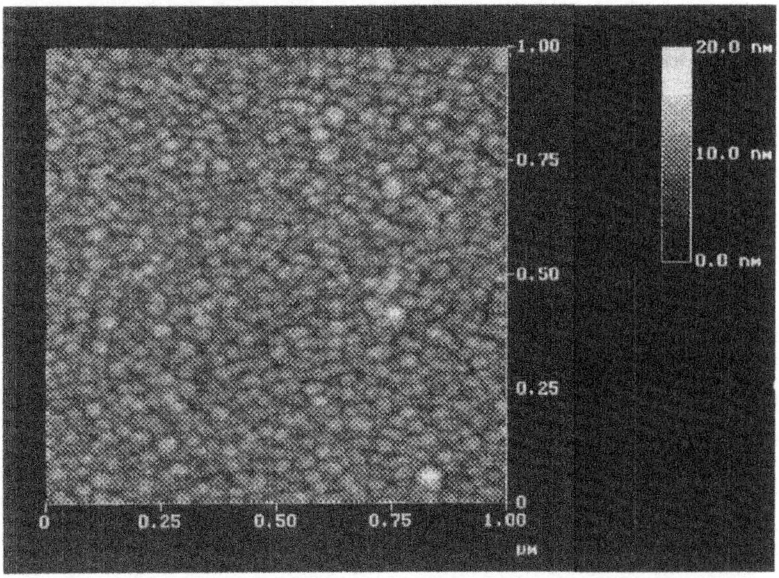

Figure 6. AFM image of carbon film deposited by ion beam (IB) method

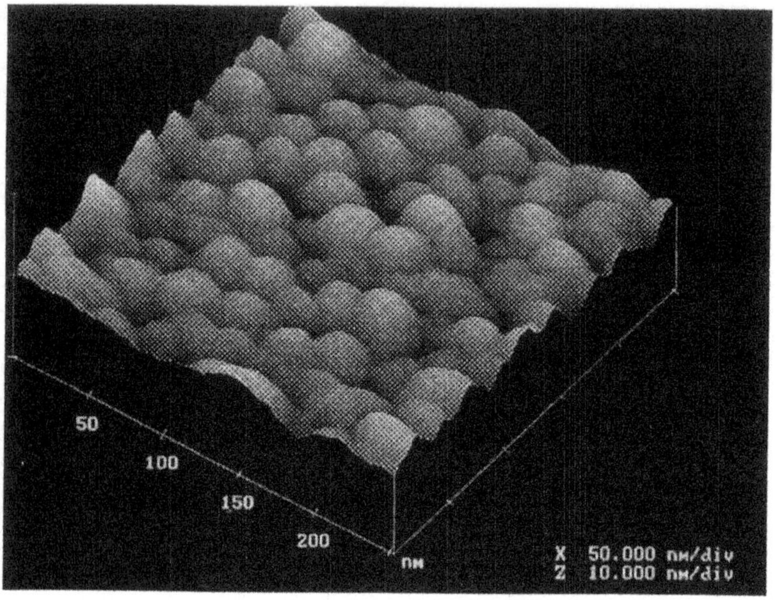

Figure 7. 3D AFM image of carbon film deposited by ion beam (IB) method

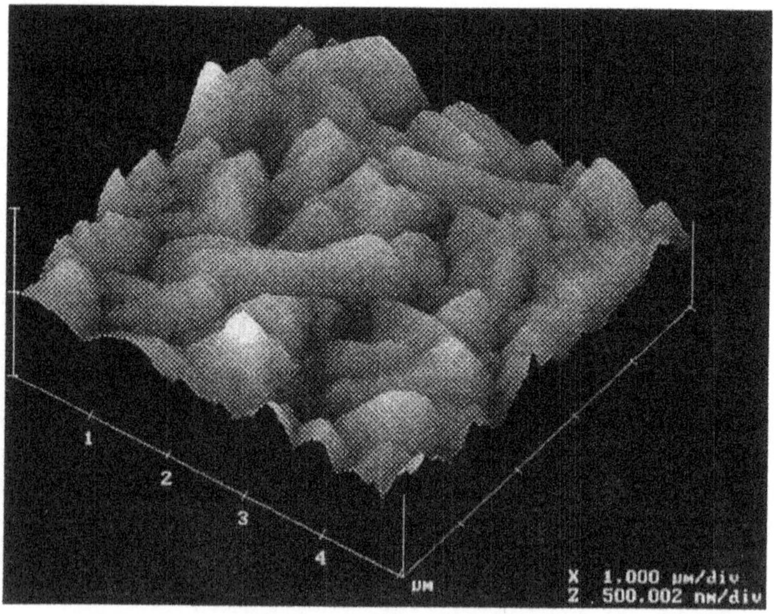

Figure 8. 3D AFM image of carbon film deposited by hf CVD method
[Courtesy Dr. Greg M. Swain (from Utah State University)]

4. Discussion

The big clusters are formed by coagulation of nuclei in the gas phase. Large packed grains were clearly viewed by AFM 3D image (Figure 3); grains average size is about 150 nm.

As the sketch suggest, the number of diamond particles connected from the gas phase seems rather small compared with typical nucleation densities observed on substrate surfaces.. Nevertheless, it is clear that their contribution to the deposition process is not negligible.

Another interesting feature provided by optical microscopy is the repartition of the clusters on the substrate surface. During deposition, the flow of nuclei passes over the substrate/layer surface. Therefore, deposition of the nuclei coagulated in the gas phase to form clusters occurs according to the direction of this flow. Another question arising: are these clusters cohesive with film ?

Topography extending between diamond clusters is studied and displayed in Figures 4 to 8. This area is the result of surface nucleation directly on the substrate. It exhibits nucleation density of a-C films. Grains average size is about 30 nm.

It is important to underline that the big nuclei from the gas phase do not spread laterally to connect with other nuclei.

5. Conclusion

The significant differences in the properties of the layers, obtained by the four methods, are explained by the critical process parameters of plasma power, gas pressure, surface temperature, and carbon species concentrations impact on the quality and microstructure of the deposited films.

In the dense RF methane plasma method, nanocrystalline diamond (NCD) films are prepared consisting essentially (~90%) of tetrahedral carbon (ta-C). While typical DLC films are constituted essentially by carbon in three main forms: diamond (~65%), graphite (~30%) and polymers (~5%).

Acknowledgments

The work was partly financed by Grant *"Manufacturing of nanocrystalline diamond - biomaterial for medical applications"* no. 7 S201 016 07, no. 8 T11B 058 08 and C/1526/95 from the Polish State Committee for Scientific Research and by Project ATP of French Government 6494 *"Couches minces diamantées"*.

References

1. Angus, J., Koidl, P. and Domitz S. (1986) Carbon Thin Films, in: J.Mort, F.Jansen (eds.), *Plasma Deposited Thin Films* CRC Press, Inc., Boca Raton.
2. Holland, L. and Ojha, S. M. (1976) *Thin solid films* **38** L17.
3. Bubenzer, A., Dischler, B. and Nyaiesh, A. (1982) *Thin solid films* **91** 81.
4. Vora, H. and Moravec, T. (1981) *J. Appl. Phys.* **52** 6151.
5. Has, Z., Mitura, S. and Wendler, B. (1983) in T. Takagi (ed.), *Proc. Int. Ion Engineering Congr. ISIAT and IPAT'83*, Kyoto, 1143.
6. Has, Z. and Mitura, S. (1985) *Thin Solid Films* **128** 353.
7. Mitura, S., Has, Z. and Gorokhovsky, V. (1991) *Surf. Coat. Technol.* **47** 106.
8. Mitura, S., Mitura, E. and Mitura, A. (1995) Manufacturing of amorphous carbon layers by RF dense plasma, *Diamond Rel. Materials* **4** 302.
9. Couvrat, P., Denis, M., Langer, M., Mitura, S., Niedzielski, P. and Marciniak, J. (1995) The corrosion tests of amorphous carbon coatings deposited by RF dense plasma onto steel with different chromium content, *Diamond Rel. Materials* **4** 1251.
10. Marciniak, J., Koczy, B., Boba, S. and Mitura, S. (1993) Einfluss von Passivirungs und Kohlenstoffschichten auf austenitischen CrNiMo-Stahlen auf die Bestandigkeit gegen Lochfrass und Spannungsrisskorrosion, *Werkstoffe und Korosion* **44** 379.
11. Tochitsky, E.I., Selifanov, O.V., Stanishevsky, A.V., Akulich, V.V. and Kapustin, I.A. (1991) *Surface and Coat.Technol.* **47** 522.
12. Tochitsky, E.I., Stanishevsky, A.V., Kapustin, I.A., Akulich, V.V. and Selifanov, O.V. (1991) *Surface and Coat.Technol.* **47** 292.
13. Rylski, A., Niedzielski, P. and Mitura, S. (1995) AES analysis of interface carbon/substrate *Journal of Chemical Vapor Deposition* **3(4)** 296.

DIRECT OBSERVATIONS OF THE ELASTIC MODULUS AND TENSILE STRENGTH OF CVD DIAMOND FILMS AND FIBERS

J. L. DAVIDSON
Vanderbilt University
Box 99-B, Nashville, TN 37235, USA

1. Abstract

We have directly measured, by bending tests of loaded microbeams of CVD diamond, the elastic modulus of polycrystalline deposited diamond. The values are in close agreement with those reported for natural diamond. By using patterning techniques developed in our work, diamond film tensile specimens were created and subjected to loading tests. Ultimate load measurements were taken leading to estimates of the strength of diamond films in orthogonal load conditions. Also, using a variety of extremely small diameter substrate fibers such as graphite and molybdenum, diamond fibers were created with diameters ranging from 50 to 300 microns. The grain structure and other topological features of these diamond fibers were examined by SEM analysis. Details of these activities will be reported in this paper.

2. Introduction

The modulus of elasticity, strength and fiber characteristics of diamond have been examined. This paper reviews those results observing that all the fundamentals necessary to achieve a diamond reinforced composite have been determined and such a composite could be expected to be significantly superior to any existing composite system. The information presented below is a review of work previously conducted by the author on these properties and characteristics of CVD diamond.

3. Elastic Modulus of Diamond

Micro-mechanical beams, cantilevers, and optically transparent membranes/windows of polycrystalline synthetic diamond were fabricated on single-crystal silicon <100> wafers [1]. The Young's modulus of polycrystalline diamond thin film was estimated using the elastic deformation cantilever beam formulation [1].

M. A. Prelas et al. (eds.), Diamond Based Composites, 229–240.

3.1. EXPERIMENTAL

The process flow developed for the selective deposition of polycrystalline diamond [2] and fabrication of beams (double side-supported), cantilever beams (single side-supported), and windows/membranes is summarized in Figure 1. Starting substrates were

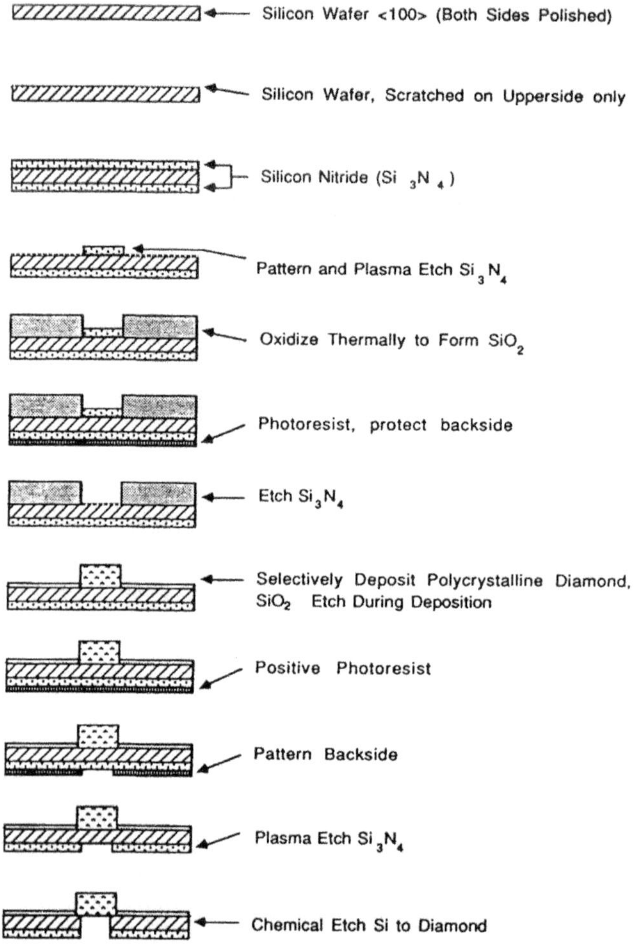

Figure 1. Schematic diagram of the process flow steps to fabricate polycrystalline diamond microstructures on a silicon substrate.

p-type silicon wafers, 5 cm diameter, <100> orientation, polished on both sides, and resistivity of 9-13 Ω-cm. They were manually polished on one side only, designated the top side, using diamond paste of 0.25 micron mean particle size. The wafers were then washed in running DI water for 15 min. rinsed with methanol, ultrasonically agitated in DI water for 15 min. and blown dry with a nitrogen jet. Silicon nitride (Si_3N_4) was then deposited (thickness = 1600A) on both sides, using low-pressure chemical vapor deposition (LPCVD) with dichlorosilane and ammonia, at a substrate temperature of

~800°C. The silicon nitride was then photolithographically patterned on the top side of the substrate, plasma etched in a 90% CF_4 and 10% O_2 mixture of gases, and the photoresist rinsed away with acetone. With the nitride patterned, the silicon wafers were thermally oxidized (using pyrogenic steam) at 1000°C for 10h. Silicon dioxide thickness was 1.2 micron. Photoresist was spin-coated on the back side of the wafer in order to protect it while the wafers were etched in buffered oxide etch (NH_4F + HF) for 30s to remove the thin surface layer of oxide from the silicon nitride on the top side. After rinsing in running DI water for 15 min. the silicon nitride was completely etched from the top side of the wafer in hot phosphoric acid (~180°C), cooled, washed with running DI water for 15 min. rinsed with acetone and methanol, and nitrogen-jet dried.

Details of the plasma-assisted CVD procedure used to grow polycrystalline diamond thin films are described elsewhere [3]. After growing the diamond, photoresist was spin-coated over the silicon nitride on the back side (i.e., opposite to the diamond film), and then exposed for patterning using an infrared mask aligner. The silicon nitride was plasma-etched using the same gas mixture described above until the silicon surface was exposed in the patterned areas. The diamond side of the wafer was attached in wax to a glass plate and placed in silicon etch of KOH + H_2O (2:1) at 60°C for ~ 12h. The silicon etched completely through the 250 micron thick wafer, creating the polycrystalline diamond bridges, cantilever beams, and windows/membranes intended by the mask's design. We are addressing cantilever beams in this review, see Figure 2.

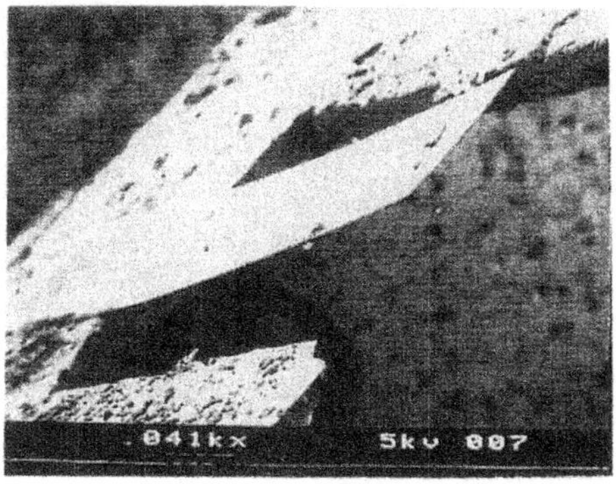

Figure 2. SEM of diamond cantilever microbeam of length 2 mm on a silicon surface.

3.2. MICRO-MECHANICAL MEASUREMENTS

After the fabrication of the cantilever beams, the silicon wafers were mounted under the optical microscope. The deflection of the poly-crystalline diamond "Microbeams" by loading was used as a means of determining the mechanical properties. The beams were deflected by applying loads manually and measuring the deflection using calibrated vertical focusing of an AO Scientific Instruments (Model Microstar) optical microscope. Using the cantilever beam formula and load deflection data, the Young's modulus of thin-film materials that comprise the cantilever beams was determined.

Beam lengths and widths were measured with a calibrated filar micrometer eyepiece on an optical microscope. Beam thickness was measured by a calibrated Alpha-Step (TM) profilometer. Small weights attached to fine wires were carefully hung, manually with the aid of tweezers, on the ends of the free-standing polycrystalline diamond cantilever beams, while in place under the microscope. The configuration is shown schematically in Figure 3. After measuring the amount of deflection, the weights and supporting wire were measured for each case on an analytical chemical balance. The amount of deflection was determined by a point-to-point focus technique, with a calibrated vertical stage optical microscope, of the loaded and unloaded beam. In this manner, the total deflection was determined to within approximately +/- 1 micron and the load to within +/- 1 microgram.

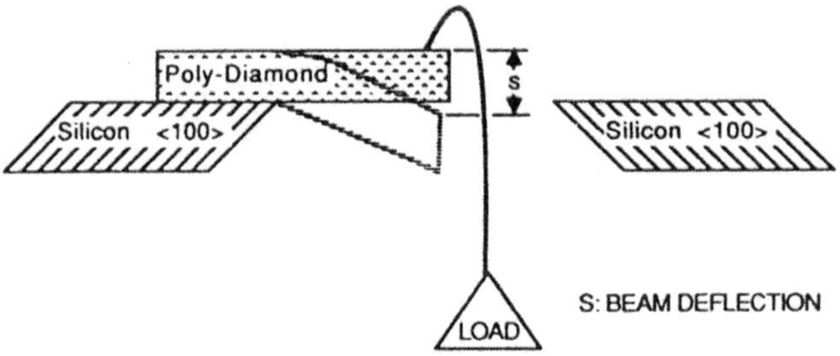

Figure 3. Schematic diagram illustrating the load application at the end of a free standing polycrystalline diamond cantilever beam (cross section view).

The deformation of a cantilever beam can be expressed in terms of the deflection of the beam from its original zero-load position. The relationship between cantilever beam deflection, S, applied load, P, at the end of a cantilever beam, and the length of the beam, L, is

$$S=(PL^3)/3EI \quad (1)$$

Where E is the Young's modulus of the thin film cantilever beam material (kg/cm^2), and I is the moment of inertia of the beam (cm^4). Specifically

$$I=(bh^3)/12 \quad (2)$$

where b and h are width and thickness of the cantilever beam, respectively [4].

From these formulas it is seen that the unknown, E, may be determined experimentally from S(P) at constant L or S(L) at constant P. Both approaches were used to obtain an estimate of E, and were defined as Case 1 and Case 2, respectively. Reiterating:

Case 1. - Data may be obtained with a single cantilever beam, variable load. Therefore, a plot of S vs. P at a constant length, width and thickness should yield a straight line with a slope of $L^3/3EI$ and Young's modulus may be determined from the slope.

Case 2. - Data may be obtained from different length cantilever beam. Therefore, a plot of S vs. L^3 at a constant width, thickness, and load should yield a straight line with a slope of P/3EI.

Figure 4 shows the typical data of deflection vs. load obtained with a single polycrystalline diamond cantilever beam. Figure 5 shows the typical data of deflection vs. length of various polycrystalline diamond cantilever beams at a constant applied load. Thus, Young's modulus was determined for both cases and the values observed were 1.20×10^7 kg/cm^2 and 1.25×10^7 kg/cm^2, for case 1 and case 2, respectively. This compares interestingly with the reported value for single crystal diamond of 1.15×10^7 kg/cm^2 [5].

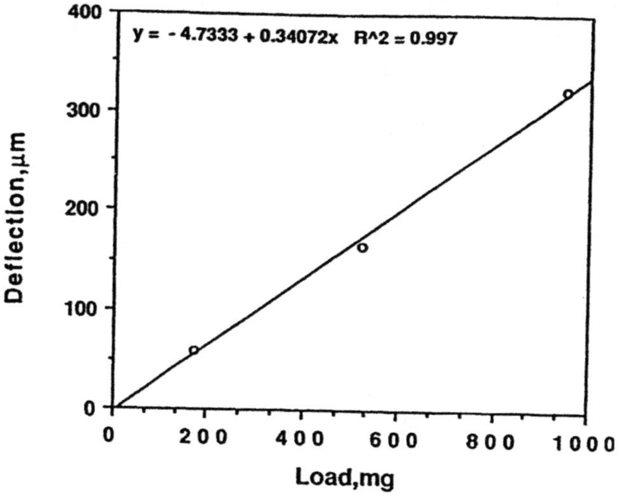

Figure 4. Deflection vs. load with a polycrystalline diamond cantilever beam on a silicon substrate.

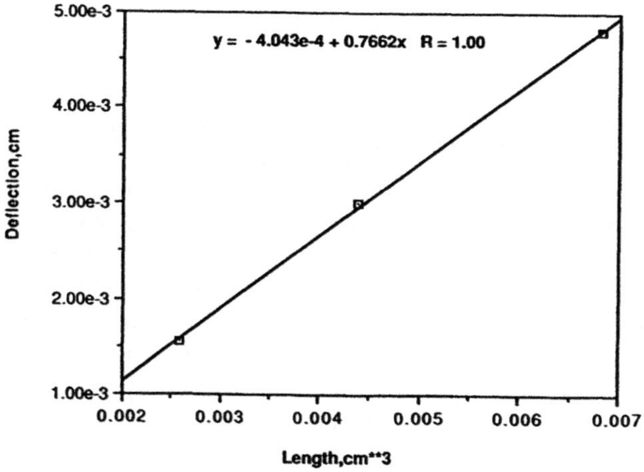

Figure 5. Typical data of deflection vs. length^3 of polycrystalline diamond cantilever beams at a constant applied load (175 mg) on a silicon substrate.

It was observed in the free-standing cantilever beams that they curl upward slightly, an indication that residual stresses may exist in the polycrystalline diamond films. It is not known at this time whether these stresses i.) arise from mismatch at the silicon-diamond interface, ii.) are distributed through the thickness of the film as it was grown, or iii.) are a combination of these effects. If there is a transition from silicon to diamond that involves a silicon carbide layer on silicon, as suggested by some [6], then this curling could be due to differences in the layers.

Thus the average Young's modulus was estimated by measuring the deflection under load of polycrystalline diamond-film cantilever beams. From the data on deflection-load at a constant length, and on deflection-length^3 at a constant load obtained from bending under load, the Young's modulus was measured and compared with the reported value for single-crystal diamond.

4. Tensile Strength of Polycrystalline Diamond Films

Although the fact that diamond is "hard" is a well-established point of fact, lay and scientific, the actual "conventional" mechanical properties such as the modulus of elasticity, tensile strength and compressibility are much less quantified. This is especially true of the interesting polycrystalline diamond films, PDF, resultant from the synthesis work with filament and plasma growth techniques of the last decade. The potential mechanical applications for diamond, which are numerous, suggest more property characterization would be useful.

As described above, the Young's modulus of PDF was found to be $1.20(10^7)$ kg/cm^2 [1] as compared to $1.15(10^7)$ kg/cm^2 for natural, single crystal diamond. As regards *tensile strength*, the experimentation that had been performed, (monitoring deflection as a function of pressure and the pressure to *rupture* PDF *membranes*) and the resulting estimates of tensile strength [7], were unclear due apparently to complications in the model selection for the membrane, (that is, should the PDF be treated as a membrane or a plate?) led to wide variations in the estimates.

Ideally, and as we attempted in our effort, a more conventional tensile test specimen of PDF could be fabricated and positioned for true orthagonally loaded tensile stress testing. In our work we "shaped" PDF tensile specimens fabricated by microwave chemical vapor deposition described elsewhere [8]. For PDF stripes, the diamond was deposited on silicon and the silicon was then etched away in a conventional silicon etch solution which had no effect on the diamond. Analytical techniques, e.g., SIMS, indicated all of the silicon was removed. For some specimens the silicon substrate was shaped to result, directly from the deposition, in a stripe of diamond suitable for tensile testing. Other diamond specimens were deposited as a film and a patterning process, described elsewhere [9], was used to achieve the desired specimen shape. Figure 6 is a photograph of a free standing PDF tensile specimen. Raman spectra and visual inspection indicate films with very low sp^2 content.

Figure 6. Photograph of patterned polycrystalline diamond tensile specimen, length = 2.5 cm.

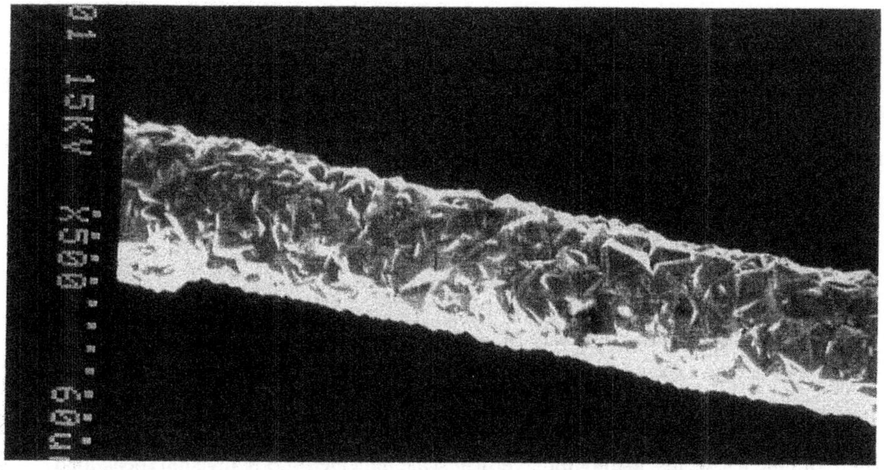

Figure 7. Polycrystalline diamond fiber, width ~ 100 microns, (Molybdenum core, diameter ~ 25 microns).

Two forms of tensile specimens were pursued, stripes, as just described, and *filaments*. Filaments were formed by depositing diamond on free standing sections of small (nominally 10-25 μ diameter) fibers of tungsten, molybdenum, graphite and aluminum oxide. The molybdenum core fibers were selected to attempt filament tensile testing, see Figure 7.

4.1. MECHANICAL TESTING

Free standing PDF structures are fragile and brittle, easily broken by macroscopic bending loads. Large compressive grips are unsuitable for gripping the films or fibers in the tensile machine. Employing an MTS model 5002 tensile test machine, specially designed connectors and a specimen mounting technique using a supporting aperture care for a pre-loading, successful tensile tests were performed.

"Successful" is defined here as pulling the tensile specimens apart and achieving a clear separation at or near the middle (gauge length) of the specimen.

A plastic frame aperture card was used in some instances, enabling placement of the tensile specimen in readiness for load. The ends of the specimen were "super glued" to the frame which had holes centrally located into which universal hooks to connect to the load cell and base of the tensile machine were attached

PDF structures were loaded to fracture and the maximum load noted. A typical load:time output is shown in Figure 8. The calibrated load cell signal as f(t) was processed through a PC based data recording system for storage, plotting, etc. Due to unquantifiable "give" in the entire arrangement, no attempt at accounting for strain in the PDF is made. The first, smaller peak in the example represents a shifting in the hooks as the load began to build up.

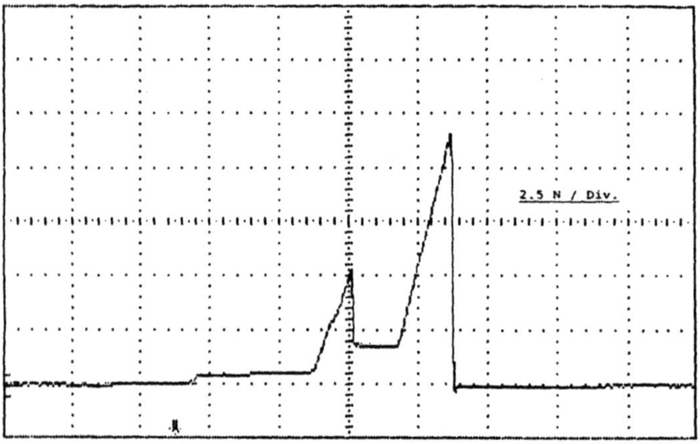

Figure 8. Example tensile test load output, vertical scale indicated, 2.5 N/division (peak load approximately 11 N), horizontal scale, time = 5 seconds per division.

By noting the maximum load sustained to failure and measuring the width and thickness (area) of the cross section that fractured under that load, the tensile strength is estimated. It is noted that only data where the PDF failed (separated) clear of the plastic frame "grips", i.e., preferably at or near the middle of the length of PDF stripe, was used in these estimates. Slightly less than half of specimen tensile tested to date fractured in a "good data" zone.

Figure 9 is an SEM of the fracture surface of the cross section of PDF that failed at maximum load. The fracture surface resembles that of other fractured PDF, characterized by Hoff [10], although loading conditions in that work may be different in application and unknown in magnitude.

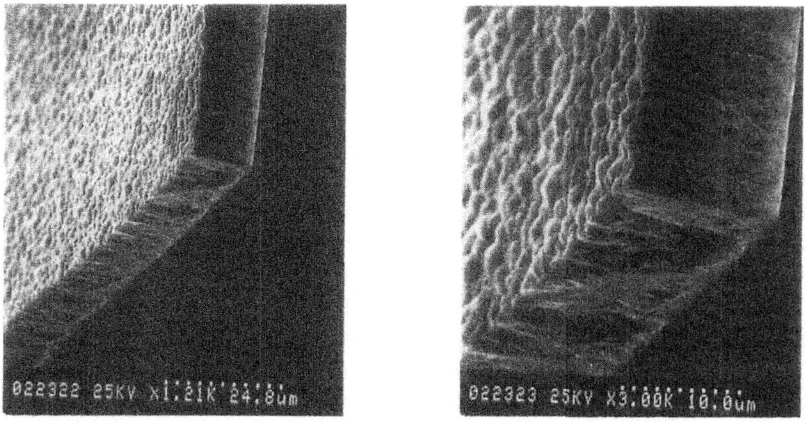

Figure 9. SEM of fractured tensile specimen (the vertical surface was the load bearing interface).

Table 1 summarizes the tensile test data. All data is from stripes [no filaments]. It was presumed that negligible plastic flow occurred (none was observed) during tensile load, and thus the area at fracture cross section measured was taken as the area that carried the load at failure.

No load data is available for poly diamond fibers. No successful handling technique is yet achieved for “zero” or start conditions of the load testing on the size instrument and fixtures used without first breaking the fiber specimens by cross loading.

TABLE 1. Tensile test summary results

SPECIMEN	w(mm.)	t(μ)	L(N)	T.S.(psiX10^5)
1	1.0	10	11.5	1.66
2	2.0	5.7	10.5	1.33
3<1>	1.0	15	3.0	.29
4	3.0	8	12.5	.75
5<2>	2.2	8	6.0	.49
6<1>	4.1	16	13.5	.30

w = width in mm. (± .1 mm.)
t = thickness in microns (± .1 μ)
L = Load in Newtons
T.S. = Tensile Strength in psi (est. ± 20%)

notes <1> broke "near" grip
<2> "hourglass" shape

4.2. DISCUSSION

Even the highest value observed for tensile strength, 166 ksi, is smaller than expected based on the very high Young's modulus of diamond and based on our own membrane rupture data, see next section. There are several possible explanations: (a) loading in these experiments is still not perfectly axial and bending moments are present, (b) microcracks (notches/stress risers) exist at the edges or surfaces of the specimens, or (c) intergranular fracture is occurring through material inferior to pure diamond. Also, the small number of samples "successfully" tested does not allow statistical assessment.

In summary, diamond films grown by microwave plasma CVD, thickness approximately 5 - 20 microns, were patterned into tensile test specimens. Using conventional tensile test techniques, the diamond was stressed in tension to fracture and the load recorded. The cross sectional area at separation was measured and the tensile strength observed to be nominally 1.5 (10^5) psi. Fracture characteristics were examined by SEM.

5. Diaphragm Rupture Stress

Information about the stress in diamond films and its rupture stress is important for composite consideration. The stress in the film may be tensile or compressive and compressive stresses could produce film wrinkling or decohesion by buckling in some situations. CVD diamond films have been reported with tensile stress of 0.84 GPa [11].

The rupture stress of a diaphragm can be defined as the stress at which the diaphragm reaches the elastic strain limit where permanent damage occurs to the diaphragm. For a circular diaphragm with a clamped edge, the maximum stress σ_{max} at the circumference is [12]

$$\sigma_{max} = \frac{3}{4} P_{max} \left(\frac{r}{t}\right)^2 \quad (3)$$

r and t are diaphragm's radius and thickness, P_{max} is pressure. To examine the rupture stress of diamond diaphragms, eight samples were tested. Their dimensions, rupture pressures and stresses are listed in Table 2.

TABLE 2. Sample data from rupture stress test

Sample number	Thickness (μm)	Radius (mm)	Rupture pressure (psi)	Rupture Stress $\times 10^{11}$ (dyne/cm^2)
1	4.5 ± 0.5	3.75 ± 0.05	14.5 ± 0.5	5.20 ± 1.20
2	8.0 ± 0.5	4.30 ± 0.05	17.5 ± 0.5	2.61 ± 0.59
3	4.5 ± 0.5	3.75 ± 0.05	12.0 ± 0.5	4.31 ± 0.98
4	3.0 ± 0.5	4.25 ± 0.05	4.0 ± 0.5	4.15 ± 1.10
5	7.0 ± 0.5	4.20 ± 0.05	13.0 ± 0.5	2.42 ± 0.55
6	8.0 ± 0.5	4.90 ± 0.05	9.5 ± 0.5	1.84 ± 0.42
7	4.0 ± 0.5	4.30 ± 0.05	6.5 ± 0.5	3.88 ± 0.92
8	6.0 ± 0.5	5.00 ± 0.05	7.2 ± 0.5	2.58 ± 0.61

It can be seen that the rupture stress of diamond diaphragms has the range of $1.84 \pm 0.42 \times 10^{11}$ to $5.20 \pm 1.2 \times 10^{11}$ dyne/cm^2 with an average of $3.40 \pm 0.79 \times 10^{11}$ dyne/cm^2. The difference in rupture stress among these samples is because of the imperfections of the diamond diaphragms, differences in thickness and boundary clamping.

Compared to the rupture stress σ_r of silicon 6.24×10^8 dyn/cm^2 [13] , diamond film has rupture stress considerably larger. Hence a diamond diaphragm is potentially more rugged than silicon and can withstand much higher pressure conditions. On the other hand, a thinner diaphragm thickness could be achieved using diamond as a diaphragm. For example, for a given pressure P_{max}, the minimum thickness t_{min} of diaphragm has the relationship:

$$t_{min} = r\sqrt{\frac{3}{4}\frac{P_{max}}{\sigma_r}} \tag{4}$$

For example, assume P_{max} to be 250 mmHg with radius r as 1 mm, the silicon diaphragm would require a minimum thickness of 20 micron while diamond diaphragm would require only 2.5 micron. A thinner diaphragm can reduce the CVD deposition time, improving sensitivity and reducing cost for real world applications.

6. References

1. Davidson, J.L., Ramesham, R., and Ellis, C. (1990) Synthetic Diamond Micromechanical Membranes, Cantilever Beams, and Bridges, *J. Electrochemical Society*, **10**, 137.
2. Davidson, J.L., Ramesham, R., and Ellis, C. (Oct. 15-20, 1989) Abstract 682, p. 973, *The Electrochemical Society Extended Abstracts*, Hollywood, FL.
3. Davidson, J.L., Ellis, C., and Ramesham, R. (1989) *J. Electron. Mater.*, **18**, 711.
4. Nash, W.A. (1972) Theory and Problems of Strength of Materials, 2nd ed., McGraw-Hill Book Company, New York, 159.
5. Moses, A.J. (1978) The Practicing Scientist's Handbook, van Nostrand Reinhold Company, New York, 596, 662, 879.
6. Williams, B.E., Glass, J.T., Davis, R.F., Kobashi, K., and Horiuchi, T. (1988) *J. Vac. Sci. Technol.*, **A6**, 1819.
7. Air Force Report AL-TR-89-044, CN F04611-88-C-0074, Edited by D. M. Jassowski, Nov. 1989.
8. Davidson, J.L., and Cao, X. (May 5-10, 1991) Ultimate Tensile Strength of Polycrystalline Diamond Films, Second International Symposium on Diamond Materials, *179th Meeting of the Electrochemical Society*, Washington, D.C.
9. Davidson, J.L. (April 1991) Diamond as an Active Microelectronic Material, *Proceedings of the Advanced Materials Symposium*, International Society of Hybrid Microelectronics, Dallas, Texas.
10. Hoff, H.A. (Sept. 1990) Fracture Characteristics of Chemical Vapor and Combustion Deposited Diamond, *Second International Conference on the New Diamond Science and Technology*, Washington, D.C.

11. Windischmann, H., Epps, G.F., Cong, Y. and Collins, R.W. (1991) Intrinsic Stress in Diamond Films Prepared by Microwave Plasma CVD, *J. Appl. Phys.*, **69**(4), 2231-2237.
12. Ko, W.H., Bao, M., and Hong, Y. (1982) A High-Sensitivity Integrated-Circuit Capacitive Pressure Transducer, *IEEE Transactions on Electron Devices*, **29**(1), 48-63.
13. Clauser, et al. (1963) *The Encyclopedia of Engineering and Processing*, London, England, Reinhold, 616.

MAGNETIC RESONANCE STUDIES OF SOLID-STATE HYDROGEN AND HYDROGEN-RELATED DEFECTS

K.M. McNAMARA RUTLEDGE*
General Electric Company, Corporate Research & Development, Schenectady, NY, USA

G.D. WATKINS AND X. ZHOU
Department of Physics, Lehigh University, Bethlehem, PA, USA

K.K. GLEASON
Department of Chemical Engineering, MIT, Cambridge, MA, USA

Abstract

Hydrogen is a detrimental impurity in many chemical vapor deposited (CVD) materials, particularly those involved in electronic or optical applications. For example, active hydrogen defects have been observed in materials such as silicon, Si, gallium arsenide, GaAs, and diamond, C, thin films. Hydrogen and its related defects can be identified, quantified, and observed using magnetic resonance techniques. These techniques allow a unique quantitative, non-destructive view of hydrogen in the solid-state. Nuclear magnetic resonance (NMR) is used to study hydrogenated defects directly, while electron paramagnetic resonance (EPR) is used to observe hydrogen associated with paramagnetic defects. These observations can enhance our understanding of the effects of hydrogen incorporation on the properties of such materials.

**Current Address: Department of Chemical Engineering, Worchester Polytechnic Institute, Worchester, MA, USA*

M. A. Prelas et al. (eds.), Diamond Based Composites, 241–259.

1. Introduction

Many of the carrier gases used to deposit thin films contain hydrogen and thus, hydrogen is incorporated as an impurity in many chemical vapor deposited (CVD) semiconductor materials. Since hydrogen can represent an electrically active impurity, it may interfere with the predicted properties and performance of devices containing materials such as CVD silicon, Si, gallium arsenide, GaAs, and diamond, C. It is important to be able to observe hydrogen in the solid-state in order to predict and control its detrimental effects.

Hydrogen and its related defects can be identified, quantified, and observed using magnetic resonance techniques. These techniques allow a unique quantitative, non-destructive view of hydrogen in the solid-state. Nuclear magnetic resonance (NMR) is used to study hydrogenated defects directly, while electron paramagnetic resonance (EPR) is used to observe paramagnetic defects, such as dangling bonds, and their associated hydrogen impurities.

Specifically discussed are the presence of hydrogen-related defects associated with lattice vacancies in chemical vapor deposited (CVD) diamond as well as the probable distribution of these and other hydrogen environments in CVD diamond. These environments include both rigidly held static hydrogen and hydrogen-containing groups with significant mobility. The relative and absolute concentrations of each defect type are discussed. Although these results consider defects in diamond, specifically, they can provide insight into phenomenon observed in other semiconductor materials, such as hydrogen activation at grain boundaries is silicon.

2. Nuclear magnetic resonance

Quantitative measurements of impurities such as ^1H, averaged over the entire sample volume, can be obtained using solid-state NMR. Proton NMR is non-destructive and particularly valuable for measuring hydrogen concentrations, since this mobile, low Z element is difficult to quantify by other spectroscopies. In cases where paramagnetic impurity concentrations are low and spin-lattice relaxation times are fast, (which we demonstrate is the case for CVD diamond films), each nuclei in the magnetic field will give rise to the same integrated signal intensity, regardless of its local bonding environment. Thus, comparison of the integrated area under the NMR lineshape with that of a known standard provides a quantitative overall hydrogen count. Typical ^1H NMR spectra yield overall hydrogen concentrations ranging from < 0.017 to ~1 at. % H for CVD films.[1]

As mentioned, NMR experiments are quantitative, regardless of the local bonding environment of the nuclei, only for samples in which the concentration of paramagnetic and ferromagnetic defects is small in comparison to the total concentration of the nuclei of interest. In samples containing such impurities, a small fraction of nuclei in close proximity to such defects will remain undetected as a result of extreme line broadening due to interaction with the unpaired electron of the paramagnetic impurity.[2] That this represents only a small fraction (< 0.05%) of the total hydrogen in CVD diamond films will be demonstrated in the EPR section below. As additional supporting evidence, the ability to quantitatively correlate infrared spectroscopy and NMR results confirms that paramagnetic defects do not shield a large number of protons from NMR detection in films examined to date.[3]

For quantitative measurements, it is also important to insure that sufficient spin-lattice relaxation occurs between signal acquisitions[4]. This relaxation is characterized by a time constant, T_1, known as the spin-lattice relaxation time constant, and determined by saturation recovery techniques and subsequent application of the Bloch equation[5]:

$$\frac{M(t) - M_{eq}}{M_0 - M_{eq}} = \exp(-t/T_1), \qquad (1)$$

where M(t) is the net magnetization parallel to the external magnetic field at time, t; M_{eq} is the equilibrium magnetization; and M_0 is the initial magnetization at t = 0. The spin-lattice relaxation for hydrogen in CVD diamond is relatively fast, between 5 ms and 1 s at room temperature [6,7].

NMR spectroscopy can also provide information on the distribution and local environment of an element. The room-temperature and low-temperature ^1H NMR spectra for a typical CVD diamond film is shown in Figure 1. Such spectra often contain two components, a narrow Lorentzian and a broad Gaussian, indicating at least two different bonding configurations for hydrogen[8]. The ratio of integrated intensities of these components shows that the majority of the hydrogen contributes to the Gaussian component, while the Lorentzian component can be attributed to only a small fraction of the hydrogen present. The actual integrated intensity under each lineshape is directly proportional to the number of nuclei in its environment. In addition, the average interproton spacing within the Gaussian component can be determined through application of the van Vleck equation[9] which reduces for a polycrystalline material with closely-space hydrogen nuclei, to,:

$$FWHM_G = 189.6\ (\Sigma\, r_{ij}^{-6})^{1/2}\ \text{Å}^3\text{kHz} \qquad (2)$$

where $FWHM_G$ is the full-width at half-maximum intensity of the Gaussian lineshape, and r_{ij} is the interproton spacing.

In addition, the dimensional distribution of hydrogen on a greater length scale can be explored by ^1H multiple quantum (MQ) NMR studies [6]. Up to 20 hydrogen atoms within a 10 Å radius were correlated in CVD diamond films despite the very low (< 0.1 at.% H) overall bulk concentrations, indicating large scale clustering of hydrogen. The initial MQ growth for the films can be compared with that for known bulk (three dimensional) hydrogen distributions, such as for CaH_2, and with known two-dimensional proton distributions, such as intentionally hydrogenated diamond powder. Such comparisons lead to conclusions on the dimensionality of clustered hydrogen in diamond films.

3. Electron Paramagnetic Resonance

Electron paramagnetic resonance (EPR) spectroscopy detects defects which have an unpaired electron. In diamond, unpaired electrons are often localized at impurity atom sites, such as substitutional nitrogen, or at lattice defects, such as vacancies. Like NMR, EPR can give quantitative and distributional information about the unpaired electron. In addition, detailed information can be obtained about the identities of the atoms and symmetry which surrounds the unpaired electron. [10] Nuclei having angular momentum which are in the proximity of the paramagnetic defect can often be identified through EPR as a result of hyperfine interactions. The strength and orientation dependence of hyperfine couplings depend on the location of the nuclear spin (or spins) with respect to the unpaired electron. Since the predominant isotope of hydrogen, ^1H, is a spin-1/2 nucleus, its interactions with nearby paramagnetic defects can be observed in this manner.

In contrast to natural and HPHT diamond, which often contain significant nitrogen impurities, dangling bond defects are often assumed to dominate the EPR spectra of CVD diamond films. Their concentration, N_S, ranges from 10^{17} to 10^{19} cm^{-3} [11-13], significantly lower than observed hydrogen contents. Previous studies have identified an S=1/2 electron paramagnetic resonance (EPR) at g=2.0028 with a pair of weaker partially resolved satellites in CVD diamond films produced from methane and hydrogen gases.[14-17] One of these studies, using samples grown from deuterated reactants, has shown that the satellites arise from the hyperfine interaction with nearby hydrogen. A second study, combining observations at 9.8 and 35 GHz, has additionally proposed that they arise from the forbidden $\Delta m\pm1$ nuclear spin flips of hydrogen during the EPR $\Delta M\pm1$ transitions. Linewidths of dangling bond defects in polycrystalline diamond range from 1.9 to 12 gauss[14-19]. The smaller linewidths are observed in films with lower dangling bond densities, consistent with the reduction of dipole-dipole interactions between randomly distributed paramagnetic centers [19].

Here, we report the study of this EPR signal at the intermediate frequencies of 14 and 20 GHz. Our results confirm in detail the $\Delta m\pm1$ origin of the satellites. In addition, we find the result that the relative intensities of the central and satellite components, as well as their overall lineshapes, can be accurately matched at all four frequencies by the assumption of a single unique defect with hyperfine interaction $A\parallel$ =+28 MHz and $A\perp$ =-6 MHz with a single hydrogen atom. From analysis of the hyperfine anisotropy, the separation between the electronic and nuclear spins is determined, leading to a microscopic model of the defect. An additional resonance is also observed in the spectra of samples containing relatively low hydrogen contents and two possible interpretations for it are discussed.

Our results at microwave frequencies n = 14 and 20 GHz are shown in Figure 2, along with the results of Holder et al. [14] at 9 and 35 GHz. Taken together, they clearly demonstrate a monotonically increasing separation and decreasing intensity of the satellites as the microwave frequency increases. That this is the characteristic signature for forbidden $\Delta m\pm1$ nuclear spin flip transitions [14,17,20] can be seen as follows: The spin Hamiltonian for an electronic spin, S=1/2, with isotropic g, coupled to a nuclear spin, I=1/2, is

$$\mathcal{H} = g\mu_B \mathbf{S}\cdot\mathbf{B} + \mathbf{S}\cdot\mathbf{A}\cdot\mathbf{I} - g_N\mu_N \mathbf{I}\cdot\mathbf{B} \tag{3}$$

from which the nuclear Hamiltonian, to first order in $A/g\mu_B B$, becomes

$$\mathcal{H}_N = -g_N\mu_N \mathbf{I}\cdot[\mathbf{B} - (\mathbf{A}\cdot\mathbf{B}M/g_N\mu_N B)] \tag{4}$$

where g and g_N are the electronic and nuclear g-values, respectively, μ_B and μ_N, the corresponding Bohr magnetrons, **S** and **I**, the corresponding spin operators, **A**, the nuclear hyperfine tensor, and $M\pm 1/2$ is the azimuthal quantum number for **S** quantized along **B**, the applied magnetic field. The bracketed term represents an effective magnetic field seen by the nucleus, and when **A** is anisotropic, its two vector components, **B** and $\pm\mathbf{A}\cdot\mathbf{B}/2g_N\mu_N B$, can point in different directions, as illustrated schematically in Figure 3(b). Thus, the quantization axis for the nucleus differs for the two quantization states, and as a consequence, the orthogonality of the $\Delta m \pm 1$ nuclear quantization states is destroyed. The intensity of the EPR transitions are proportional to

$$I(+1/2, m <-> -1/2, m') \propto |\langle +1/2|S+|-1/2\rangle|^2\, |\langle m|m'\rangle|^2\, H_1^2 \tag{5}$$

where $|\langle m|m'\rangle|^2 = \cos^2(\theta/2)$ for the $\Delta m=0$ transitions and $\sin^2(\theta/2)$ for the $\Delta m=\pm 1$ transitions and θ is the angle between the nuclear quantization axis for $M=-1/2$, and that for $M=+1/2$, and H_1 is the amplitude of the microwave magnetic field. As illustrated in the figure, θ can depart substantially from zero when the anisotropy in **A** $\sim 2g_N\mu_N B$, leading to significant intensity in the forbidden $\Delta m=\pm 1$ transitions.[21] As B increases, θ decreases as B^{-1}, the forbidden intensities decrease, therefore, as B^{-2}, where $B \cong B_o = h\nu/g_B$, and the satellite positions, given by equation (1), approach field values of $\pm(g_N\mu_N/g\mu_B)B_o$ to either side of the central line.

4. Experimental

More than ten samples were produced for this work by a proprietary chemical vapor deposition technique, using a microwave plasma of hydrogen with a low concentration of a carbon-containing precursor. Deposition conditions were varied to produce diamond films with a range of hydrogen content (measured by NMR) and thermal conductivity, as these are two commonly used indicators of diamond film quality (Table I). The samples produced were free standing films with thicknesses from 100-500 mm, after removal of the substrate by a proprietary method. The films were cleaned in boiling acid to eliminate surface contamination before being examined by NMR and EPR. The EPR spectra were obtained at room temperature in absorption, using balanced bolometer spectrometers operating at 14 and 20 GHz. Low microwave power (~5 mW) and magnetic field modulation amplitude (~0.5 G peak to peak) were used to avoid saturation and lineshape distortion of the spectra, respectively. The NMR spectra were also obtained at room temperature, unless otherwise indicated, using a 270 MHz home-built NMR spectrometer with quadrapolar detection.

5. Results and Discussion

The total hydrogen content determined for a number of diamond films, whose ^{1}H NMR spectra contain both Lorentzian and Gaussian components, as shown in Figure 1, is listed in Table 1. A "hole-burning" NMR experiment, where magnetization is transferred from the Gaussian environment to the Lorentzian environment (Figure 4), shows these two hydrogen environments are located within 5 Å of each other.[1] In addition, the low temperature NMR experiment at 100 K (Figure 1) shows that the Lorentzian component is broadened, indicative of reduced molecular motion[3,22]. It is unlikely that this peak is a result of trapped H_2 as observed in amorphous silicon, since the motion of H_2 would not be significantly reduced at 100K. The Gaussian component remains unchanged at 100 K, indicating rigidly held hydrogen [8]. As the hydrogen content increases, it can sometimes become difficult to resolve the narrow feature from the dominant broad line[23].

Although the average hydrogen concentration in the CVD films is low, the proton homonuclear dipolar line broadening in the Gaussian component is large (~60 kHz). This discrepancy indicates locally high hydrogen concentrations exist, requiring significant segregation of hydrogen in polycrystalline diamond [8,24]. Randomly dispersed CH or CH_2 groups would provide too little homonuclear broadening to account for the

TABLE I. Estimated defect concentration, measured by NMR and EPR.

Sample	Total H content (at %)	H1 (10^{16} cm^{-3})	H2 (10^{16} cm^{-3})
1	0.021	0	3.2
2	0.025	0	4.8
3	-----	0	≤ 0.4
4	0.027	0	≤ 0.5
5	0.031	≤ 3	0
6	0.043	50	0
7	0.045	16	0
8	0.085	≤ 6	0
9	0.150	67	0
10	0.290	119	0
11	0.320	92	0
12	0.330	78	0

observed linewidth. However, A fit to the van Vleck equation (2) shows that the broadening is consistent with the areal densities for hydrogen passivated diamond surfaces, typically 1-3 x 10^{15} H/cm^3.

Also, consistent with this interpretation is correlation of the observed NMR hydrogen contents with the absorption of the films in the CH-stretch region (2750-3050 cm^{-1}) of the infrared spectra (Figure 5). This agreement suggests that the majority of hydrogen in CVD diamond is covalently bonded to the lattice, with absorption coefficients similar to long-chain alkanes and paraffinic hydrocarbons.[24] Finally, MQ-NMR experiments show correlation growth rates for CVD diamonds which are similar to those observed for a two-dimensional hydrogen surfaces on natural diamond powder. Thus it is likely that the majority of the hydrogen in CVD diamond films is involved in surface passivation. Both internal surface area (for example grain boundaries) as well as that of the top growth surface are potential sites for such passivation. Internal voids may also contribute to the surface area [23] and influence the thermal conductivity of diamond films [25].

Quantitative values for the concentration of two dangling bond-hydrogen defects, H1 and H2, respectively, are also listed in Table I.[21] Note these values are one to two orders of magnitude smaller than the total hydrogen content observed by NMR. To simulate the EPR spectra, axial symmetry is assumed for **A** with a single hydrogen atom, which, for any angle, ϕ, between the symmetry axis and **B**, gives four lines, two for the $\Delta m=0$ transitions, and two for the $\Delta m=\pm 1$, whose intensities are given by equation (5). This is illustrated in figure 3(a). These are convoluted with a random distribution function for the defect axis orientation, $N(\phi)d\phi = (1/2)\sin\phi d\phi$, and then convolute again with the derivative of a Lorentzian lineshape vs. B. There are thus only three adjustable parameters, A||, A^, (or, the isotropic part $a=\{A_{\parallel} + 2A_{\perp}\}/3$, the anisotropic part $b=\{A_{\parallel}-A_{\perp}\}/3$, and the Lorentzian linewidth, which must give a good fit to the experimental derivative spectra at all four microwave frequencies.

In Figure 2, the fit for all four frequencies using the values of a = 5.5 MHz and b = 11.0 MHz is shown. [The Lorentzian used for the best match shown in the figure for the sample used in our 14 and 20 GHz results had a peak-to-peak derivative width (PPDW) = 2.8 G. For the sample used by Holder et al. [14] at 9.8 and 35 GHz, the PPDW was taken to be 3.6 G. Studies reported in previous work display different degrees of resolution of the satellites consistent with sample-dependent breadth difference[14-17], and such an adjustment, which is one of resolution only, is therefore reasonable]. By adjusting a and b, we conclude that the excellent fit at all four frequencies shown in the figure requires b = +11.0 MHz and a = +5.5 MHz, with an uncertainty of ±2.5 MHz for each. The relative intensities of the satellites are determined primarily by b, {i.e. $\sim b^2$}, and the shape of the central line, which arises from the normally allowed hyperfine structure, depends strongly on the ratio of a to b. Because of this, the range of values for a and b beyond which a satisfactory fit cannot be obtained is so narrowly defined. The absolute signs of a and b cannot be determined but have been taken as those appropriate for dipole-dipole origin of the anisotropy.

The narrow allowable range of the hyperfine parameters strongly suggests that the resonance results from a single well-defined defect. The isotropic part of the hyperfine interaction, a=5.5 MHz, is small, corresponding to <0.5% of the atomic value (1420 MHz), so the hydrogen atom must be considered only a neighbor of the paramagnetic site. We treat, therefore, the anisotropic part, b=+11.5 MHz, as arising from the dipole-dipole interaction between separated electronic and nuclear dipoles, $b=g\mu_B\mu_N/r^3$, which gives for the separation r=1.9Å (and determines the sign of b to be positive). This is an interesting result, the distance between the nearest C-C diamond lattice separation, 1.54Å, and the next-nearest separation, 2.51Å.

These are similar to what one might expect for distances between a carbon atom next to a lattice vacancy and a hydrogen atom bonded to one of the other carbon neighbors of the vacancy (1.74Å with a typical C-H bond distance of 1.09Å and the carbon atoms in their normal unrelaxed lattice sites. Upon breaking the weak next nearest neighbor reconstructed bond, we expect relaxation backward and this distance should increase.) Many vacancy-like sites are expected in such polycrystalline films -- in the bulk at dislocation cores, but particularly at or near the many grain boundaries, which, of necessity, contain vacancy-like reconstructed bonds adjacent to the extra atom planes that support the angle between the lattice planes between two crystalline grains. Such stretched bonds are vulnerable to single hydrogen atoms which can enter and break the bond, attaching to one of the carbon atoms, and activating a vacancy-like dangling bond on the other as the two relax backward. A schematic of the defect is shown in Figure 6. If we had a hydrogen in an isolated bulk vacancy, we would expect to see easy hopping of the dangling bond amongst the three remaining carbon neighbors as well as motion of the hydrogen itself, as is commonly observed for such defects.[26-28] For all of the EPR active defects observed here, however, no such motion is observed in experiments from 1.5K to room temperature. We therefore expect to find these highly-distorted vacancy-like defects at grain boundaries, surfaces, and other areas of misfit resulting from the polycrystalline growth. The results of molecular modeling are in agreement with this interpretation, predicting r=1.96Å.

Finally, let us mention that in a few samples that contained significantly lower hydrogen contents, as determined by nuclear magnetic resonance and infrared spectroscopy, the H1 signals described above is missing and a weaker but distinguishably different hydrogen-related center, H2, which has not previously been reported, is observed. It is narrower and with relative intensity factors of ~2-3 weaker than the H1 defect. For it, we estimate b~6.7 MHz, which corresponds to a hydrogen separation of ~2.3Å. It is shown elsewhere that this is a distinct defect from the H1 defect, and that its EPR Gaussian lineshape supports the notion of nearby regions of extended hydrogenated structural defects, such as the hydrogen terminated grain boundaries and surfaces indicated by previous NMR results.[1]

6. Summary

Hydrogenated defects in polycrystalline diamond films have been quantitatively observed using magnetic resonance techniques. Results show that the overall hydrogen content in these materials is typically ≤ 0.2 atomic %. The majority of this hydrogen (> 99%) is spatially removed from paramagnetic sites, covalently bound, and rigidly held. A small fraction (< 1%) of the hydrogen observed is associated with covalently bound motional groups, such as rotating surface methyl groups, while < 0.0005 at. % is associated with dangling bond defects. Evidence indicates that all of these defects are likely to be located within the same region in the sample, specifically at highly distorted defects found at grain boundaries, surfaces, and other areas of misfit resulting from polycrystalline growth. Because of their potential electrical activity, even very low concentrations of such defects are expected to effect the electrical, thermal, and optical performance of these materials.

7. Acknowledgments

The authors would like to acknowledge the contributions of B.E. Scruggs, R.P. Messmer, and S. Chawala, for assistance and helpful discussion. The author would also like to acknowledge the Corporate Research and Development Center of General Electric Company for samples studied. The EPR studies were performed at Lehigh University and supported by USONR under grant no. N00014-94-1-0117. The remainder of the work was performed with the support of General Electric Corporate Research and Development and Superabrasives Divisions.

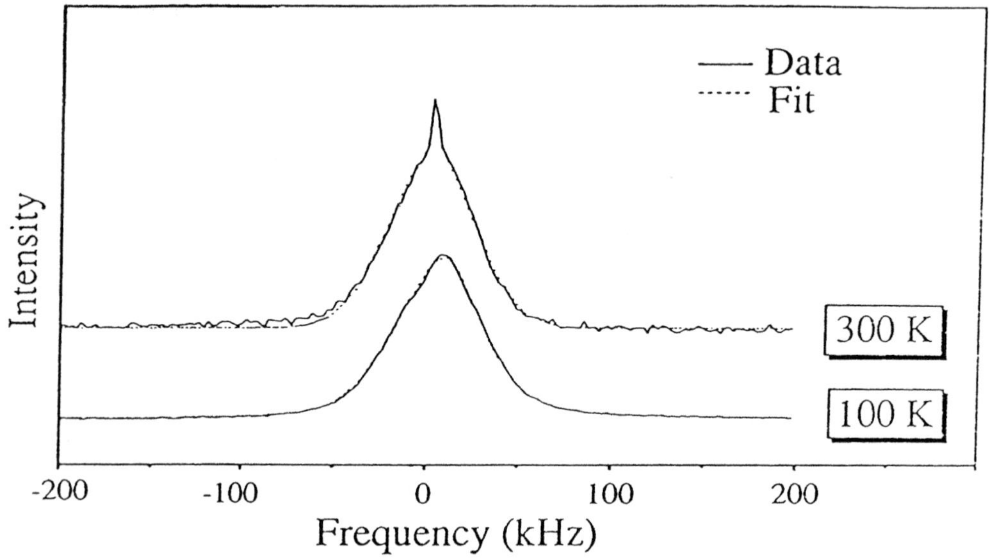

Figure 1. Proton NMR spectra of diamond film at 300 K showing two components. Upon lowering the temperature to 100 K, the broad Gaussian remains the same while the narrow Lorentzian broadens.

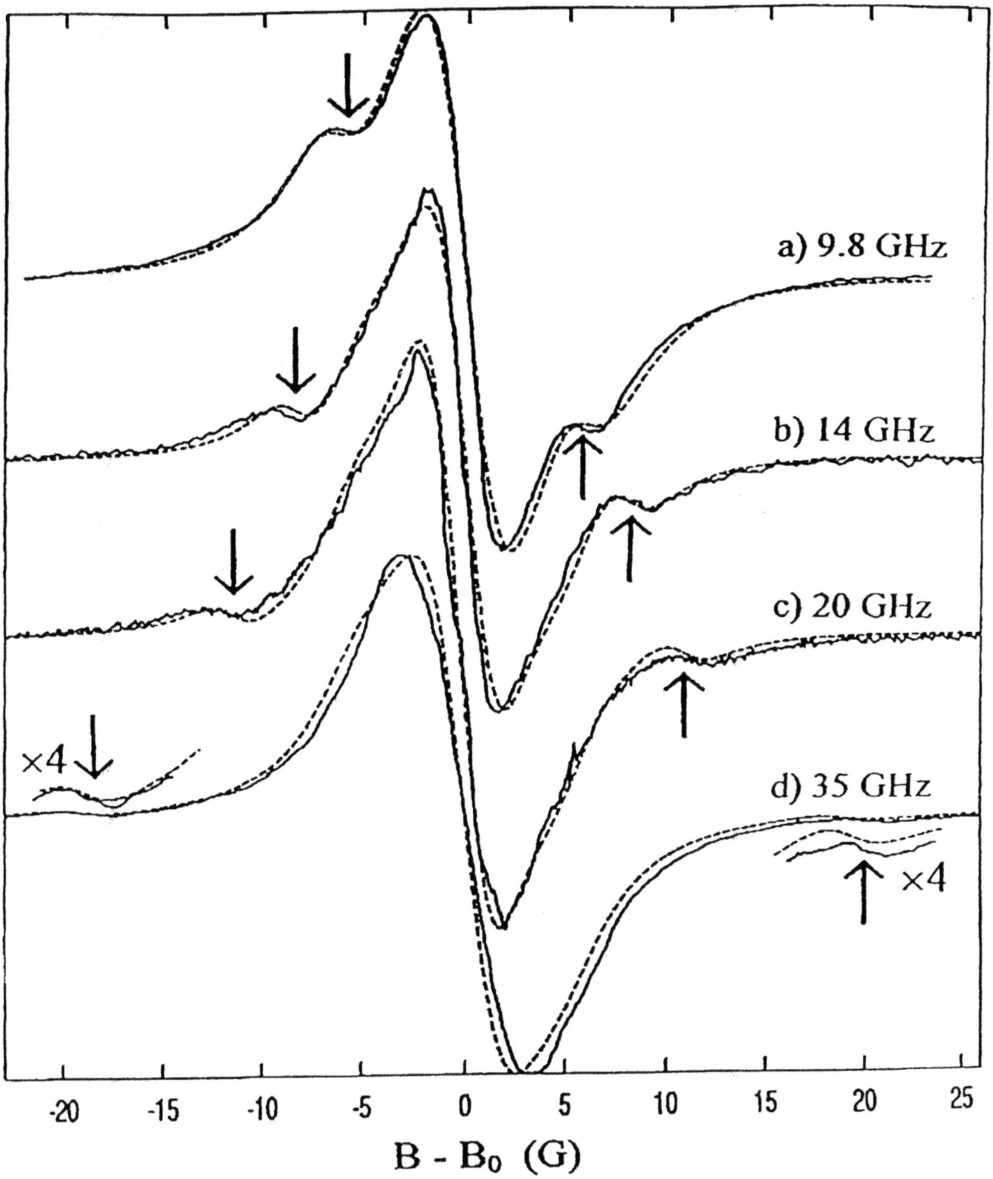

Figure 2. Magnetic field dependence of the EPR spectra of a dangling bond defect in a polycrystalline film. The shoulders on the central feature which are quite strong at 9.8 GHz (top), and decrease in intensity as the frequency increases, being barely resolved at 35 GHz (bottom)[Zhou 1995].

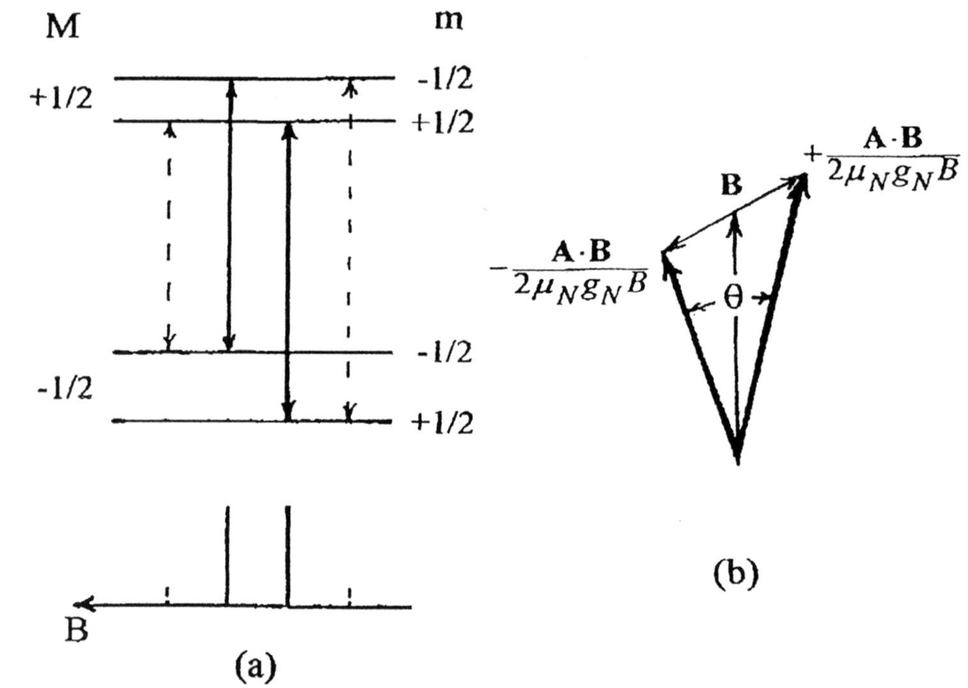

Figure 3. (a) Energy level diagram and EPR transitions for an S=1/2 electron spin coupled to an I=1/2 nucleus. (b) The effective magnetic field seen by the nucleus for the two M states.

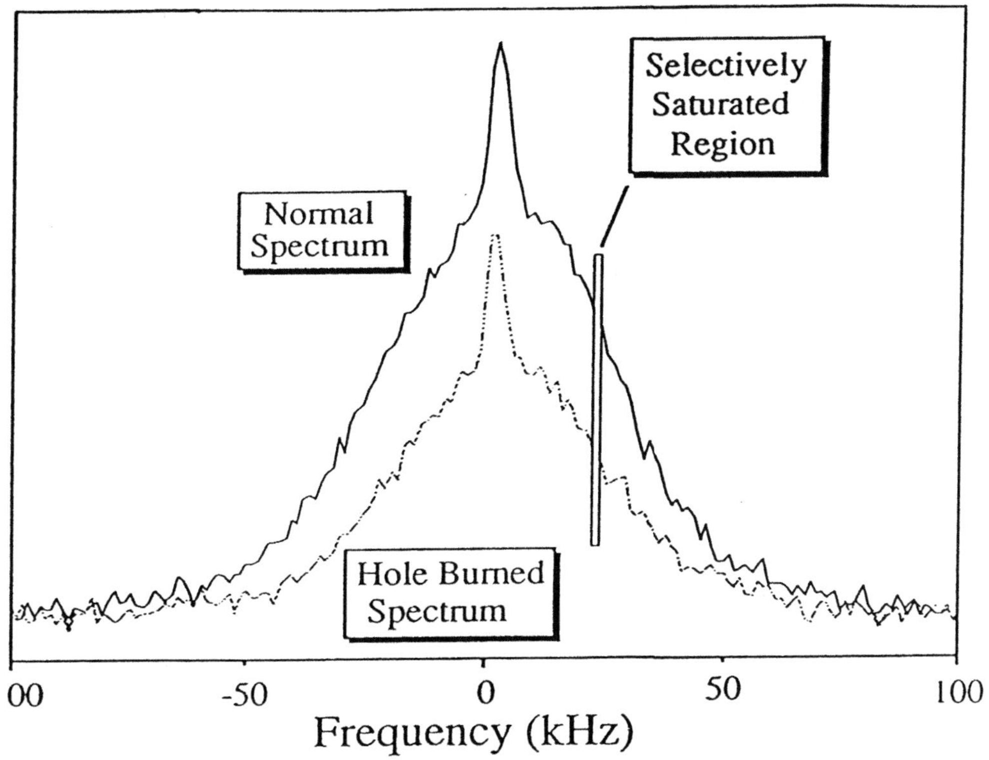

Figure 4. NMR spectrum resulting from a hole-burning experiment (dashed line), where the indicated portion of the Gaussian component is irradiated and magnetization is allowed to transfer to the Lorentzian component. The original signal (solid line) is shown for comparison.

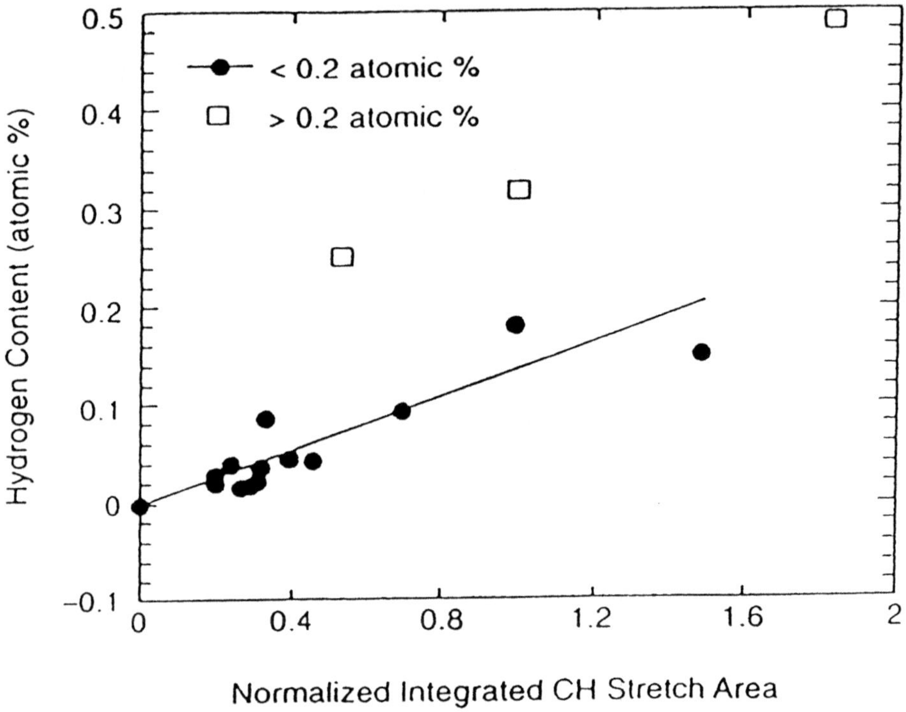

Figure 5. Correlation of the normalized absorption in the CH-stretch region of the IR spectra with the total hydrogen content, as measured by NMR spectroscopy.[McNamara, 1994]

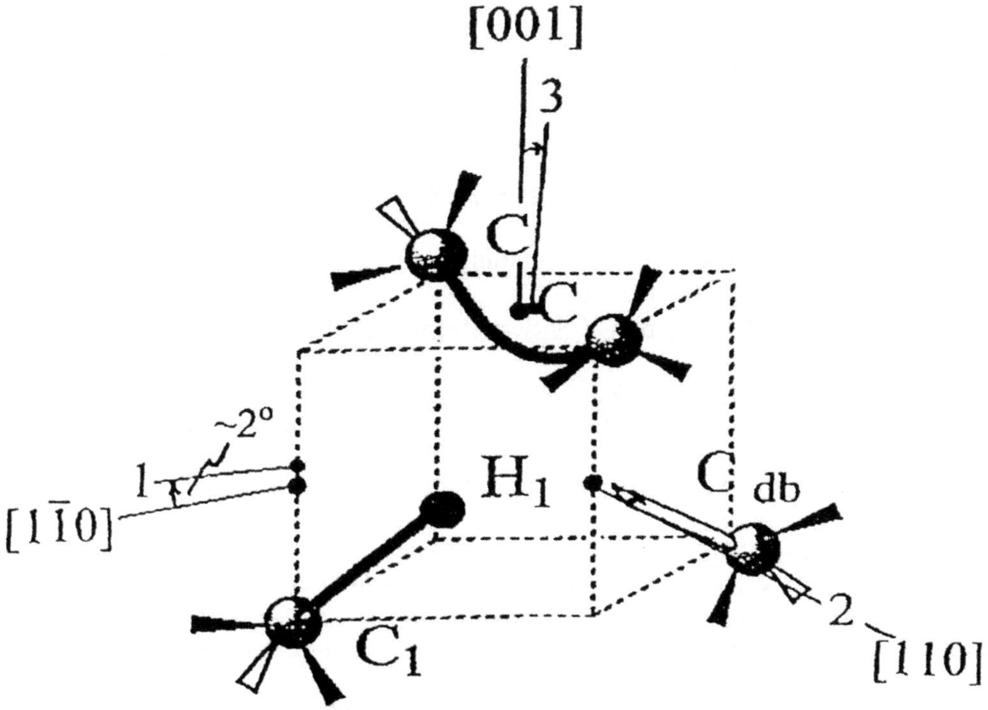

Figure 6. Model showing approximate atom positions for a single hydrogen atom in a vacancy, and the
principal axes for the hydrogen hyperfine interaction.

8. References

1. K.M. McNamara Rutledge and K.K. Gleason, *Chem. Vap. Deposition*, **2**(2), 37 (1996).
2. Henrichs, P.M., Cofield, M.L., Young, R.H. and Hewitt, J.M., *J. Magn. Reson.*, **58,** 85-94 (1984).
3. McNamara, K.M., Williams, B.E., Gleason, K.K. and Scruggs, B.E., *J. Appl. Phys*., 1994, **76**, 2466.
4. McNamara, K.M., Gleason, K.K., and Butler, J., and Vestyck, D.J., *Diamond and Related Materials,* 1992, **1**, 1145-1155.
5. Abragham, A., The Principals of Nuclear Magnetism, Oxford University Press, NY (1983).
6. Levy, D.H. and Gleason, K.K., *J. Phys. Chem.*, **31**, 8125-8131 (1992).
7. Lock, H., Wind, R.A., Maciel, G.E., and Johnson, C.E., *J. Chem. Phys.,* **99**, 3363-3373 (1993).
8. McNamara, K.M., Levy, D.H., Gleason, K.K. and Robinson, C.J., 1992, *Appl. Phys. Letts.,* **60**, 580-582.
9. J. H. van Vleck, *Phys. Rev*., **74**, 1168 (1948).
10. Loubser, J.H.N. and van Wyk, J.A., *Rep. Prog. Phys*., **41**, 1201-1248 (1978).
11. Watanabe, I. and Sugata, K., *Jap. J. Appl. Phys*., **27**, 1808-1811 (1988).
12. Fabisiak, K., Maar-Stumm, M., and Blank, E., *Diamond and Related Materials*, **2** , 722-727 (1993).
13. Fanciulli, M. and Moustakas, T.D., *Phys. Rev. B* , **48**, 14982-14988 (1993).
14. S.L. Holder, I.G. Rowan, J.J. Krebs, *Appl. Phys. Lett*., **64**, 1091 (1994).
15. Hoinkis, M., Weber, E. R., Landstrass, M.I., Plano, M.A., Han, S., and Karris, D.R., *Appl. Phys. Lett*., **59** 1870 (1991).
16. Portis, A.M., *Phys. Rev*., **91**, 1071 (1953).
17. G.T. Trammell, H. Zeldes, and R. Livington, *Phys. Rev*., **110**, 630 (1958).
18. Zhang,W., Zhang,F., Wu, Q. and Chen,G.,*Materials Letters* , **15**, 292-297 (1992).
19. Erchak, D.P., Ulyashin, A.G., Glefand, R.B., Penian, N.M., Zaitsev, A.M., Varichenko, V.S., Efimov, V.G., and Stelmakh, V.F., *Nuclear Instruments and Methods in Physics Research*, **B69**, 271-276 (1992).
20. L.E. Halliburton, B.D. Perlson, R.A. Weeks, J.A. Weil, and M.C. Wintersgill, *Sol.St. Comm*., **30**, 575 (1979).
21. Zhou, X., Watkins, G.D., McNamara Rutledge, K.M., Messmer, R.P., Chawala,S., *Submitted to Phys. Rev*., May, 1996.
22. McNamara, K.M. and Gleason, K.K., *Chem. of Materials*, 1994, **6**, 39-43.
23. Pruski, M., Lang, D.P., Hwang, S.-J., Jia, H., and Shinar, J., *Phys. Rev. B,* 1994, **49**, 10635-10642.

24. McNamara, K.M., Gleason, K.K., and Robinson, C.J., *J. Vac. Sci. Technol. A*, 1992, **10**, 3143-3148.
25. McNamara, K.M., Scruggs, B.E. and Gleason, K.K., *J. Appl. Phys.*, 1995, **77**, 1459-1462.
26. Watkins, G.D., and Corbett, J.W., *Phys. Rev.*, **134**, A1359 (1964).
27. Elkin, E.L., and Watkins, G.D., *Phys Rev.*, **174**, 881 (1968).
28. Watkins, G.D., and Corbett, J.W., *Phys. Rev.*, **138**, A543 (1965).

PHOTOTHERMAL DETERMINATION OF THE THERMAL CONDUCTIVITY OF SUPERHARD THIN FILMS

J. MAZUR, J. BODZENTA, Z. KLESZCZEWSKI
Institute of Physics
Silesian Technical University
ul. Krzywoustego 2
44-100 Gliwice
POLAND

The nondestructive method is proposed for determination of the thermal conductivity of transparent, thermally thin coating on opaque, thermally thick substrates. The sample is heated by intensity modulated, unfocused laser beam. The photodeflection detection is used for signal registration and the "mirage" signal dependence on modulation frequency is measured.. The method is based on 1-D model of heat propagation in layered structure. The light interference in coating layer is taken into account. Two geometries of photothermal experiments are considered. The modulated light may illuminate the coated surface of sample or the opposite one. It is shown that the ratio of signals measured in these geometries does not depend on parameters of the probe beam. The proposed method is verified experimentally for silicon wafers coated with DLC. Obtained results are very promising.

1. Introduction

Thermal properties of diamonds and diamond films were investigated by many authors using different experimental techniques - e.g. thermocouple method [1], optically induced transient grating experiment [2]. In the last few years photothermal measurements of thermal properties of diamond layers became popular [3-7].

Potentialities of photothermal measurement techniques strongly depend on method using for signal detection. The mirage effect or the photothermal beam deflection (PBD) method is used for signal detection in our laboratory. Intensity modulated laser beam illuminates the sample surface and causes local changes of temperature distribution in the sample and air above it. The probe beam which passes parallel to the sample surface is deflected by refractive index gradient caused by temperature gradient in the gas. Two different dependencies of PBD signal can be measure - dependence on distance between the probe and the power beam (spatial dependence), and dependence on modulation frequency.

Boudina and co-workers [6], and Bachmann and co-workers [7] had measured spatial dependence of PBD signal and then used rather complicated multiparameter fitting methods for determination of thermal parameters of polycrystalline diamond films (PCD) layers. Moreover, the measurements were performed for free standing

M. A. Prelas et al. (eds.), Diamond Based Composites, 261–267.

PCD membranes, which required special preparation of samples. It may be difficult to determine thermal properties of thin diamond-like carbon (DLC) coatings using this method.

In this paper new measuring technique is proposed for thermal conductivity determination of thin DLC layer on thermally thick substrate. The proposed method bases on PBD signal frequency dependence and simple one dimension (1D) theoretical model. The method is nondestructive and noncontact.

2. Theory

Let us consider a layered structure which consists of gas over the sample, coating layer of thickness w, substrate with thickness d, and gas below the sample. Thermal properties of each layer are characterized by its thermal conductivity κ_i and thermal diffusivity β_i (where $i = s, c, g$ denotes the substrate, the coating layer and the gas respectively). Optical properties of coating and substrate are described by their complex refractive indexes. The sample is illuminated by intensity modulated light (the angular frequency of modulation is equal ω). Two orientations of the sample are considered (Fig. 1). The light beam illuminates either the coated surface of the sample ("top illumination") or the opposite ("bottom illumination"). The probe beam is passing parallel to the sample surface at height h above it.

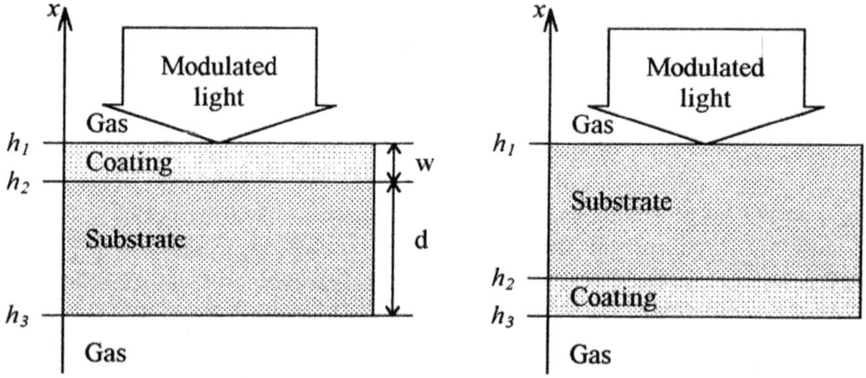

Figure 1. "Top" and "bottom " illumination of the sample

In PDB detection the measured signal is proportional to the temperature gradient in the gas over the surface of sample [8]

$$S(\omega) = K \left. \frac{\partial T_g(x,\omega)}{\partial x} \right|_{x=h} \tag{1}$$

where K is a constant which depends on thermooptical properties of gas and sensitivity of detector.

As one can see the signal is completely determined by distribution of temperature in gas over the sample. The following section is devoted to description of the temperature field in considered system.

2.1 DESCRIPTION OF TEMPERATURE FIELD IN GAS AND SAMPLE

2.1.1 Fourier - Kirchhoff equation and boundary conditions

The distribution of temperature in each layer of investigated sample and gas is described by Fourier - Kirchhoff heat transfer equation. After Fourier time-frequency transformation this equation can be written in the form

$$\frac{\partial^2 T_j(x,\omega)}{\partial x^2} - \frac{i\omega}{\beta_j} T_j(x,\omega) = \Lambda_j(x,\omega), \tag{2}$$

where $j = g$ (gas over the sample), s (substrate), c (coating) or gb (gas below the sample), $\Lambda_j(x, \omega)$ is the volume density of heat source in j-th medium and $T_j(x, \omega)$ is the Fourier component of temperature field on frequency ω.

Above differential equations must be completed with following boundary conditions

- temperature field continuity

$$T_j(h_j,\omega) = T_{j+1}(h_j,\omega); \tag{3a}$$

- conservation of heat flux

$$\kappa_j \left.\frac{\partial T_j(x,\omega)}{\partial x}\right|_{x=h_j} = \kappa_{j+1} \left.\frac{\partial T_{j+1}(x,\omega)}{\partial x}\right|_{x=h_j} - Q_j(x,\omega), \tag{3b}$$

where $Q_j(x, \omega)$ is a heat flux source on j - j+1 interface.

It is also assumed that

$$T_g(+\infty,\omega) = 0, \; T_{gb}(-\infty,\omega) = 0. \tag{3c}$$

The solutions of equation (2) can be written as

$$T_j(x,\omega) = \lambda_j(x,\omega) + A_{j,1}(\omega)\exp\left(-\sqrt{\frac{i\omega}{\beta_j}}x\right) + A_{j,2}(\omega)\exp\left(\sqrt{\frac{i\omega}{\beta_j}}x\right), \tag{4}$$

where $\lambda_j(x, \omega)$ is the inhomogeneous solution caused by volume heat source, and $A_{j,n}$ are functions of ω. In particular for gas above the sample we have

$$T_g(x,\omega) = A_{g,1}(\omega)\exp\left(-\sqrt{\frac{i\omega}{\beta_g}}x\right), \tag{5}$$

and then we obtain from (1) an expression for PDB signal

$$S(\omega) = -K\sqrt{\frac{i\omega}{\beta_g}} A_{g,1}(\omega)\exp\left(-\sqrt{\frac{i\omega}{\beta_g}}h\right). \tag{6}$$

The function $A_{g,1}(\omega)$ depends on thermal and optical properties of substrate and coating.

2.1.2 *Heat sources in the sample*

In photothermal experiments heat sources arise as a result of absorption of energy from light beam. In case of opaque media energy is absorbed near illuminated surface. Resulting heat source can be involved to the theoretical model as additional heat flux in flux -continuity condition (Q_j in equation (3b)). For transparent media absorption of energy leads to volume heat sources (Λ_j in equation (2)).

As it is mentioned above the sample is illuminated in two ways. For the "top illumination" one should consider two heat sources. The first one is a volume heat source in coating. Proper description of this source requires analysis of light interference in the layer. It can be shown that general form of the source is

$$\Lambda_1 = \frac{\alpha I_1}{\kappa_c}\left(B_1 \exp(\alpha x) + B_2 \exp[-\alpha(2w - x)] + B_3 \cos\left(\frac{4\pi n_c}{\lambda} x - \phi\right)\right) \tag{7}$$

where I_1 - intensity of light entering the coating, B_1, B_2, B_3, ϕ - constants depending on optical properties of gas, the coating and the substrate, α - absorption coefficient of coating layer, n_c - refractive index it, λ - light wavelength. The second source is placed on coating-substrate interface and is proportional to the energy absorbed in substrate

$$Q_2 = I_2 / \kappa_s \,, \tag{8}$$

where I_2 - energy flux entering the substrate.

For the "bottom illumination" only one plain source arises on the bottom surface of sample. This source depends on the intensity I_3 of light illuminating the sample

$$Q_1 = I_3 / \kappa_s \,. \tag{9}$$

2.2 ANALYSIS OF EXPERIMENTAL DATA

It follows from equation (6) that the ratio of signals measured for "top" and "bottom" illuminations is of the form

$$R(\omega) = \frac{S^{top}(\omega)}{S^{bottom}(\omega)} = \frac{A_{g,1}^{top}(\omega)}{A_{g,1}^{bottom}(\omega)}, \tag{10}$$

where $A_{g,1}^{top}$ and $A_{g,1}^{bottom}$ can be derived from equations (3a,3b,3c) and (4) with volume heat source (7) and heat flux sources (8) and (9).

The analytical formula for $R(\omega)$ obtained from Symbolic Computation System is very complicated and the following subsection is devoted to discussion of some simplifications.

2.2.1 *Simplifications*

The first simplification is based on assumption that the thermal conductivity of gas is much smaller then thermal conductivities of coating and substrate. In this case formula for $R(\omega)$ has following form

$$R(Z)=\frac{iC_1\alpha I_1\beta_c\left[\gamma_s\tanh(\Gamma_s dZ)+\gamma_c\tanh(\Gamma_c wZ)\right]\Gamma_s}{I_3\kappa_c\left[\gamma_s+\gamma_c\tanh(\Gamma_c wZ)\tanh(\Gamma_s dZ)\right]Z}+$$
$$-\frac{iC_2\alpha I_1\beta_c\sqrt{2\pi i\kappa_s}\tanh(\Gamma_s dZ)}{I_3\kappa_c\beta_s\cosh(\Gamma_c wZ)\left[\gamma_s+\gamma_c\tanh(\Gamma_c wZ)\tanh(\Gamma_s dZ)\right]Z}+ \tag{11}$$
$$+\frac{C_3\alpha I_1\left[\gamma_c+\gamma_s\tanh(\Gamma_c wZ)\tanh(\Gamma_s dZ)\right]}{I_3\gamma_c\left[\gamma_s+\gamma_c\tanh(\Gamma_c wZ)\tanh(\Gamma_s dZ)\right]}+$$
$$+\frac{\left(C_4\alpha I_1+I_2\kappa_c/\kappa_s\right)}{I_3\sqrt{\beta_s}\cosh(\Gamma_c wZ)\left[\gamma_s+\gamma_c\tanh(\Gamma_c wZ)\tanh(\Gamma_s dZ)\right]}$$

where $\Gamma_j=\sqrt{2\pi i/\beta_j}$, $\gamma_j=\kappa_j/\sqrt{\beta_j}$, and $Z=\sqrt{\omega/2\pi}$. Subsequent simplifications can be done with assumption that for considered range of modulation frequency, coating layer is thermally thin. Then

$$\cosh(\Gamma_c wZ)\cong 1+\tfrac{1}{2}\Gamma_c^2 w^2 Z^2$$
$$\tanh(\Gamma_c wZ)\cong\Gamma_c wZ, \tag{12}$$
$$C_1\approx C_2.$$

Additionally, when the coating is transparent

$$\alpha C_3\to 0. \tag{13}$$

Above assumptions lead to the more convenient formula for signals ratio

$$R_f(Z)=\frac{\alpha I_1(C_4-2\pi C_1 w)+I_2\frac{\kappa_c}{\kappa_s}-Z\pi\alpha I_1 C_1 w^2\frac{\kappa_s}{\kappa_c}\Gamma_s\tanh(\Gamma_s dZ)}{I_3\left(\kappa_s+Z\rho_c c_c w\frac{\Gamma_s}{\beta_s}\tanh(\Gamma_s dZ)\right)}. \tag{14}$$

3. Experimental verification

The measurements were carried out for two silicon wafers coated with thin DLC layers. Optical parameters of the coatings and their thickness have been determined from independent optical measurements. Results are collected in Table 1. Thermal parameters of silicon substrates have been measured for non-coated silicon wafer (reference sample) using method described in [9]. The determined thermal diffusivity of silicon is 0.95cm^2/s.

The PBD signal dependence on square root of frequency has been measured for top and bottom illumination in frequency range 100Hz-40kHz For each sample. Then the ratio of signals obtained for "top" and "bottom" illumination has been calculated. Results are shown in the Figure 2.

Theoretical curves have been fitted to the experimental data using least square method. Fitted curves are shown in Fig. 2 and Fig. 3 together with experimental data. Preliminary fitting results show that the thermal conductivity of DLC layer is three orders of magnitude smaller than for pure diamond.

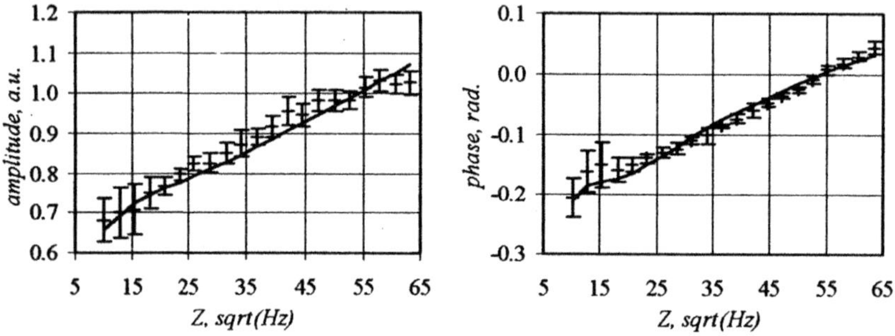

Figure 2. Amplitude and phase of experimental data and fitted curves for sample #1

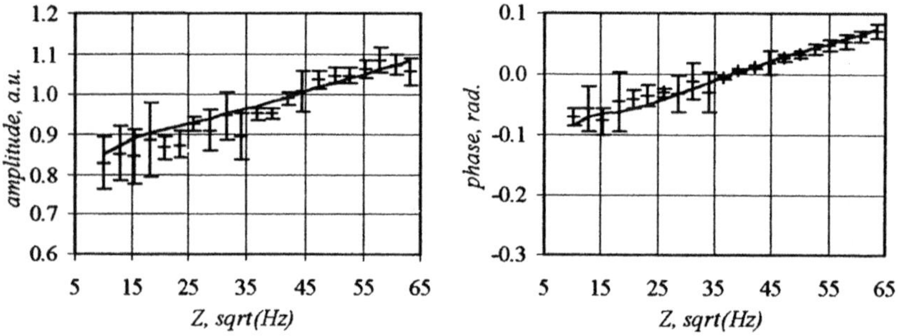

Figure 3. Amplitude and phase of experimental data and fitted curves for sample #2

TABLE 1. Parameters of investigated samples

Sample	Substrate	Coating layer	
	thickness d[μm]	thickness w[μm]	refractive index n
1	480±5	1.32±0.06	2.52±0.02
2	465±5	0.65±0.01	2.46±0.02
Reference	480±5		

4. Conclusions

The new method for nondestructive determination of thermal conductivity of thermally thin layer on thermally thick substrates is proposed. An analytical formula for description of measured signals ratio has been derived. Experimental results confirm possibility of using the method for investigation of DLC carbon layers.

Values of thermal conductivity of investigated DLC layers on silicon substrates are much smaller than the thermal conductivity of pure diamond. Probably so small value of thermal conductivity of thin diamond-like carbon layers is caused by small grains size, which restricts phonon free pass. Since the phonon free path at room temperature are much larger than grain dimensions in DLC films, which would be expected to scatter phonons. Moreover, estimated thermal conductivity values should be treated as effective values for diamond and interface amorphous material from grain boundaries.

5. Acknowledgment

We gratefully acknowledge the Polish State Committee for Scientific Research (KBN) for support under grant No. 2 P302 084 06.

6. References

1. Graebner, J.E., Mucha, J.A., Seibles, L., and Kamlot, G.W. (1992) The thermal conductivity of chemical-vapor-deposited diamond films on silicon, *J. Appl. Phys.* **71**(7), 3143-3146.
2. Tokmakoff, A., Banholzer, W.F., and Fayer, M.D. (1993) Thermal diffusivity measurements of natural and isotopically enriched diamond by picosecond infrared transient grating experiments , *Appl. Phys.* **A56**, 87-90.
3. Visser, E.P., Versteegen, E.H., and van Enckevort, W.J.P. (1992) Measurement of thermal diffusion in thin films using a modulated laser technique: Application to chemical-vapor-deposited diamond films, *J. Appl. Phys.* **71**(7), 3238-3248.
4. Chen, Z., and Mandelis, A. (1992) Thermal-diffusivity measurements of ultrahigh thermal conductors with use of scanning photothermal rate-window spectrometry: Chemical-vapor-deposition diamonds, *Phys. Rev.* **B 46**(20), 13526-.
5. Petrovsky, A.N., Salnick, A.O., Mukhin, D.O., and Spitsyn, B.V. (1992) Thermal conductivity measurements of synthetic diamond films using the photothermal beam deflection technique, *Mater. Sci. Eng.* **B11**, 353-354.
6. Boudina, A., Fitzer, E., Netzelmann, U., and Reiss, H. (1993) Thermal diffusivity of diamond films synthesized from methane by arc discharge plasma jet CVD, *Diamond Relat. Mater.* **2**, 852-858.
7. Bachmann, P.K., Hagemann, H.J., Lade, H., Leers, D., Wiechert, D.U., Wilson, H., Fournier, D., and Plamann, K. (1995) Thermal properties of C/H-, C/H/O-, C/H/N- and C/H/X-grown polycrystaline CVD diamond, *Diamond Relat. Mater.* **4**, 820-826.
8. Murphy, J.C., and Aamodt, L.C. (1980) Photothermal spectroscopy using optical probing: Mirage effect, *J. Appl. Phys.* **51**(9), 4580-4588.
9. Bodzenta, J., Mazur, J., Bukowski, R. and Kleszczewski, Z. (1995) Photothermal measurement for plates, *Proc. SPIE* **2643**, 286-292.

EFFECT OF LIGHT ON THE PERFORMANCES OF CVD DIAMOND NUCLEAR DETECTORS

C.MANFREDOTTI, F.FIZZOTTI, P.MUZZANI, P.POLESELLO, E.VITTONE
Università di Torino, Dipartimento di Fisica Sperimentale
Via P.Giuria 1, I-10125 Torino, Italy

Abstract

CVD (Chemical Vapour Deposited) diamond samples have been characterized by using Am-241 alpha particles in order to look for specific differences of the transport properties between electrons and holes. The effects of "priming" or exposing the samples to large doses of x-rays and to day light and to monochromatic light have been analyzed in some details with respect to charge collection efficiency and to counting efficiency values and their decay as a function of time. It turns out that the effect of priming is slightly larger for holes than for electrons and that only wavelengths in the interval between 350 and 550 nm are effective in changing the collection length. In this interval, hole collection efficiency is decreasing, while for electrons it increases. A possible interpretation of all these effects is presented and discussed, taking also into account luminescence and photoconductivity data.

1. Introduction

It is well kown that diamond nuclear particles detectors, natural but particularly CVD, are affected by polarization effects [1, 2]: both counting rate and pulse height are decaying as a funcion of time. Polarization can be eliminated, at least partially, by priming the detectors [3] with apparently any kind of radiation. Irradiation generally increases the collection efficiency [3, 4], but it is not known if this happens for electrons, holes or for both. It is also known that by exposing the detector to monochromatic light or to light in a certain wavelength interval, collection efficiency can decrease or even increase and a particular model has been even introduced [3], but with no clear distinction between electrons and holes. ^{241}Am α particles, being absorbed in about 14 μm from the sample surface, may help in shining light into this problem, particularly if, as in our case, detector thickness is large and also charge collection length is large with respect α particles range in the detector. Recently [2, 5], data were reported concerning polarization effects as investigated by α particles and

M. A. Prelas et al. (eds.), Diamond Based Composites, 269–275.

the contribution of electrons and holes to collection length, which was reported to be not much different. Also the effect of X-rays and light exposure was examinated in some details [2]. In this paper more conclusive data are reported and discussed, particularly in connection to a model for recombination in CVD diamond, which was introduced in order to explain luminescence and photoconductivity results.

2. Experimental

^{241}Am α spectra were recorded with a standard ORTEC electronic chain connected to a PC by means of a standard ADC. The CVD diamond sample was 400 μm thick and it was equipped with standard Ti/Au electrodes of 1 cm^2 surface area. The spectra were recorded in separated time steps in order to avoid or to reduce at a minimum during the measurement the polarization effects, which may show either a reduced count rate or a displacement of the spectrum towards low energies as a function of time. This trick was also used during the observation of the decay of some parameters, as for instance the collection efficiency, as a function of time. In order to evaluate the collection efficiency, a comparison was carried out with the response of a Si surface barrier detector, which was assumed to have a 100 % collection efficiency, by taking into account the different pair creation energies of Si (3.6 eV) and diamond (13.2 eV). The collection length was calculated directly from the product between collection efficiency and sample thickness: assuming the validity of Hecht's relationship [2], this can be considered as good approximation for low collection efficiencies (below 30 %).

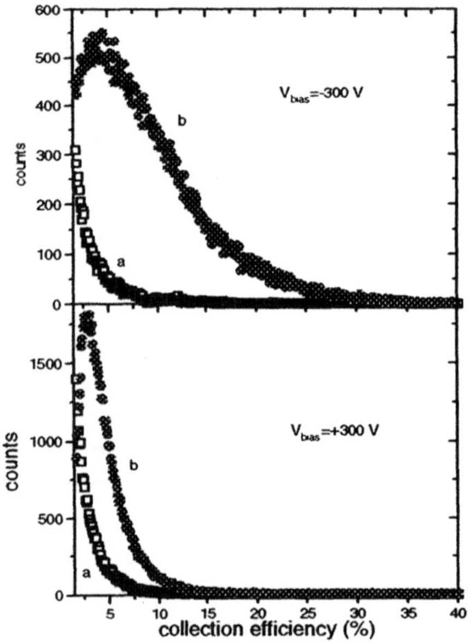

Figure 1. ^{241}Am alpha particle spectra before (a) and after (b) 35 Gy x-ray irradiation or "priming". Bias voltage is -300 V (upper figure) or +300 V (lower figure).

The bias voltage was applied to the contact at the substrate side, while α particles hit the growth side of the sample, which had been accurately polished. Since the penetration depth of α is 14 μm and the collection length, which can be considered as the summed average of the hole and electron path before recombination, is generally larger (more than 20 μm and up to about 50 μm), it has been assumed that the main contribution to the collection length is coming from the carrier which flows at the opposite electrode. That means that one apparently measures electron collection length for positive biases and hole collection length for negative ones.

3. Results and discussion

Fig. 1 shows α spectra taken for 300 s before (curve a) and immediately after (curve b) a priming with a totale dose of 35 Gy of X-rays (150 KV, 10 mA), with the assuption that the upper spectra, which refer to a bias voltage of -300 V, are mainly due to the holes contribution, while the lower ones, which were obtained with + 300 V, are to be ascribed to electrons.The effect of priming is very clear: both counting rate and collection efficiency increase dramatically. It is also evident that, while (a) curves are similar, (b) curves are quantitatively different: the peak for holes occurs at the channel 30, while the peak for electrons occurs at the channel 12. This means that the collection length for holes, at least in the first 50 μm of thickness, is more than twice the value for electrons. This fact can be even more appreciated in Fig. 2, which shows the time behaviour of the collection length for both holes and electrons before (curve a) and after (curve b) X-ray irradiation. In the "unprimed state" the collection length is larger for holes than for electrons, but only very slightly, while in the "primed state" is more than the double. In the primed state, the decay of the collection length with time is less important, in particular way for electrons. The same conclusion can be reached for the counting rate, which is shown in Fig. 3 and it decays much less strongly in the primed state than in the unprimed one. Two facts are even more evident: 1) in the unprimed state the decay of the counting rate tends to very low values, particularly for electrons, while the primed state is almost stationary; 2) in the primed state the counting efficiency increases by a

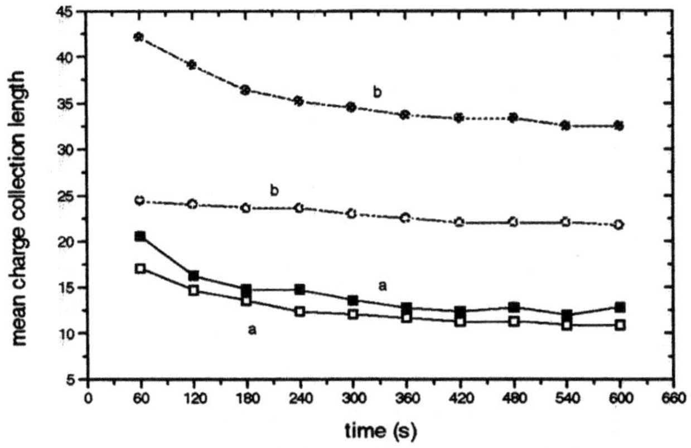

Figure 2. Behaviour of charge collection length as a function of time before (a) and after 35 Gy x-ray priming (b). Bias voltage is -300 V (solid symbols) and +300 V (open symbols).

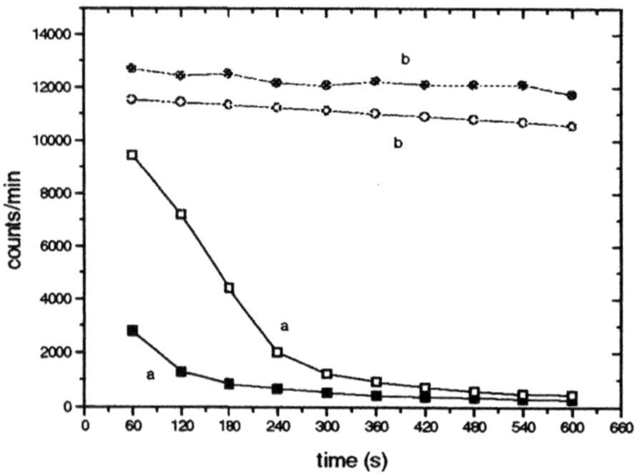

Figure 3. Behaviour of the counting rate as a function of time before (a) and after 35 Gy x-ray "priming" (b). Bias voltage is -300 V (solid symbols) or +300 V (open symbols).

factor of more than ten (18 for electrons in the quasi-stationary state after 8 minutes). That means that X-rays are not only saturating traps, with the consequent increase of almost a factor 2 in the collection length, but are also eliminating space-charge regions from the bulk of the material, i.e. low electrical field regions in which recombination between electron and holes is dominant. The "priming" procedure was accurately checked to be a reliable methods for eraising any previous "history" of the sample (particularly as for as light exposure is concerned) and for reproducing a "starting state" on which any kind of reproducible measurement can be performed. For istance, α spectra exactly similar to those ones reported in the curves (b) of Fig.1 can be obtain for sample exposed to white day light or to white light after a single exposure to a X-ray dose of more than 20 Gy.

The effects of light exposure on CVD diamond performances are even more interesting, since they clarify the mechanisms of trapping and recombination according to what is described in the second part of this discussion. Light illumination, particularly if monochromatic, clearly evidences the opposite effects on holes and electrons, as displayed in Figs. 4 and 5, which report α spectra before (a) and after (b) an exposure to a monochromatic light. In this case the tungsten lamp light was used in conjunction with an f = 0.25 m Jobin-Yvon monochromator in order to obtain a transmitted band of 1 nm. Illumination times were controlled in order to get, at each wavelenght, a photon density of 10^{16} ph/cm^2 over the sample, as measured by a photon detector Tektronix J16. Times were of the order of few minutes. The procedure was the following one for each wavelength: a) exposure to x-ray for 12 minutes (35 Gy dose); b) recording of α spectrum for 5 minutes keeping the sample in dark; c) exposure to monochromatic light; d) same spectrum as before, always in dark. Illumination is from the side but, since the absorption is very low in the whole investigated wavelength interval, the bulk may be assumed to be uniformly illuminated. For holes (Fig. 4) the "quenching" effect is maximum at λ=400 nm, (the peak disappears and it is substituted by a long exponential tail after illumination) less

strong at λ=480 nm and apparently at λ=700 nm (note that at a bias voltage of -200 V the peak is less evident than at -300 V, as it can be seen in Fig. 1).
For electrons (Fig. 5) the effect is opposite: at λ=400 nm and at λ=480 nm the spectra increase after illumination, while at λ=700 nm the spectra are coincident, as for holes. Quantitatively this effects can be even more appreciated in Fig. 6, which shows the behaviour of d_{after}/d_{before} (ratio of the collection lengths after and before the illumination) as a function of wavelength, for electrons (curve a) and holes (curve b). Under light illumination, the collection length for electrons increases below λ=550 nm, reaching a factor of 2 approximately with respect to the "primed" state, while for holes, in the same conditions, it decreases, also in this case by a factor of 2 below λ=550 nm. In this case in which both electrons and holes are contributing, the total effect should be negligeable, at least in approximation in which the collection length is proportional to the collection efficiency. This would be the case of minimum ionizing particles.

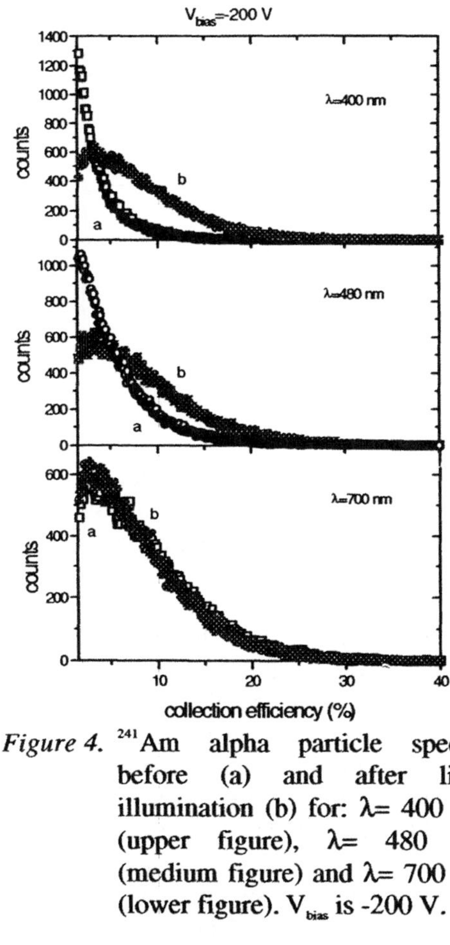

Figure 4. [241]Am alpha particle spectra before (a) and after light illumination (b) for: λ= 400 nm (upper figure), λ= 480 nm (medium figure) and λ= 700 nm (lower figure). V_{bias} is -200 V.

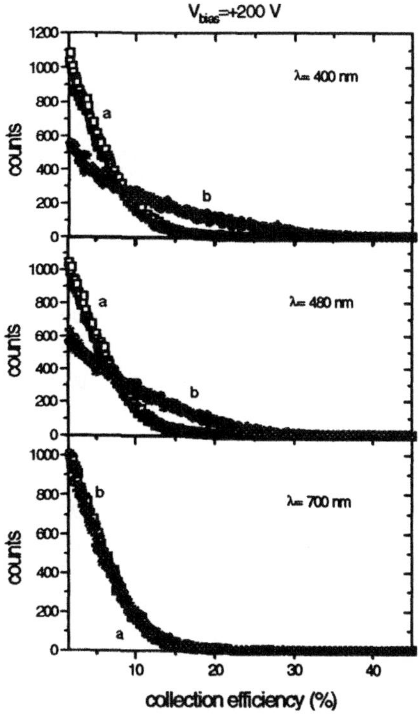

Figure 5. Same as Fig. 4, but with a 200 V applied bias

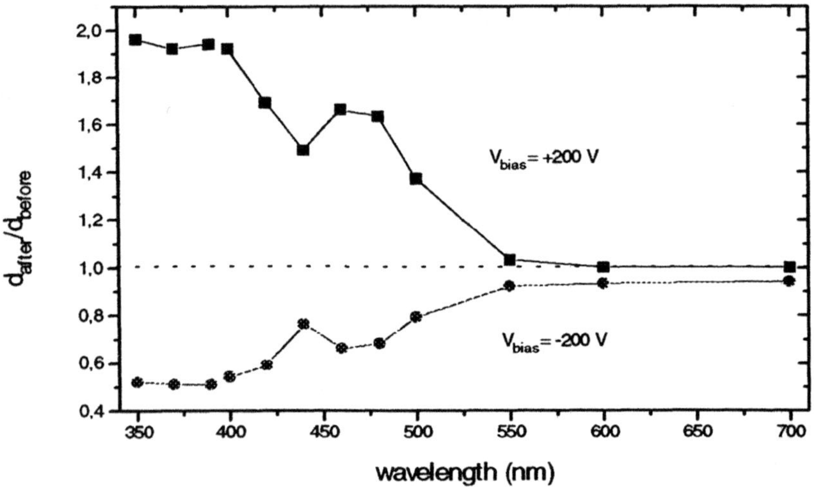

Figure 6. Wavelength dependence of the ratio of collection lengths after (d_{after}) and before (d_{before}) a monochromatic illumination (giving in all cases a 10^{16} photons/cm^2 flux) for a positive bias (+200 V, upper curve) and a negative bias (-200 V, lower curve).

The model [6] which has been introduced in order to explain both electroluminescence and photoconductivity can be adopted here with some minor revision. A recombination center exists at about 3 eV from the valence band in CVD diamond. This center can capture both electrons and holes and, if conveniently filled with electrons, for instance by a double injection mechanism, hole recombination at the center is radiative and it gives a strong luminescence at 3 eV, i.e. exactly at the same energy at which the "quenching" effect of light is maximum for holes. That means that, at λ=400 nm, the center is filled with electrons by making hole recombination easier and electron trapping less likely. This effect is less evident at λ=500 nm, where energy supplied by photons is not enough to fill the traps with electrons. The electron transition to the conduction band cannot be of any importance here, since electrons are minority carriers and recombine fast with holes, which are assumed to be majority carriers [6].

Even if the model is very rough, and it evidently needs more experimental data to be defined in details, it explains all the available data. It explains, also, with cathodoluminescence (or ionoluminescence) is much stronger in regions in which collection lenghts, as obtained locally by micro-IBIC, is shorter: hole recombination, which is radiative, is in evident competition with hole movement and it represents, in our view, the main mechanism for the reduction of the hole collection length.

4. Conclusions

X-ray irradiation can be considered as a reliable method not only for improving the detector performances of CVD diamond, but also for obtaining it in a suitable, reproducible state. X-ray exposure, by filling all the traps, improves mainly the collection length for holes, which are the majority carriers, and less the collection length for electrons, which are the minority carriers. Polarization effect, particularly the the decay of counting efficiency, are strongly reduced by x-ray "priming". CVD diamond seems to be strongly affected by the creation of spaxce charge due to hole trapping: the creation of both electrons and holes, connected to the electron trapping, should help in reducing the extension of this positive space charge.
Light illumination at 3 eV and up to 2.5 eV is very likey a suitable mechanism for detrapping holes from this center (the transition is strongly radiative, as evidenced by electroluminescence data), reducing this way the hole collection length. In the same time, electron-hole recombination at the center is strongly inhibited, and as a consequence electron collection is likely increased.
About the nature of the center, nothing sure can be said: it could be either a defect associated with the whole texture of the grain surfaces (and, in this sense, a "surface" state) or even a defect associated in nitrogen incorporation, which can be very high also in CVD diamond, because of the large fluxus of hydrogen, which contain always a certain amount, at ppm level, of nitrogen. In the gas phase, nitrogen could be present at 0.1 % atomic fraction: since nitrogen incorporation is likely in diamond, this could result in a 10^{19} cm^{-3} nitrogen content, as in IIa natural diamond. And natural diamond performances are not very much different from CVD diamond ones.

5. References

1. Kozlov S.F., Stuck R., Hage-Ali M., Siffert P. (1975), Preparation and characteristics of natural diamond detectors, *IEEE Transactions on Natural Science* **NS-22**, 160-170
2. Manfredotti C., Fizzotti F., Vittone E., Polesello P., Wang F. (1996), A study of polycrystalline CVD diamond by nuclear techniques, *physica status solidi (a)* **154**, 327-350.
3. Fallon P.J., Nam T.L., Keddy R.J. (1992), Trapping levels in pulse-counting synthetic diamond detectors, *Diamond and Related Materials* **1**, 1185-1189
4. Bauer C. et al. (the RD42 Collaboration) (1995, in press), *Wire Chamber Conference (13-17 Feb. 1995), Vienna*
5. Manfredotti C., Fizzotti F., Boero M., Vittone E., Polesello P. (1995), Characterization of CVD diamond films by nuclear techniques with α particles, *Diamond and Related Materials* **4**, 517-520
6. Manfredotti C., Wang F., Fizzotti F., Vittone E., Polesello P., Scacco A. (1995), Blue-violet electroluminescence and photocurrent spectra from polycrystalline chemical vapour deposited diamond film, *Applied Physics Letters* **67**, 3376-3378

OPEN-CIRCUIT MOBILITY MEASUREMENTS IN DLC THIN FILMS

M. DŁUŻNIEWSKI, P. STĘPNIK, E. STARYGA, G.W. BĄK
Institute of Physics, Technical University of Łódź, ul. Wólczańska 219, 93-005 Łódź, Poland

S. MITURA
Institute of Materials Science and Engineering, Technical University of Łódź, ul. Stefanowskiego 1, 90-924 Łódź, Poland

Abstract

Drift mobility in DLC thin films proves to be of the order of 10^{-6} cm^2/Vs and its activation energy is about 0.035 eV. This suggest that hopping transport is a dominant mechanism of charge transport. Computer simulation of charge transport in thin films under open-circuit conditions for volume charge generation has been carried out. The computer simulation is to improve the interpretation of results obtained by means of generation with an electron gun.

1. Introduction

Diamond-like carbon (DLC) films obtained by the plasma methods have a number of interesting properties that predestine this material to technical and industrial applications. This is because its unique thermal and chemical stability and comparatively high thermal conductivity which is close to 20 W/cmK. Semiconducting diamond has already been used for high-temperature thermistor, light detectors and fast detectors of infrared radiation. It has also been used in such electronic devices like high-temperature Schottky diode and metal-oxide-semiconductor field-effect transistor (MOSFET) [1]. Recently, much cheaper DLC layers are considered as a substitute for the traditional semiconductors in microelectronics [2-4]. It is evident that application of DLC layers depends on their properties, particularly on the electronic transport properties and that is the reason for the recent wide interest in basic physical properties of DLC thin films.

The carrier drift mobility is an important parameter characterizing the electronic properties of insulators and semiconductors because it refers directly to the charge transport mechanism.

In this paper, we consider the effect of trapped space charge on open-circuit surface-potential decay measurements.

M. A. Prelas et al. (eds.), Diamond Based Composites, 277–284.

2. Drift Mobility Measurements

There are several methods of the drift mobility determination. In case of classical semiconductors such as silicon or germanium, the Hall effect experiment is the commonly applied method of carrier mobility measurements. Unfortunately, this is not a useful method in relation to the high-ohmic semiconductors and insulators. In such materials, these studies involve measurements consisting in injecting a thin sheet of charge (via light generation or electron beam excitation) into a sample near surface and observing the transport of the sheet moving through the bulk under the action of an externally applied electric field. This method is widely known as the time of flight (TOF) technique. The carrier drift mobility is determined by measuring the carrier time of flight t_{tr} through the sample of the thickness L.

Conventional time-of-flight technique [5] is commonly used to measure the drift mobility in high-resistivity materials. Since in this technique the sample is sandwiched between two metallic electrodes, it is sometimes not possible to apply a high electric filed due to the presence of flaws in some real insulators (like thin diamond-like carbon films, for instance).

The xerographic-mode transient charge technique originated from the theory of photo-induced discharge of a high-resistivity photoconductive material [6]. It is in fact an electrodeless technique, capable of observing the charge transport phenomena in an insulator. Due to the absence of a top electrode (i.e. in open-circuit configuration), it is possible to apply a high electric field to it by charging the top free surface. In the open-circuit configuration, the DLC layer can retain charge on its top surface despite the presence of flaws in it.

Batra *et al.* [6] have made xerographic discharge measurements on corona-charged amorphous selenium. The assumption of the theory is that the non-metallized sample face is charged at the very surface(i.e. that the depth of injection of charge equals zero). They used a short intense flash of strongly absorbed light to inject all of the surface charge into a sample. The injected charge drifts under its own self-field toward the grounded electrode, resulting in a decrease in surface potential. In xerographic discharge, the intense light flash is thought to produce hole-electron pairs, which dissociate at the surface of the specimen, electrons neutralizing the positive ions (in the case of positive corona), leaving free holes available for conduction. Injection is assumed to take place in a short time compared with the transient time of the charge across the sample.

In the open-circuit technique, the free surface of the sample is charged to the voltage V_0 using the corona discharge device. According to the theory [6] the time-dependence discharge of the sample due to charge carrier transport may be described by the voltage time dependence $V(t)$ and its derivative $\mathrm{d}V/\mathrm{d}t$:

$$V(t) = L^2/2\mu_d\,(1-t/2t_{tr}) \tag{1}$$

$$\mathrm{d}V/\mathrm{d}t = -\,V_0/2t_{tr} \tag{2}$$

for $t < t_{tr}$, and

$$V(t) = (L^2/2\mu_d)(1/t) \tag{3}$$

$$dV/dt = -(L^2/2\mu_d)(1/t^2) \tag{4}$$

for $t > t_{tr}$,
where t_{tr} is the transit time of the leading charge front, L is the thickness of the sample.

The relation between the drift mobility μ_d and the transit time t_{tr} is given by the following equation:

$$\mu_d = L^2/t_{tr}V_0. \tag{5}$$

The above equations are based on the assumption that the leading front of carriers moves in a constant field V_0/L as it traverses the insulator. There are a modifications [7, 8] of Batra's theory including instantaneous partial injection and field-dependent mobility.

Using an electron pulse gun Mycielski *et al.* [9,10] charged the surface of DLC film with a low-energy (≈ 1 keV) electron beam . The theory described by Batra *et al.* [6] was applied to the case of low energy electron beam charging and the surface-potential decay characteristics were analyzed according to equations (1) - (4). It was shown [10] that phenomena observed in investigated DLC films can be explained by the same model.

The apparatus was tested using amorphous layers of selenium, which is very convenient for xerographic measurements. The obtained electron drift mobility at room temperature is in a very good agreement with this one reported by other authors. The mean value of the electron mobility in selenium is 6.2×10^{-3} cm^2/Vs [11,12] whereas the result obtained by Mycielski *et al.* [10] was 7.0×10^{-3} cm^2/Vs. Since the data obtained from this system are very close to those published values which were measured by the conventional time-of-flight technique, it can be suggested that the present xerographic-mode transient charge equipment works well for drift mobility measurements.

A typical characteristic V(t) and derivative dV/dt for the DLC films presented in ref. 10 correspond to the theoretical predictions of the Batra's model. It was confirmed by a detailed numerical analysis of the $V(t)$ and dV/dt dependencies. The electron drift mobility at room temperature for DLC films prepared by plasma decomposition at low autopolarization voltage are very low ($1.6\times10^{-6} \div 8.0\times10^{-6}$ cm^2/Vs [10]). Such low values of carrier mobility suggest that the electron transport has a hopping nature. In order to confirm the suggestion concerning the hopping transport of charge in diamond-like carbon films, the measurements of electron drift mobility at different temperature were made [13].

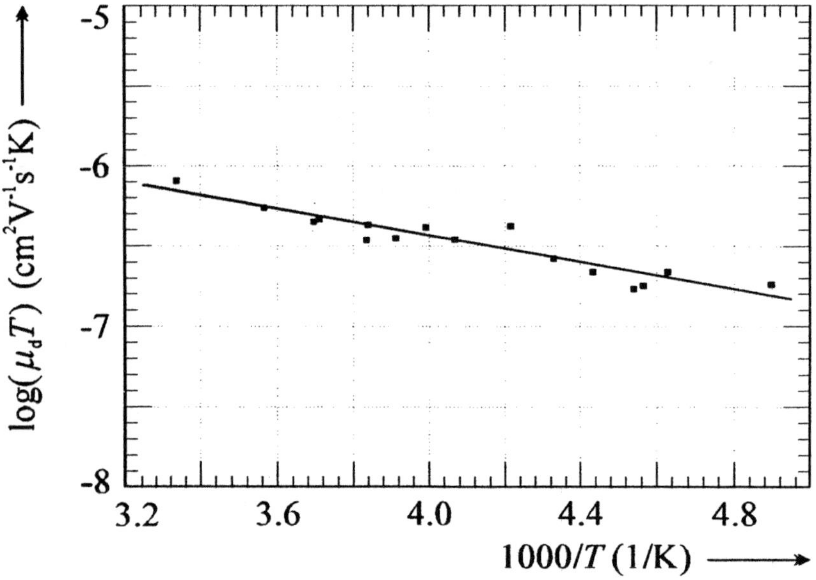

Figure 1. The dependence log ($\mu_d T$) = f(1/T) for DLC films obtained by decomposition of methane in the rf electric field at autopolarization voltage V_e = 250 V [13].

It has been found (see *Figure 1*) that mobility weakly increases with increasing temperature (the activation energy ≈ 0.035 eV [13]). The hopping mechanism in DLC films was previously suggested by some authors [14, 15] on the basis of dc-conductivity studies. The results of drift mobility measurements (presented in ref. 13) make the above hypothesis the more credible since the drift mobility relates directly to charge carrier transport.

3. Charge Transport in Thin Films under Open-Circuit Conditions: Computer Simulation

It is assumed in the theories describing the charge transport under open-circuit conditions that for the time t = 0 the charge is accumulated at the very surface of a sample. The assumption is right when a sample is corona-charged but it is not quite correct if it is charged with an electron gun as it has been done. The depth of injection may be about a few tenths of micrometers which is comparable to the sample thickness (see *Figure 2*).

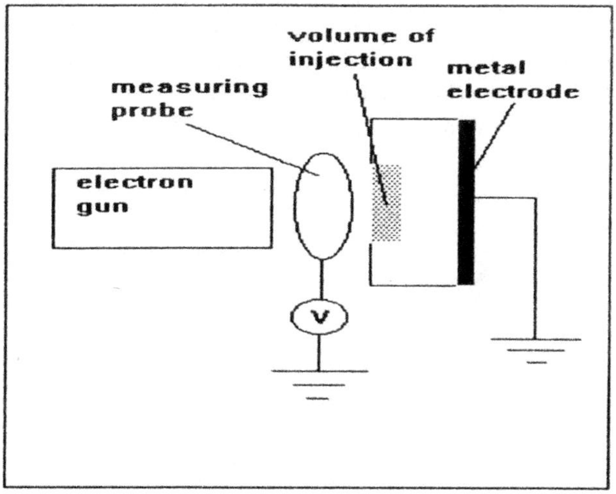

Figure 2. Schematic diagram of open-circuit mobility measurement with volume injection of charge carriers.

The analytical solutions presented above need not describe the charge transport quite correctly in case of volume charging. In order to solve the problem of charge transport in thin films under open-circuit conditions for a volume generation of charge a computer simulation has been carried out. The assumptions of the simulation were as follows:

1. Hopping transport in a narrow band of localized states is dominant. The assumption is a consequence of the experimental results obtained so far [10, 13].
2. The localized states constitute a regular lattice. In the next step a simulation for a disordered spatial distribution of localized states will be carried out.
3. The Gaussian distribution describes the energetic distribution of the localized states. The standard deviation σ_E was assumed to be 0 eV, 0.005 eV, 0.01 eV and 0.02 eV. So small values of the standard deviation result from a small activation energy of the measured mobility.
4. Density of the probability of a single jump between the localized states was assumed to be given by:

$$p_{il} = \nu_0 \exp(-\alpha r_{il}) \exp\left(\frac{-(E_l - E_i - eE(z_l - z_i))}{kT}\right) \quad (6)$$

for $E_l - E_i - eE(z_l - z_i) > 0$, and

$$p_{il} = \nu_0 \exp(-\alpha r_{il}) \quad (7)$$

for $E_l - E_i - eE(z_l - z_i) < 0$

where ν_0 is the frequency factor (equal to 10^{12} s^{-1}), α is the decay parameter of the localized state wave function, r_{il} is the distance between the states,

E is the field intensity, $z_l - z_i$ is the distance between the states along the electric field, e is the electron charge, k is the Boltzmann constant and T is the temperature.

5. The simulation was carried out for a sample 0.5 μm thick, its area was 0.15 μm × 0.15 μm. The distance between localized states was assumed to be 30 Å. The depth of injection was 0.1 μm. The solutions for other depths of injection are going to be obtained.

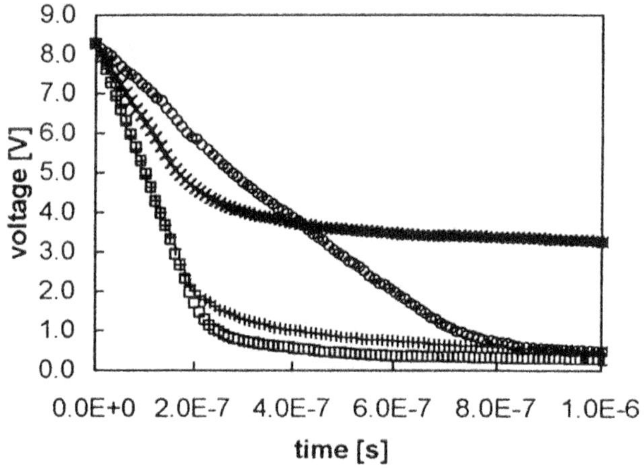

Figure 3. Voltage pulses for various values of the standard deviation of the energetic Gaussian distribution: o 0 eV, □ 0.005 eV, + 0.01 eV, x 0.02 eV.

According to equations (1) and (2) the drift mobility may be found either by estimation of the time of flight (which is assumed to be equal to the time corresponding the voltage $0.5 \times V_o$) or from the slope of the linear part of the voltage curve. The first method may not be valid for the volume charging as the distance covered by the moving carriers is difficult to be estimated in this case. However there is no reason to suppose that the second method cannot be used. In case of the initial surface charging the charge spreads in a sample with elapsing time and becomes a volume charge. This means that the dependence between the slope of the linear part of the voltage curve and the drift mobility should be similar for both the surface and the volume charging at the time $t = 0$. The value V_0 in eqn.(2) is the voltage corresponding to charging at the very surface, so the voltage $V(t = 0)$ from *Figure 3* cannot be taken for the calculations of the drift mobility. In order to get the correct value of the voltage the calculations have been made and the value 9 V has been obtained. In Table 1 the values of the drift mobility for the various standard deviations are presented.

TABLE 1. The drift mobility μ_d for various standard deviation σ_E.

σ_E [eV]	0	0.005	0.01	0.02
μ_d [cm^2/Vs]	7.0×10^{-4}	2.0×10^{-3}	2.0×10^{-3}	1.2×10^{-3}

As we see the energetic disorder of the localized states gives rise to an increase in the drift mobility from the value 7.0×10^{-4} cm^2/Vs up to 2×10^{-3} cm^2/Vs, but then the drift mobility decreases again. Moreover a remarkable „tail" becomes visible for the increasing standard deviation of the energetic distribution (see *Figure 3*). It may be supposed that the tail results from „trapping" of charge carriers in these fragments of a sample which are isolated from the rest of a sample with very „difficult" jumps, i.e. with jumps of comparatively small probability of execution.

4. Conclusions

1. Diamond-like carbon films obtained by plasma decomposition are low-mobility materials. The value of the drift mobility is of the order of 10^{-6} cm^2/Vs. Its activity energy is about 0.035 eV. The above results suggest that hopping transport of charge carriers in the narrow band of localized state dominates in investigated DLC layers.
2. The computer simulation presented above may give correct results of open-circuit measurements on thin low-conducting films dominated by hopping transport. The results suggest that the energetic disorder is equivalent to trapping of charge in comparatively deep traps giving rise to a long tail of voltage pulse.
3. The experimental results suggest that DLC films are a low-mobility hopping system. The values of the drift mobility obtained from the simulation is typical of low-mobility hopping systems. The results obtained for the simulation are by about two orders of magnitude higher than the experimental, but it may be supposed that it is possible to get a final agreement between the experimental value of the drift mobility and the simulation results by fitting the average distance between the localized states.
4. The simulation is to be developed by taking into account both the spatial disorder of the localized states and the varying depth of injection. It may be expected that the development will enable to fit the theoretical and experimental curves and to understand the charge transport in DLC thin films better.

Acknowledgments

This work was supported by Polish Committee for Scientific Research, grant No. 8T11B 058 08.

References

1. Gildenblat, G.S., Grot, S., Wronski, C.R., Badzian, A.R., Badzian, T. and Messier, C.R. (1988) Electrical characteristics of Schottky diodes fabricated using plasma-assisted CVD diamond films, *Appl. Phys. Lett.*, **53**, 586-588.
2. Gildenblat, G.S., Grot, S.A., and Badzian, A. (1991) Electrical properties and device application of homoepitaxial and polycrystalline diamond films, *Proc. IEEE Electron Dev.*, **79**, 647.
3. Chan, K.K., Silva, S.R.P., and Amaratunga, G.A.J., (1992) Electronic properties of semiconducting diamond-like carbon-diamond, *Thin Solid Films*, **212**, 232-239.

4. Seth, J., Chaudhry, M.I., and Babu, S.V., (1992) Electrical characteristics of plasma-deposited diamondlike carbon/silicon metal-insulator-semiconductor structures, *J. Vac. Sci. Technol. A*, **10**, 3125-3130.
5. Spear, W.E., (1957) *Proceeding of the Physical Society* (London), **B70**, 669.
6. I.P. Batra, I.P., Kanazawa, K.K., and Seki, H. (1970) Discharge characteristics of photoconducting insulators, *J. Appl. Phys.*, **41**, 3416-3422.
7. Sonnonstine, T.J., and Perlman, M.M., (1975) Surface-potential decay in insulators with field-dependent mobility and injection efficiency, *J. Appl. Phys.*, **46**, 3975-3981.
8. Perlman, M.M., and Sonnonstine, T.J., (1976) Drift mobility determinations using surface-potential decay in insulators, *J. Appl. Phys.*, **47**, 5016-5021.
9. Mycielski, W., Staryga, E. and Lipiński, A. (1993) Electron drift mobility in DLC thin films, *Thin Solid Films*, **235**, 13-14.
10. Mycielski, W., Staryga, E., Lipiński, A., Mitura, S. and Sokołowska, A. (1994) Open-circuit mode drift mobility measurements in DLC films, *Diamond and Related Materials*, **3**, 858-860.
11. Spear, W.E. (1969) Drift mobility techniques for the study of electrical transport properties in insulating solids, *J. Non-Cryst. Solids*, **1**, 197-214.
12. Hartke, J.I., (1962) Drift mobilities of electrons and holes and space-charge limited currents in amorphous selenium films, *Phys. Rev.*, **125**, 1177-1192.
13. Mycielski, W., Staryga, E., Lipiński, A. and Dłużniewski, M. (1994) Electron transport in diamond-like films, *Electron Technology*, **27**, 29-35.
14. Robertson, J. (1991) Hard amorphous (diamond-like) carbon, *Prog. Solid St. Chem.*, **21**, 199-320.
15. Dasgupta, D., Demichelis, F. and Tagliaferro, A. (1991) Electrical conductivity of amorphous hydrogenated carbon, *Phil. Mag.B*, **63**, 1255/

NATIVE AND LIGHT INDUCED DEFECT STATES IN WIDE BAND GAP HYDROGENATED AMORPHOUS SILICON-CARBON(a-Si_{1-x} C_x :H) ALLOY THIN FILMS

MEHMET GÜNEŞ
İzmir Institute of Technology, Faculty of Science, Department of Physics, Gaziosmanpaşa Bulv. No:16, Çankaya, İzmir, 35210-TURKEY.

1. Abstract

In this study, wide band gap a-Si_{1-x} C_x :H alloy thin films prepared with and without hydrogen dilution of (SiH_4 +CH_4) were characterized using optical absorption, dark conductivity, steady-state photoconductivity, sub-bandgap absorption obtained with both photothermal deflection spectroscopy (PDS) and dual beam photoconductivity (DBP), and electron spin resonance (ESR) techniques. Experimental results of steady-state photoconductivity and sub-bandgap absorption for different generation rates were analyzed using a detailed numerical model based on Simmons-Taylor statistics. The densities, energy location and nature of the native and light induced defect states in diluted and undiluted a-Si_{1-x} C_x :H alloy thin films were derived from the best fits to the experimental data. The extracted parameters for defect states were compared with those of a-Si:H films both in the annealed and light degraded states.

2. Introduction

High quality, wide band gap hydrogenated amorphous silicon-carbon (a-Si_{1-x} C_x :H) alloy thin films are desired for use as an intrinsic layer in the top cell of multijuntion solar cells[1] and as p-type layers of these and other electronic devices. It is well known that these materials can be prepared using glow discharge decomposition of silane and hydrocarbons such as methane(CH_4) [2,-5] or using sputtering techniques [5]. However, these wide band gap a-Si_{1-x} C_x :H alloy materials were not so photosensitive. As the concentration of carbon in the film increases, dark and photoconductivity decrease, the density of localized defect states increases so do the optical gap and inverse slope of valence band tails, E_{ov} [2-6]. Alternatively, it was reported that the wide band gap a-Si_{1-x} C_x :H alloys prepared using H_2 -dilution of (SiH_4 +CH_4) gas mixture significantly improved the optoelectronic properties [7]. Later, this was supported by the studies of Baker et al.[8] and Li et al[9], who also reported that hydrogen dilution improved both dark and photoconductivity by several orders of magnitude. Recently, Lu et al. carried out more detailed studies on the process-property relations of a-Si_{1-x} C_x :H alloys deposited in PECVD system as a function of H_2 -dilution ratio [10,11]. They reported that an increase in H_2 -dilution resulted in an increase in surface smoothening, an increase in

M. A. Prelas et al. (eds.), Diamond Based Composites, 285–299.

the electron $\mu\tau$ products, and a decrease in the sub-bandgap absorption, a decrease in the inverse slope of valence band tails, E_{ov}, and a decrease in the void fraction. They obtained optimum optoelectronic and microstructural properties of a-$Si_{1-x}C_x$:H alloys for dilution ratios of 20 to 24. Moreover, p-i-n homojuntion solar cells prepared using diluted undoped a-$Si_{1-x}C_x$:H alloy layers showed high open-circuit voltages(V_{oc}) and high fill factors [9].

Even though there were improvements in the optoelectronic properties, the defect states in a-$Si_{1-x}C_x$:H alloy thin films have not been studied and well characterized as extensively as those in a-Si:H thin films. Generally, electron spin resonance (ESR) [5], constant photo current method (CPM) [12], and photothermal deflection spectroscopy (PDS) [6,8,9] have been used. The density of defect states were derived either from the integration of CPM and PDS spectrum using the same integration constants, which were derived for a-Si:H thin films not for a-$Si_{1-x}C_x$:H alloys, or directly from ESR spins [3,5]. Due to higher density of surface/interface defect states, both ESR and PDS cannot be used for a reliable characterization of the bulk native defects in the annealed state. Alternatively, an approach of deconvolution of PDS and CPM spectrum was also used to obtain the distribution of defect states in the band gap[13]. However, this method ignores conduction band tails and does not take into account the presence of different defect states with different nature.

In addition, wide band gap a-$Si_{1-x}C_x$:H alloys also have intrinsic property of light induced degradation known as the Staebler-Wronski effect[14]. This has not been studied in detail yet and it is essential to determine long term stability of a-$Si_{1-x}C_x$:H based solar cells. Therefore, detailed characterizations of native and light induced defect states are necessary to understand the origins of these defects and mechanism for the Staebler-Wronski effect in wide band gap a-$Si_{1-x}C_x$:H alloys.

In this paper, wide band gap hydrogenated amorphous silicon-carbon alloy thin films prepared with and without H_2-dilution of (SiH4 + CH4) were studied using optical spectroscopy, dark and photo conductivity, and sub-bangap absorption obtained with dual beam photoconductivity (DBP) technique[15] both in the annealed and light degraded states. The experimental results of photoconductivity and sub-bandgap absorption spectra were analyzed using a detailed numerical model described elsewhere[16].

3. Experimental Details

Steady-state photoconductivity and sub-bandgap absorption measurements were carried out on around 1µm thick undoped a-$Si_{1-x}C_x$:H alloy thin films deposited on 7059 glass substrate. Deposition conditions for these films were discussed elsewhere[9-11]. Optical absorption in the visible region of the spectrum was obtained from the transmission(T) and reflection(R) measurements and used to normalize sub-bandgap photoconductivities. The optical gaps were derived from the Tauc relation[17]. The ESR measurements were carried out on films codeposited onto quartz substrates. The annealed state was obtained by heating the samples at 200 °C for 12h in N_2 ambient and stabilized light soaked state was obtained by illuminating through both sides of the samples for more than 300h with 1W/cm^2 white light from an ELH light source filtered with both an IR reflector and IR absorber. Light soaking temperature was maintained around 40°C for high intensity light soaking by air cooling through both sides of samples. Steady-state photoconductivities were measured in the ohmic regime with generation rate $G=10^{15}$ to 10^{20} $cm^{-3}s^{-1}$ using volume absorbed light of λ=620 nm. The sub-bandgap photoconductivities, measured using dual beam photoconductivity (DBP) [15,16] technique with volume generation rates from $G=10^{16}$ to 10^{18} $cm^{-3}s^{-1}$, were also carried out in the ohmic regimes of photocurrents.

4. Results

4.1. ANNEALED STATE

Undoped a-$Si_{1-x}C_x$:H alloy thin films were deposited at 300°C from decomposition of CH_4 and SiH_4 and dilution ratio (H_2/(CH_4 + SiH_4)) was 0:1 and 20:1. The optical gap obtained from the Tauc relation is 1.90 eV for both diluted and undiluted a-$Si_{1-x}C_x$:H films, which are higher than that of intrinsic a-Si:H film. The activation energies are 0.90 eV for undiluted and 0.98eV for diluted a-$Si_{1-x}C_x$:H film which indicate that both films have the Fermi level close to middle of the gap.

The experimental results of the electron $\mu\tau$ products in the annealed state are presented as a function of generation rate with open symbols in Figure 1. The generation rate dependence of σ_{ph} ($\propto G^{\gamma}$) is similar to that observed in a-Si:H films. γ is 0.86 for diluted, 0.80 for undiluted a-$Si_{1-x}C_x$:H, and 0.90 for intrinsic a-Si:H. It is clearly observed that undiluted a-$Si_{1-x}C_x$:H film has the electron $\mu\tau$ products almost two orders of magnitude lower than those of intrinsic a-Si:H film. However, H_2-dilution technique developed by Matsuda[7] improves the optoelectronic properties significantly [7-11]. The values of the electron $\mu\tau$ products of diluted a-$Si_{1-x}C_x$:H film increased and became very close to those of intrinsic a-Si:H, which are shown in Figure 1. This has been attributed to a decrease in the densities of defect states in the bandgap.

The defect states in a-$Si_{1-x}C_x$:H films prepared with and without H_2-dilution were investigated using the sub-bandgap absorption measured by DBP technique at different generation rates. These results are presented in Figure 2 for both films. The α(hv) measured using PDS on the same a-$Si_{1-x}C_x$:H films in the annealed state are also shown in this Figure. It is important to note that PDS results are much higher than those measured by DBP and it also overestimates the E_{0V} because PDS detects all possible optical transitions to and from the defect states and those due to surface/interface defect states. The E_{ov} is 68 meV for diluted and 70 meV for undiluted film as determined from PDS measurements and corresponding DBP results are 55 and 60 meV for diluted and undiluted films, respectively.

As expected, the sub-bandgap absorption due to bulk defect states measured by low generation rate DBP is higher for undiluted a-$Si_{1-x}C_x$:H, i.e. α(1.3eV) is 1.1 $cm^{-1}$and decreased to 0.35 cm^{-1} for diluted a-$Si_{1-x}C_x$:H, which is very close to that of intrinsic a-Si:H film. The generation rate dependence of sub-bandgap absorption, $\Delta\alpha$(1.1eV) [$\Delta\alpha$(1.1eV)=α(1.1eV) measured by high G - α(1.1eV) measured by low G], is 0.11 cm^{-1} for undiluted film and 0.14 cm^{-1} for diluted a-$Si_{1-x}C_x$:H film. These are almost similar to those

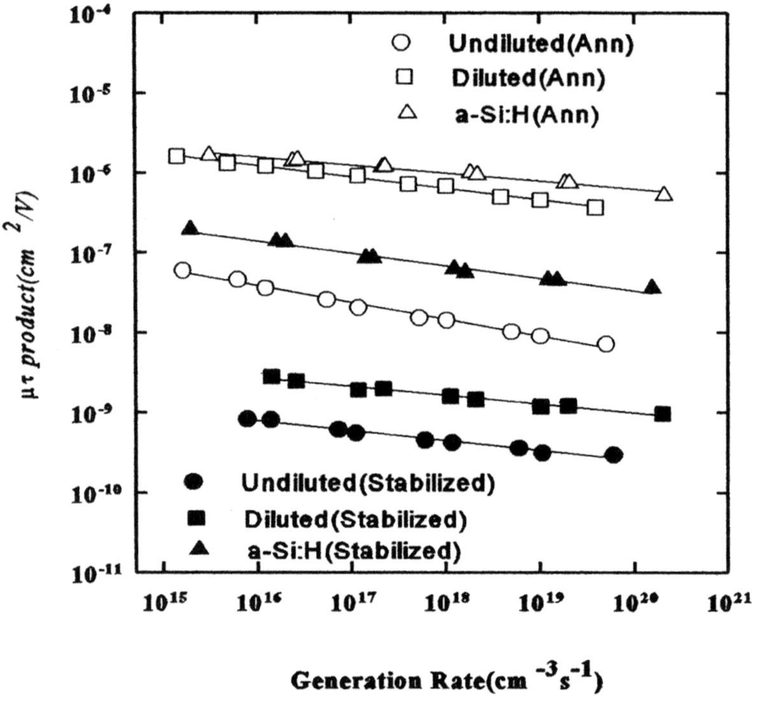

Figure 1. The electron $\mu\tau$ products versus generation rates for diluted and undiluted $a\text{-}Si_{1-x}C_x$:H films and for an intrinsic a-Si:H film in the annealed and stabilized light soaked states.

measured in intrinsic a-Si:H films[18, 19] and much lower than those of non-intrinsic a-Si:H films, which have large densities of charged defect states[20]. Although undiluted film has a factor of 3 higher value of α(1.3eV) than diluted $a\text{-}Si_{1-x}C_x$:H, there is a factor of 35 differences between the electron $\mu\tau$ products of diluted and undiluted films. This indicates that there could be possible differences in either the nature of defect states or free carrier mobilities or both in diluted and undiluted $a\text{-}Si_{1-x}C_x$:H films.

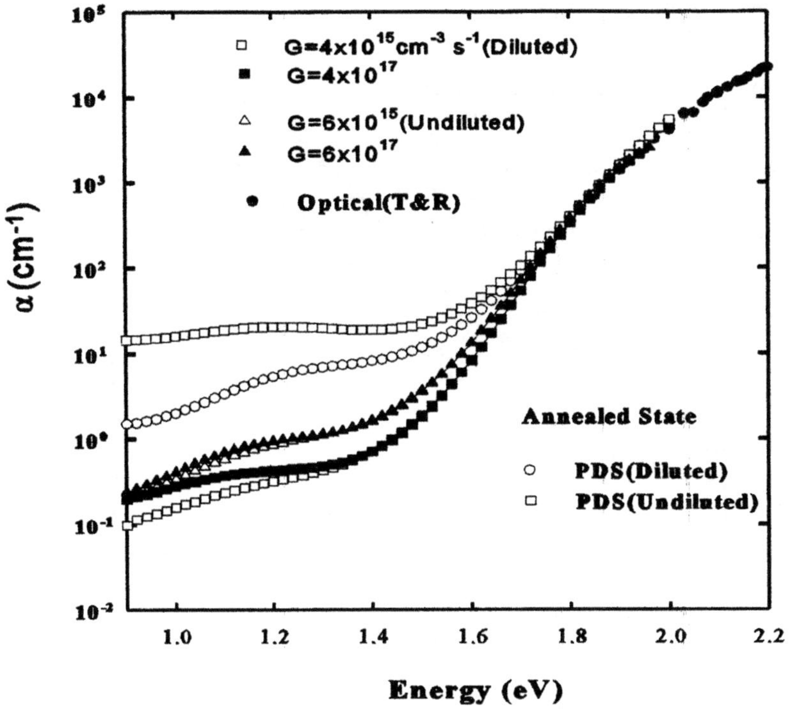

Figure 2. The $\alpha(h\nu)$ versus energy measured by DBP at two generation rates and by PDS for diluted and undiluted a-$Si_{1-x}C_x$:H alloy thin films in the annealed state.

4.2 LIGHT SOAKED STATE

Although hydrogenated amorphous silicon-carbon alloys exhibit similar light induced instability, the Staebler-Wronski effect [14], they are less stable than a-Si:H films. In this study, a-Si_{1-x} C_x :H alloy films were light soaked up to the stabilized soaked state using high intensity, 1 W/cm^2, ELH white light source. Experimental results of the electron $\mu\tau$ products in the stabilized soaked state are shown as filled symbols in Figure 1 and those of sub-bandgap absorption are shown in Figures 3. The results are also summarized in Table 1 for diluted and undiluted a-Si_{1-x} C_x :H alloys as well as those of an intrinsic a-Si:H film.

Even though the electron $\mu\tau$ products measured at $G=10^{16}$ $cm^{-3}s^{-1}$ were similar for diluted a-Si_{1-x} C_x :H and intrinsic a-Si:H films in the annealed state, it decreased more drastically in diluted a-Si_{1-x} C_x :H. As seen in Figure 3, both diluted and undiluted a-$Si_{1-x}C_x$:H films exhibit much lower electron $\mu\tau$ products than intrinsic a-Si:H in the stabilized soaked state. However, γ changes to around 0.85, which is similar to those observed in a-Si:H films

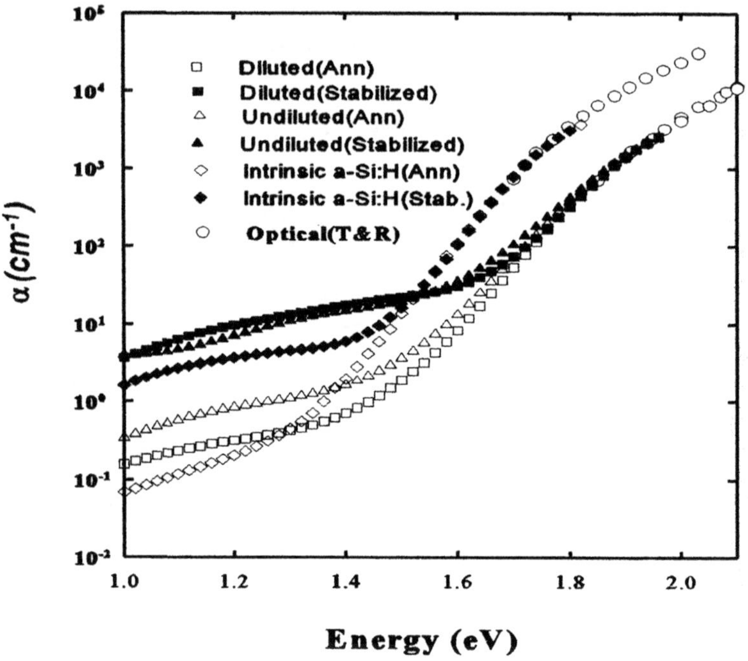

Figure 3. The sub-bandgap absorption spectra measured using DBP at low G for diluted and undiluted a-$Si_{1-x}C_x$:H films as well as those for intrinsic a-Si:H film both in the annealed and stabilized soaked states.

in the light soaked state [20]. This indicates that the distributions of the light induced defect states are similar in a-$Si_{1-x}C_x$:H alloys and a-Si:H films but the densities of light induced defect states must be significantly higher in a-$Si_{1-x}C_x$:H films to compensate the large degradation in the electron $\mu\tau$ products.

Changes in the densities of light induced defect states were characterized using the sub-bandgap absorption in the stabilized soaked state. The results of $\alpha(h\nu)$ for diluted and undiluted a-$Si_{1-x}C_x$:H and that for intrinsic a-Si:H are shown in Figure 3 both in the annealed and stabilized light soaked states. Although there is no detectible change in the slope of the valence band tail absorption for diluted a-$Si_{1-x}C_x$:H and a-Si:H film, there is slight increase in E_{0V} for undiluted film. It increased from 60 meV to 70 meV. The values of α(1.3eV) for diluted and undiluted a-$Si_{1-x}C_x$:H films are comparable in the stabilized state but they are a factor of 4 to 5 higher than that of intrinsic a-Si:H. In addition, the $\Delta\alpha$(1.1eV) for diluted a-$Si_{1-x}C_x$:H increased from 0.14 cm^{-1} to 1.7 cm^{-1}, which is also higher than those measured in a-Si:H films. This indicates that the densities of light induced defect states in a-$Si_{1-x}C_x$:H

Table 1. The experimental results of diluted and undiluted a-$Si_{1-x}C_x$:H films and those of intrinsic a-Si:H film both in the annealed and stabilized light soaked states. Where N_{DOS}= α(1.3eV)x3x10^{16} for a-$Si_{1-x}C_x$:H and N_{DOS} = α(1.2eV)x3x10^{16} for a-Si:H.

Films	Undiluted a-$Si_{1-x}C_x$:H	Diluted a-$Si_{1-x}C_x$:H	Intrinsic a-Si:H
E_{opt}(eV)	1.90	1.90	1.72
E_A(eV)	0.90	0.98	0.90
γ(Ann)	0.80	0.86	0.90
γ(Soak)	0.85	0.85	0.85
μτ(G=10^{16} cm^{-3}s^{-1}) (cm^2/V)(Ann)	3.6x10^{-8}	1.2x10^{-6}	1.5x10^{-6}
μτ(G=10^{16} cm^{-3}s^{-1}) (cm^2/V)(Soak)	8x10^{-10}	2.8x10^{-9}	1.5x10^{-7}
α(1.3eV) (cm^{-1}) (Ann)	1.1	0.35	α(1.2)=0.16
α(1.3eV) (cm^{-1}) (Soak)	11.2	13.3	α(1.2)=3.7
N_{DOS}(cm^{-3}) (Annealed)	3.3x10^{16}	1.05x10^{16}	4.8x10^{15}
N_{DOS}(cm^{-3}) (Soaked)	3.4x10^{17}	4.0x10^{17}	1.1x10^{17}
N(ESR) (cm^{-3}) (Soaked)	4.1x10^{17}	4.5x10^{17}	1.0x10^{17}
E_{0V} (meV)(Ann.)	60	55	49
E_{0V} (meV)(Soaked)	70	56	49
Δα(1.3eV)(cm^{-1}) (Annealed)	0.11	0.14	0.06
Δα(1.3eV) (cm^{-1}) (Soaked)	----	1.7	0.8

films, below as well as above the Fermi level, are higher than those in a-Si:H films. This was supported by the dark ESR spin density measurements carried out in the stabilized soaked state on both a-Si_{1-x} C_x :H and intrinsic a-Si:H films. It was found that the dark ESR spin signal comes from the same defect states with the same g number, 2.0055 [21]. The density of bulk defect states were significantly higher than that of surface/interface defect states present in the annealed state[18]. The absolute density of ESR spins(after subtracting the density of surface/interface defect states) is $1.0x10^{17}$ cm^{-3} for a-Si:H, $4.5x10^{17}$ cm^{-3} for diluted, and $4.1x10^{17}$ for undiluted a-Si_{1-x} C_x :H. The densities of defect states below the Fermi level can also be obtained from sub-bandgap absorption using the values of α(1.3 eV)[18]. It is $3.5x10^{17}$ cm^{-3} for undiluted, $4.0x10^{17}$ cm^{-3} for diluted a-Si_{1-x} C_x :H alloy and $1.1x10^{17}$ cm^{-3} for a-Si:H film. These values agree very well with those measured directly using ESR. Even though the densities of the D^0 states are factor of 4-5 higher in a-$Si_{1-x}C_x$:H alloys, the electron $\mu\tau$ products in the stabilized state were lower by a factor of 54 for diluted and 187 for undiluted a-Si_{1-x} C_x :H. These large differences in the degradation of the electron $\mu\tau$ products can be attributed to the increases in the densities of other type defect states such as charged Si dangling bonds as in a-Si:H films or possibly any carbon related light induced defect states. This can only be possible through a detailed analysis of the experimental results both in the annealed and light soaked states.

5. Native Defect States in a-Si_{1-X} C_X:H Alloys

The defect states in hydrogenated amorphous silicon- carbon alloys have not been studied as extensively as those in a-Si:H materials. In 1982, Morimoto et al.[5] reported that ESR spins were due to silicon and carbon dangling bonds depending on the carbon content in the film where for x ($x=C_{Carbon}/(C_{carbon}+C_{Si})$) less than 0.2, g is 2.0055, the same as the neutral Si dangling bonds. The same group also found that the density of charged defect states in a-Si:H obtained from LESR decreased with carbon alloying but increased with O and N alloying[12].

In 1992, Robertson reported a theoretical study of defect states in a-Si_{1-x} C_x :H alloys using the tight binding approximation [22]. It was found that for $x < 0.5$, Si dangling bond defects and $x > 0.6$ carbon dangling bond defects dominate in the alloy, but no charged defect states resulted from these calculations, The nature of tail states were like those of a-Si:H, i.e. the valence band tails are wider than the conduction band tail states.

Since a-Si_{1-x} C_x :H alloy films studied here were Si rich and they were treated like a-Si:H films. The concentration of carbon in these alloys determined from electron probe microanalysis was less than 0.1 at.% [11]. Preliminary results of the analysis on diluted and undiluted a-Si_{1-x} C_x :H films were presented assuming the same distribution of the gap states and charged defect states as used in a-Si:H films[18]. The distribution used in the modeling consists of parabolic extended states, exponential valence and conduction band tail states, two Gaussian distributions of neutral and negatively charged states located below the Fermi level and a Gaussian distribution of positively charged states located above the Fermi level [18]. The results of photoconductivities and sub-bandgap absorption were analyzed using sub-bandgap absorption model (SAM) [16,19]. This model is based on Simmons-Taylor statistics [23], which was developed for a continuous distribution of gap states in the bandgap of insulators. In the modeling, the parameters for the extended states, tail states, and carrier capture cross-sections of the defect states were assumed to be the same as those in a-Si:H films. Experimental values of the optical gap, the Fermi level position, and E_{0V}(meV) were important inputs in the modeling. The mobility gaps were assumed to be 0.18 eV larger than optical gap. Because the electron $\mu\tau$ products in diluted a-Si_{1-x} C_x :H and intrinsic a-Si:H film were virtually the same in the annealed state, the same free carrier mobilities were assumed as those used for a-Si:H. There was no any priory assumptions for the densities and energy location of the native defect states. These were derived from the self-consistent fits to the results of both σ_{ph} and sub-bandgap absorption in the annealed state.

The results(symbols) of the sub-bandgap absorption and σ_{ph} (in the inset)and the best fits (solid lines) are shown in Figure 4 for diluted and in Figure 5 for undiluted a-Si_{1-x} C_x :H films. It was found that the distribution of the charged defect states can be successfully used to fit the results of the diluted a-Si_{1-x} C_x :H. Even though it is possible to obtain equivalent fits

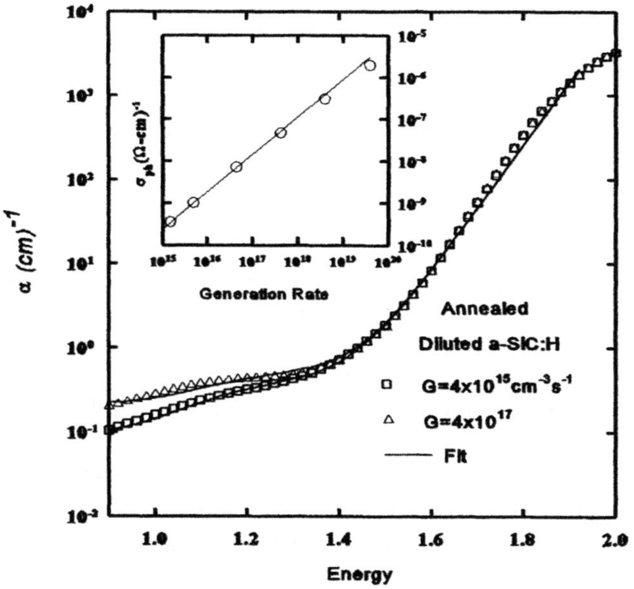

Figure 4. The sub-bandgap absorption spectra and σ_{ph} (inset) for diluted a-Si_{1-x} C_x :H film in the annealed state. The symbols are the experimental results and solid lines are the best fits to the data.

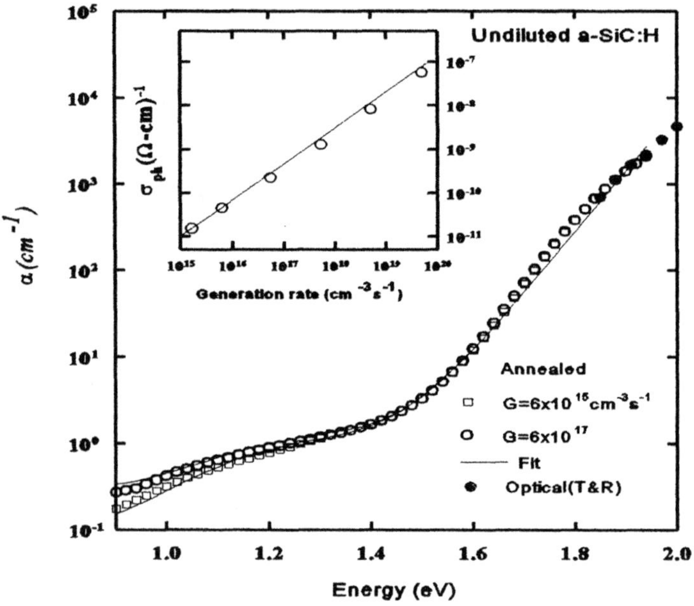

Figure 5. The sub-bandgap absorption spectra and σ_{ph} (inset) for undiluted a-Si_{1-x} C_x :H film in the annealed state. The symbols are the experimental results and solid lines are the best fits to the data.

to the sub-bandgap absorption spectra for two different generation rates, the values of steady-state photoconductivity in undiluted a-Si_{1-x} C_x :H alloy film cannot be matched using the same parameters of the gap states as used in diluted a-Si_{1-x} C_x :H . This requires to use a factor of 25 lower free electron and hole mobilities than those used for diluted alloy or to introduce different states, such as carbon related defect states with higher capture cross-sections or both to fit photoconductivities in the analysis. The densities and energy locations of the neutral and charged defect states in diluted film as well as those in intrinsic a-Si:H film are summarized in Table 2.

The densities of the neutral and charged defect states in diluted a-Si_{1-x} C_x :H film are higher than those of intrinsic a-Si:H film. This results in higher sub-bandgap absorption as well as higher $\Delta\alpha(1.1eV)$ for diluted a-Si_{1-x} C_x :H. The energy position of the defect states are found to be further away from the valence band edge in wide band gap a-Si_{1-x} C_x :H alloy. The ratio of the charged to neutral defect states, R= ([N_{D+} + N_{D-}]/N_{D0} , is 1.0 for diluted a-Si_{1-x} C_x :H film, which is slightly lower than those obtained in intrinsic a-Si:H films studied earlier[19]. In conclusion, the distribution of charged defect states can be successfuly applied to self-consistent analysis of steady-state photoconductivity and sub-bandgap absorption spectra for diluted a-Si_{1-x} C_x :H alloys.

Table 2. The densities and energy locations of the native defect states derived from the analysis of σ_{ph} and $\alpha(h\nu)$ spectra in diluted a-Si_{1-x} C_x :H and those for intrinsic a-Si:H film in the annealed states. $N_{DOS}=3x10^{16}x\ \alpha(ref.)$, where *ref.* is 1.2 eV for a-Si:H and 1.3 eV for a-Si_{1-x} C_x :H films. The extended state parameters and capture cross-sections for defect states are the same as those in ref [19].

Parameters	Diluted a-Si_{1-x} C_x :H Film	Intrinsic a-Si:H
E_{opt}(eV)	1.90	1.72
E_F(eV)(from CB)	0.98	0.90
E_{0V}(meV)	55	49
E_{0C}(meV)	30	21
E_{D0}(eV) from VB	0.90	0.78
Width(eV)	0.13	0.13
$N_{D0}(cm^{-3})$	$9x10^{15}$	$4.5x10^{15}$
E_{D+}(eV) from VB	1.46	1.28
Width(eV)	0.1	0.08
$N_{D+}(cm^{-3})$	$4.5x10^{15}$	$3x10^{15}$
E_{D-}(eV) from VB	0.60	0.50
Width(eV)	0.1	0.08
$N_{D-}(cm^{-3})$	$4.5x10^{15}$	$3x10^{15}$
$N_{DOS}(cm^{-3})$	$1.1x10^{16}$	$4.8x10^{15}$
$R=(N_{D+}+N_{D-})/N_{D0}$	1.0	1.4
$\mu_n(cm^2/Vsec)$	10	10
$\mu_p(cm^2/Vsec)$	1	1

6. Light Induced Defect States in a-Si_{1-x} C_x :H Alloys

It was shown that hydrogenated amorphous silicon-carbon alloys were less stable than a-Si:H films. Since they exhibited more degradation in σ_{ph} and large increase in $\alpha(h\nu)$ in the stabilized soaked state. However, the degradation of σ_{ph} was not directly correlated with the increase of sub-bandgap absorption which is due to the increases in the densities of the dangling bond states below Fermi level. In order to understand the light induced degradation in these alloy films, the numerical analysis of σ_{ph} and $\alpha(h\nu)$ were extended to the results of diluted and undiluted a-Si_{1-x} C_x :H films in the stabilized soaked state.

In the previous section, it was shown that the distributions of the charged defect states can be used to characterize the native defect states only in diluted a-Si_{1-x} C_x :H film but not for undiluted film. The same distribution of defect states is now used in the modeling of the light induced defect states where only changes in the densities of midgap states were assumed to occur with light soaking. The energy locations and capture cross-sections of defect states were maintained the same as those in the annealed state. The densities of the light induced defect states were derived from the best fits to the results of the σ_{ph} and $\alpha(h\nu)$. The experimental results(symbols) of σ_{ph} and $\alpha(h\nu)$ and the best fits(solid lines) are shown in Figure 6 for diluted a-Si_{1-x} C_x :H film both in the annealed and stabilized soaked state. The densities of the neutral Si dangling bond states, the D^0, are 4.9×10^{17} cm^{-3}, which is close to the value of 4.5×10^{17} cm^{-3}, the density of ESR spins measured on the same film in the stabilized state. It is factor of 4 to 5 higher than those measured in intrinsic a-Si:H. Although equally

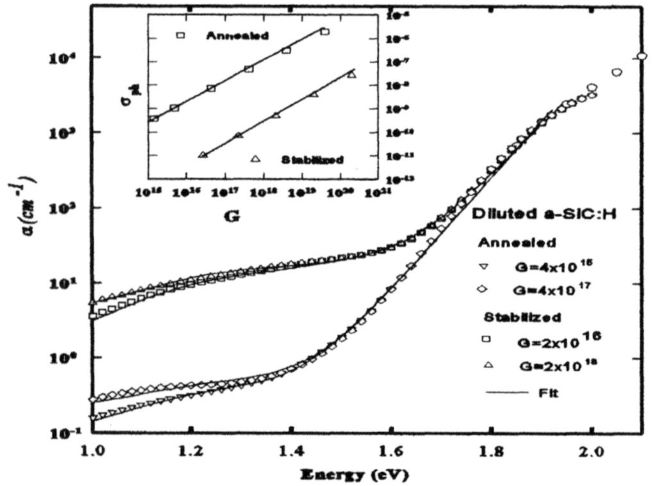

Figure 6. The sub-bandgap absorption spectra and σ_{ph} (inset) for diluted a-Si_{1-x} C_x :H film in the annealed and stabilized states. The symbols are the experimental results and solid lines are the best fits to the data.

good fits to the generation rate dependence of the sub-bandgap absorption were obtained using the densities of $1.7x10^{17}$ cm^{-3} for both the D^+ and the D^- states, these were not sufficient to fit the large decrease in the steady-state photoconductivities. In order to obtain the best fits to σ_{ph} for wide range of generation rates, a factor of 25 decrease in the free carrier mobilities were required in the stabilized light soaked state. In addition, even if the results of undiluted $a\text{-}Si_{1-x}C_x:H$ film were fitted in the annealed state using the same distribution of charged defect states and with certain free carrier mobilities, it is also impossible to match the values of photoconductivities in the stabilized soaked state without decreasing free carrier mobilities by more than a factor of 40. This indicates that light induced defect states in diluted and undiluted $a\text{-}Si_{1-x}C_x:H$ aloys are not the same as those in a-Si:H films. Even though the neutral Si dangling bond states, the D^0, created by light are the same both in a-Si:H and Si rich $a\text{-}Si_{1-x}C_x:H$ alloys[21], different type of defect states with higher capture cross-sections could be created by light soaking in $a\text{-}Si_{1-x}C_x:H$ alloys. Further studies are necessary to investigate the light induced defect states in $a\text{-}Si_{1-x}C_x:H$ alloys and any possible changes in the transport properties with light soaking.

7. Conclusions

Preliminary results on device quality, one diluted and one undiluted, $a\text{-}Si_{1-x}C_x:H$ film show that the distribution of charged defect states used in the analysis of a-Si:H films can be used to characterize the defect states in diluted $a\text{-}Si_{1-x}C_x:H$ film in the annealed state. The densities of the neutral and charged defect states in diluted $a\text{-}Si_{1-x}C_x:H$ are comparable with those in a-Si:H films. The ratio of the charged to neutral defect states is around 1.0 and slightly lower than those obtained in intrinsic a-Si:H films and much lower than those of undoped (non-intrinsic) a-Si:H films studied previously [20]. The lower values of R in $a\text{-}Si_{1-x}C_x:H$ film are consistent with the experimental results of ESR and LESR carried out on a-Si:H and $a\text{-}Si_{1-x}C_x:H$ materials reported by Shimizu et al.[12]. However, the results of undiluted $a\text{-}Si_{1-x}C_x:H$ film in the annealed states cannot be analyzed using the same distributions and gap states parameters of diluted film. It was inferred that different type of defect states with higher capture cross-sections could be present in undiluted film and/or possible lower free carrier mobilities.

In the light soaked state, it was found that both diluted and undiluted $a\text{-}Si_{1-x}C_x:H$ films are less stable than a-Si:H films. Self-consistent analysis of the results in the light soaked state was impossible even for diluted film using the same distribution of gap states and parameters used in the annealed state. Even though almost the same densities of the D^0 states measured by ESR were used as input to fit low G sub-bandgap absorption spectrum, this was not sufficient to account for the light induced degradation in photoconductivities. Different type of light induced defect states which are not present in a-Si:H film and/or lower free carrier mobilities or both are necessary to fit lower photoconductivities in both diluted and undiluted $a\text{-}Si_{1-x}C_x:H$ alloys. Further research on the native and light induced defect states in

Si rich a-Si_{1-x} C_x :H alloys are necessary to understand the Staebler-Wronski effect in these materials.

8. Acknowledgements

Experimental part of this study was carried out at the Pennsylvania State University. It was supported by NREL under Subcontract No. XG-1-10063. The author would like to thank Drs. C.R.Wronski and R. W. Collins at the Pennsylvania State University for scientific collaborations and helpfull discussions, Drs. T.J. McMahon and R. Crandall of NREL for ESR and electron probe microanalysis experiments, and Dr. Yuan-Min Li of Solarex for PDS measurements. The author would also like to thank Turkish Government, Ministry of National Education for the partial Ph.D scholarship during his stay in the USA.

9. References

1. Kuwano, Y. (1986) *Plasma Deposited Thin Films*, edited by J.Mort and F. Jansen, CRC, Boca Raton, p. 161.
2 Paul, W. Paul, D. K., Von Roedern, B., Blake, J., and Oguz, S. (1981) *Phys. Rev. Let.* **46**, 1016 .
3 Morimoto, A., Miura, T., Kumeda, M., and Shimizu, T. (1981) *Jpn. J. Appl. Phys.* **20**, L833.
4 Schmidt, M.P., Bullot, J., Gauthier, M., Cordier, P., Solomon, I., and Tran-Quoc, H. (1985) *Phil. Mag.* **B51**, 581.
5 Morimoto, A., Miura, T., Kumeda, M., and Shimizu, T. (1982) *J. Appl. Phys.* **53**, 7299.
6 Boulitrop, F., Bullot, J., Gauthier, M., Schmidt, M.P., Catherine, Y. (1985) *Solid State Comm.* **54**, 107.
7 Matsuda, A., and Tanaka, K. (1987) *J. Non-Cryst. Solids* **97&98**, 1367.
8 Baker, S. H., Spear, W. E., and Gibson, R. A. G. (1990) *Phil. Mag.* **B62**, 213.
9 Li, Yuan-Min., Catalano, A., Fieselmann, F. B. (1992) *Mat. Res. Soc. Symp. Proc.* **258**, 923.
10 Lu, Y., An, I., Gunes, M., Wakagi, M., Wronski, C.R., and Collins, R.W. (1993) *Appl. Phys. Lett.* **63**, 2228.
11 Lu, Y., Kim, S., Gunes, M., Lee, Y., Wronski, C.R., and Collins, R.W. (1994) *Mat. Res. Soc. Symp. Proc.* **336**, 595.
12 Shimizu, T., Kidoh, H., Morimoto, A., Kumeda, A. (1989) *Jap. J. Appl. Phys.* **28**, 586.
13 Demichelis, F., Crovini, G., Giorgis, F., Pirri, C.F., Tresso, E., Amato, G., Herremans, H., Grevendonk, W., Rava, P. (1993) *J. Non-Cryt. Solids* **164-166**, 1015.

14 Staebler, D.L., Wronski, C.R. (1977) *Appl. Phys. Lett.* **31**, 292.

15 Lee, S., Kumar, S., Wronski, C.R., and Maley, N. (1989) *J. Non-Cryst. Solids* **114**, 316.

16 Lee, S., Gunes, M., Wronski, C.R., Maley, N., and Bennett, M. (1991) *Appl. Phys. Lett.* **59**, 1578.

17 Tauc, J. (1972) *in Optical properties of Solids*, edited by F. Abeles, North Holland, New York, p.279.

18 Gunes, M., Wronski, C.R., and McMahon, T.J. (1994) *J. Appl. Phys.* **76**, 2260.

19 Gunes, M., Collins, R.W., and Wronski, C.R. (1994) *Mat. Res. Soc. Symp. Proc.* **336**, 413.

20 Gunes, M. (1995) *Ph. D.Thesis*, The Pennsylvania State University,University Park PA, USA.

21 McMahon, T.J. (1995) Private communication.

22 Robertson, J. (1992) *Phil. Mag.* **B66**, 615(1992).

23 Simmons, J.G., and Taylor, G.W. (1971) *Phys. Rev.* **B4**, 502.

GENESIS OF DEFECTS SUITABLE FOR HIGH-TEMPERATURE SPECTRAL HOLE BURNING IN DIAMOND

A. OSVET and I. SILDOS
Institute of Physics,
Riia 142, EE2400 Tartu; Estonia

Abstract

The present research is devoted to the characterization of nitrogen-rich diamonds of different origin from the point of view of creation of defects possessing persistent spectral hole-burning properties at room temperature. A set of natural diamond crystals was subjected to neutron irradiation at different doses in the range of 10^{17} - 10^{19} neutrons/cm^2 followed by isochronal annealing at different temperatures. After each step of annealing photoluminescence of all samples in the red spectral region was investigated and optimal conditions for the genesis of different kinds of photochromic defects were found.

Keywords: radiation defects, neutron-irradiated diamond, photo-luminescence, spectral hole burning

1. Introduction

The creation of materials where persistent spectral hole burning is feasible not only at cryogenic temperatures but also at ambient conditions, is crucial for the optoelectronic applications of this phenomenon. In our previous papers [1-4], we have introduced a number of radiation-created defect aggregates having zero-phonon lines in the red spectral region of photoluminescence and demonstrated the possibility of using neutron-treated diamond as a prospective material for information storage operating at room temperature.

M. A. Prelas et al. (eds.), Diamond Based Composites, 301–308.

In spite of extensive investigations carried out to study the defects in natural diamonds still numerous defects have remained unidentified and the origin of the defects giving rise to the ZPL-s observed in luminescence spectra at 649, 681, 723, 731, 774, 813 nm ZPL-s is also unknown.

The aim of this work is to get some knowledge of the origin and of the conditions needed to create the above-mentioned defects. We intended to vary systematically the irradiation doses and annealing temperatures of the samples and to gain information about the genesis of the defects (defect aggregates) giving rise to the ZPL-s in the red spectral region. by investigating the photoluminescence spectra of the samples subjected to different conditions of preparation. Namely these ZPL-s are interesting because of their persistent photochromic properties. Especially the 774 nm and 813 nm lines with their inhomogeneous widths exceeding the homogeneous ones at room temperature by a factor of 4 and where spectral holes burnt in and registered at RT have been demonstrated [2].

2. Experimental Details and Samples

A piece of a natural Yakutian-origin IaB-type diamond was cut, polished and the pieces were irradiated with fast neutrons (E >1 MeV) at the doses of 10^{17}, 10^{18} and 10^{19} neutrons/cm^2, respectively. This treatment was followed by a 30-minute isochronal annealing at the temperatures up to 1100°C with the intervals of 200°C. Also a diamond originating from Sierra Leone with the irradiation dose of 10^{18} neutrons/cm^2 was subjected to similar heat treatment. After each annealing cycle (performed in vacuum to prevent surface grafitization) the samples were cooled down to 6 K in the vapors of liquid He and the photoluminescence spectra of the samples were measured in the spectral region of 720 - 830 nm. An Ar^+-ion-laser-pumped cw dye laser operating at 704 nm was used to excite photoluminescence. The spectra were recorded with a 0.8 m double monochromator, the RCA C31034 photomultiplier tube and photon counting electronics for photoluminescence detection.

The nitrogen content of the Yakutian samples was estimated from the infrared absorption measurements by using the [5] relation between absorption coefficients and the nitrogen content. As a result we obtained the values of $4 \cdot 10^{19}$ cm^{-3} and $8.5 \cdot 10^{19}$ cm^{-3} for the nitrogen present in A (pairs) and B1 (aggregates of four atoms) forms respectively. In addition the samples contain the B2-aggregates (platelets). The samples contain also N3 defects (three N atoms and V) and single substitutional nitrogen atoms. The number of the latter is much smaller than that of larger aggregates and the defects can be recognized from the presence of the weak 637 nm line belonging to the N-V defect and appearing in the spectra after annealing of the irradiated samples.

3. Results and Discussion

In the following we lay out experimental results of photoluminescence measurements of differently treated samples. The alteration of the photoluminescence spectrum during the isochronal annealing of the Yakutian IaB-type crystal irradiated with the dose of 10^{18} neutrons/cm^2 is shown in Fig. 1. In general three stages can be distinguished: from RT to 500°C the spectra are dominated by the GR1 line at 741 nm stemming from the neutral vacancies. At the second stage where the samples have been annealed at temperatures higher than 500°C lines at 730, 774 and 813 nm appear simultaneously with the reduction of the strength of the GR1 line. Indeed, according to G.Davies et al [6] the energy of migration of vacancies is 2.3 eV and fast decrease of the vacancy-induced lines begins at 550°C. The decrease of the inhomogeneous width of the GR1 line from 2.3 nm to 1.0 nm between annealing temperatures of 500°C and 700°C can be explained either by the removal of disorder in the lattice during annealing or, as in [6] by the fact that vacancies are preferentially created in the vicinity of nitrogen vacancies. In the latter case those vacancies, strongly disturbed by nitrogen and contributing to the wings of the ZPL get trapped first when migration of the vacancies starts. At the very early stage of annealing self-interstitials move to the vacancies and part of the latter are destroyed quickly. At higher temperatures vacancies move to A and B1 aggregates and form H3 and H4 defects with ZPL-s at 503 and 496 nm, respectively. Part of vacancies join to form bivacancies, some are trapped by dislocations and impurities but as in natural crystals the concentration of nitrogen exceeds the concentration of vacancies and other impurities the vacancies recombine mainly with nitrogen aggregates. When the annealing temperature exceeds 700°C, the GR1 line disappears completely and the lines at 731, 734, 739, 774 and 813 nm become dominant. We suppose that the defects are radiation defect + nitrogen

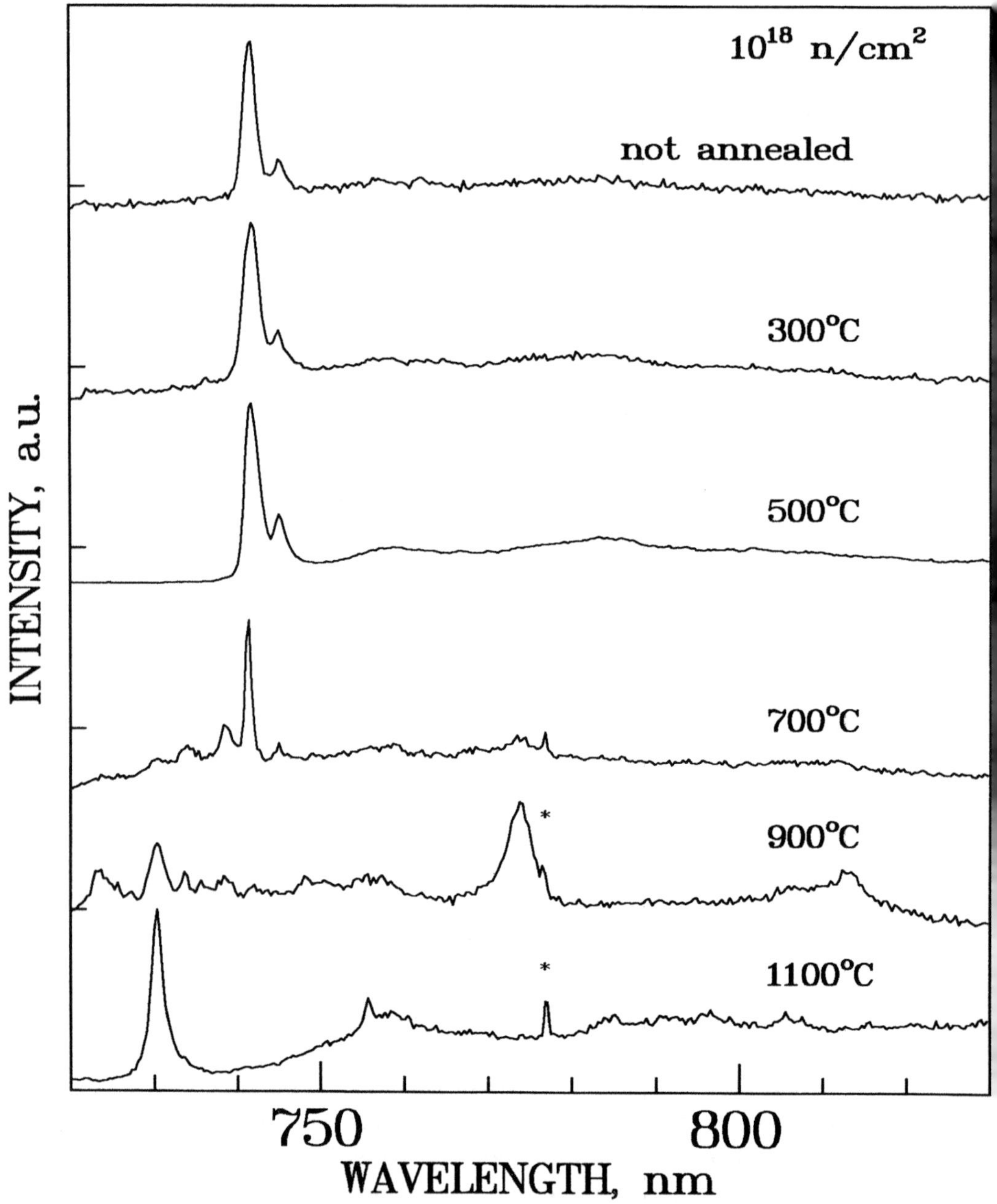

Figure 1. The effect of annealing temperature to the luminescence spectra of neutron–irradiated IaB–type diamond. Measured at T= 6 spectral resolution 0.1 nm. Excitation 704 nm, 40 mW/mm^2.
* marks the 1332 cm^{-1} Raman line of diamond.

aggregates, but in this case they cannot be the N-V-N (H3) or V-N-N-V (H4) defects. It is hard to propose a definite structure of the defect. It is also possible that some A or B1 aggregates are destroyed during annealing and new forms of defects are created. Nisida et al propose [7] that decomposition of nitrogen pairs and vacancy-assisted migration of nitrogen starts at temperatures as low as 800°C. After annealing at T=900°C the intensities of the 774 nm, 813 nm, 734 nm and 723 nm lines reach their maxima. In case of the latter strong self-absorption should be taken into account. Coincidence of the saturation of these lines with the destruction of the GR1 line could refer to the fact that defects are formed by vacancies being trapped by some impurities. It is clear from the annealing data that for production of the 774 and 813 nm lines temperature of 900°C is the most suitable. The annealing over 900°C leads to the weakening of 723, 734, 739, 774 and 813 nm lines, while the line at 731 nm starts to be dominant after annealing at 1100°C. From a similar behavior of the 774 and 813 nm lines we could suppose that these lines belong to different electronic transitions of one and the same defect. This assumption is also consistent with the spectral hole-burning data.
Figure 2 shows the dependence of defect genesis on the irradiation doze. We can see that up the doses of 10^{19} n/cm^2, the increase of the radiation doze causes the increase of the number of defects, responsible for 723, 731, 739, 774 and 813 nm lines. We are not able to estimate the threshold of saturation, as much higher doses of irradiation are needed. We can also conclude that much higher doses are needed to create the 731 nm ZPL-s in the spectra.
In figure 3 the photoluminescence spectra of the samples with different origin are compared after similar treatment. Both samples have been irradiated with fast neutrons up the doses of 10^{18} neutrons/cm^2 and annealed for 30 minutes at 900°C. The ZPL-s at 774 and 813 nm are present in both spectra but the lines in the 720 - 740 nm region are completely absent in the spectrum of the diamond from Sierra Leone. From this fact and from the different influence of the annealing temperature we can conclude that different defects are responsible for the indicated group of lines and 774 nm lines.

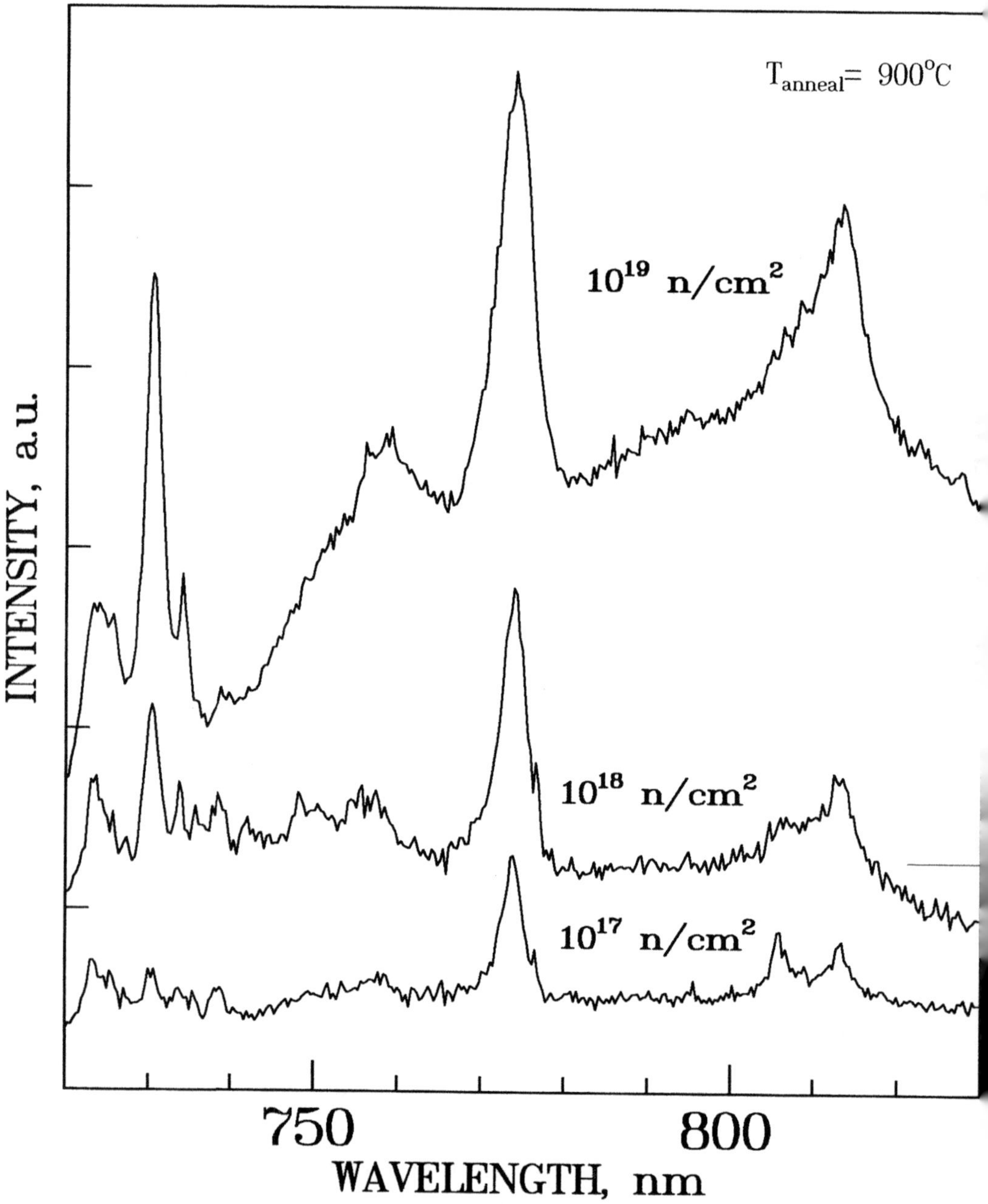

Figure 2. The effect of irradiation doses to the spectra of photo–luminescence of natural IaB–type diamond. Excitation and registration as in Fig.1.

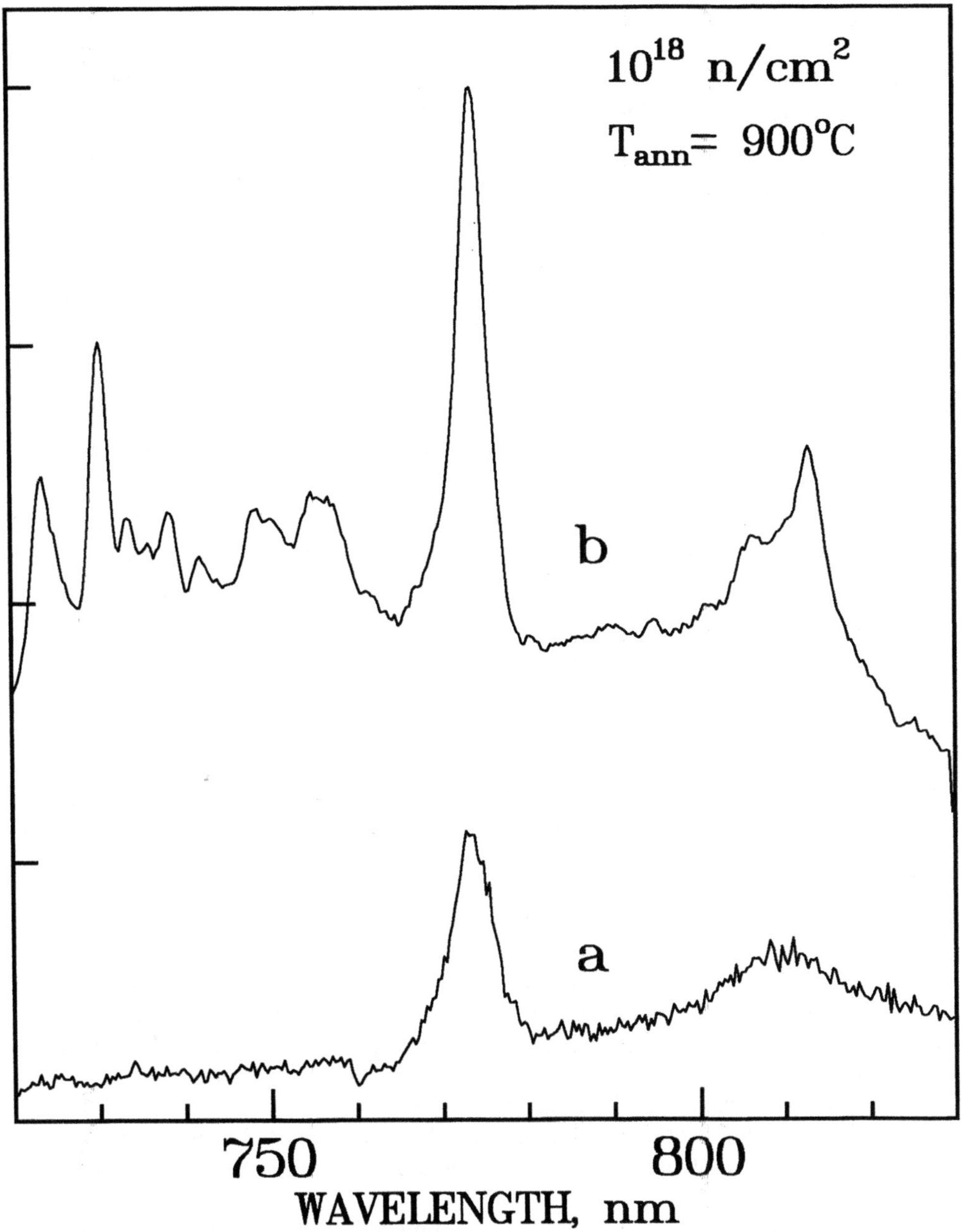

Figure 3. Photoluminescence of neutron–irradiated diamond samples of different origin:
a – Sierra Leone diamond, b – Yakutian diamond.
Spectra measured at T= 6 K. Excitation at 704 nm.

4. Conclusions

From the results of this investigation we can conclude that it is possible to control the number of defects giving rise to the 723, 731, 774 and 813 nm ZPL-s in the luminescence spectra of nitrogen-rich IaB-type diamonds. The most favorite conditions for the production of the room-temperature photochromic defects is annealing at 900°C - 950°C.

Acknowledgment

This work was partly supported by the International Science Foundation, grant No LL 9100 and by the Estonian Science Foundation, grant No 2277.
We are indebted to Dr. U. Bogner for providing pieces of diamond originating from Sierra Leone and for the help in polishing the samples. We are grateful to Prof. N. Mironova for the reactor treatment of the samples.

References

1. Osvet, A., Suisalu, A., Sildos, I. (1992) Persistent High-Temperature Spectral Hole Burning in a Neutron-Colored Diamond, *Proc.Estonian Acad. Sci. Phys. Mat.* **41**, 222.
2. Bauer, R., Osvet, A., Sildos, I., Bogner, U. (1993) Room-Temperature Persistent Spectral Hole Burning in Neutron-Irradiated Iab-type Diamond, *J.Lumin.* **56**, 57-60.
3. Sildos, I., Osvet, A. (1994) Spectral Hole-Burning Study of a Neutron-Irradiated Type IaB Natural Diamond, *Diamond and Related Materials* **3**, 725 - 727.
4. Sildos, I. Osvet, A. (1994) Spectral Hole Burning in Reactor-Treated Diamond Materials, *Proceedings of the Fourth International Conference on the New Diamond Science and Technology*, Kobe, Japan, 395 - 398.
5. Woods, G.S., Purser, G.C., Mtimkulu, A.S.S. and Collins, A.T. (1990) The Nitrogen Content of Type Ia Natural Diamonds, *J.Phys. Chem.Solids* **51**, 1191 - 1197.
6. Davies, G., Lawson, S.C., Collins, A.T., Mainwood, A. and Sharp S.J. (1992) Vacancy-related Centers in diamond, *Phys Rev B* **46**, 13157 -13170.
7. Nisida, Y., Mita, Y., Mori, K., Okuda, S., Sato, S., Yazu, S., Nakagawa, M., and Okada, M. (1989) Color Centers in Annealing of Neutron-Irradiated Type Ib and Ia Diamonds, *Materials Science Forum* **38-41**, 561 - 565.

RAMAN SPECTROSCOPY OF AMORPHOUS DIAMONDLIKE CARBON FILMS PRODUCED WITH A MASS-SEPARATED ION BEAM AND PULSED ARC DISCHARGE

L. Yu. KHRIACHTCHEV and M. RÄSÄNEN
Laboratory of Physical Chemistry, University of Helsinki, P.O. Box 55, FIN-00014 Helsinki, Finland

R. LAPPALAINEN and M. HAKOVIRTA
Department of Physics, University of Helsinki, P.O. Box 9, FIN-00014 Helsinki, Finland

Raman spectra of amorphous hydrogen-free diamondlike films prepared by two different methods, namely, by using a mass-separated carbon ion beam, and pulsed arc-discharge deposition, were investigated. For the samples prepared with both deposition methods, clear correlation between the carbon deposition energy and such parameters of the Raman spectra as the intensity ratio $R=I_{500}/I_{1550}$ and the average slope $S=(I_{1300}-I_{1100})/I_{1100}$ was established. Also, the data show a reproducible dependence between the parameters S and R, and the region corresponding to the best diamondlike films is well-defined in the S-R plot. The extremes of the introduced parameters R and S indicate the optimal deposition conditions with respect to the film hardness, electrical resistance, etc. In the case of mass-separated ion-beam deposition, the data obtained provide a straightforward dependence between the sp^3 fraction and the introduced numerical parameters.

M. A. Prelas et al. (eds.), Diamond Based Composites, 309–321.

1. Introduction

Mass-separated ion beams and pulsed plasma arc discharge are two particular methods allowing amorphous hydrogen-free diamondlike carbon films of high quality to be prepared [see, for example, Refs. 1-5]. The structure of such films has been proved to consist of an extremely disordered network of sp^2- and sp^3-bonded carbon atoms with the fraction of the sp^3 coordination as high as 80%, at least for the better controlled case of the ion-beam deposition. Studies of this so called tetrahedral amorphous carbon (ta-C) are very popular now because of its quite unusual and promising properties. Amorphous diamondlike carbon films are found to be extremely hard, optically transparent and chemically inert, which makes them perspective for coating technology, and a wide defect-free optical gap of about 2.5 eV establishes interest in this material as a semiconductor. From the fundamental point of view, this new allotrope of carbon is also extremely interesting, and the physical modeling of its microstructure and growth is still in progress.

Raman spectroscopy, among other methods of diagnostics, appears to be promising for structural analysis and characterization of amorphous diamondlike carbon material [6,7, and references therein]. Raman spectra of amorphous hydrogen-free diamondlike films differ essentially from those of the hydrogenated materials [7-9], in particular, they show relatively intensive low-frequency scattering in the 400 to 800 cm^{-1} region. Some systematic behavior of the shape of Raman spectra for the hydrogen-free material was noticed [10,11], and the influence of radiation interference on the signals was discussed [12].

Theoretical simulations of ta-C structure were performed for the materials with the sp^3 fraction of 80-90% [13,14]. The vibrational properties of this theoretically generated diamondlike structure are essentially characterized, in agreement with experimental results, by the unlocalized low-frequency modes in the 200 to 1300 cm^{-1} region, and the additional modes with frequencies above 1300 cm^{-1} are associated with well-localized geometrical anomalies in the network. As a result, in a very simple structural model consistent with many observed properties, the material is composed of small sp^2 clusters embedded in a more transparent and less scattering sp^3 host network.

In the presented work, the experimental features of Raman spectra of amorphous hydrogen-free diamondlike films prepared with two different methods, namely, a mass-separated carbon ion beam and pulsed plasma arc-discharge deposition, are investigated. We study the Raman spectra of a series of samples prepared by using the same set-ups but with different energies of carbon ions in comparison with other parameters of the material, in particular, with the sp^3 fraction.

2. Experiment

The first part of measurements was performed with amorphous hydrogen-free diamondlike films prepared by using mass-separated carbon ion deposition onto precleaned crystalline <100> silicon substrates, with the energies of carbon ions varying from 5 to 5000 eV for different samples. The typical area of deposition was about 0.1 cm^2, the thickness of the films varied from 100 to 500 nm, and all coatings were deposited with approximately the same carbon ion flux. The details of the deposition technique are described elsewhere [2]. Previous ESCA (Electron Spectroscopy for Chemical Analysis) measurements have shown that the sp^3 fraction for these films reaches the maximum of about 80% at the deposition energies around 100 eV, and relatively high fraction of sp^3 coordinated carbon remains up to the energy of 1 keV [15]. The data obtained by using ESCA generally correspond to known EELS (Electron Energy Loss Spectroscopy) measurements of similar coatings [4], and there exists some quantitative disagreement between them only in the region of high deposition energies, which can be associated with the surface-sensitive character of ESCA diagnostics.

Also, we investigated the second set of samples prepared with a curved pulsed arc discharge method. In the pulsed arc discharge method, a plasma pulse is obtained from an arc created between a graphite cathode and anode in vacuum by discharging a capacitor bank. The plasma pulse is typically deflected by 90^0 in order to filter out macroparticles and droplets. In order to achieve good adhesion, the deposition is typically started using high plasma ion energies corresponding to the discharge voltage of about 4.5 kV, and the rest of the film can be, in general, deposited with lower energies to obtain better diamondlike properties. It should be mentioned that extremely high adhesion can be achieved using this set-up [16]. Typically, no pre-treatment of the samples, e.g., etching or cleaning with Ar-ion sputtering is necessary. We used single crystal <100> Si wafers as substrates. Film thickness varied in the range 0.1-2.1 μm. The films are hydrogen-free (‹ 0.1 at. % measured by ERDA method) and the overall purity is better than 99 at. % measured by RBS method. The details of the deposition technique and the effect of the deposition parameters are described elsewhere [2,3,17]. For the second set of samples, the deposition conditions such as discharge voltage, deposition rate and filtering were found to affect the resulting sp^3 fraction, hardness and the amount of macroparticles in the structure. In particular, here we report the results obtained by varying the deposition voltage.

Raman spectroscopic measurements were carried out using a 1-meter Jarrell Ash double spectrometer (8 cm^{-1} resolution) equipped with a cooled low-noise photomultiplier tube. The excitation radiation of an argon ion laser (typically 200

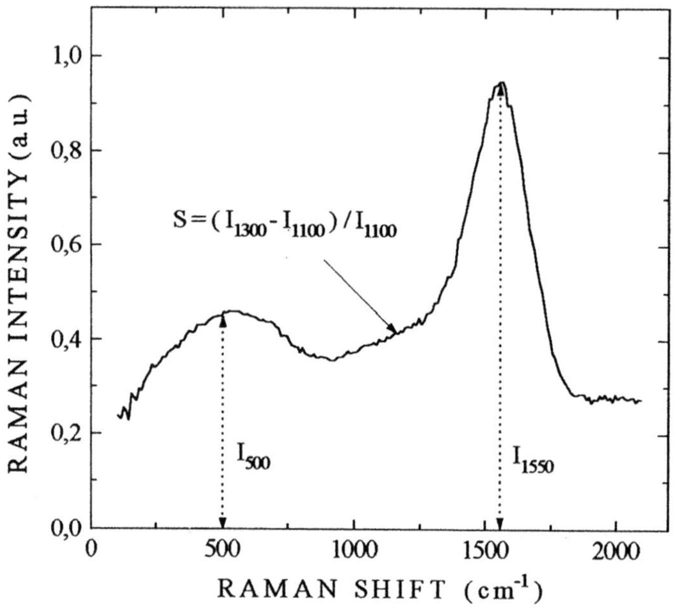

Figure 1. The background-subtracted Raman spectrum of an amorphous hydrogen-free diamondlike film (ion-beam deposition, E=1000 eV) and notations used in the text.

mW at 514.5 nm) was focused to a round spot without visible degrading the coatings. The spectra were recorded in 200 points for the 100-2100 cm^{-1} region, and then a silicon-substrate background was carefully subtracted. Because we did not notice any clear difference between the results in various pressure and temperature conditions, most of the spectra were recorded under normal laboratory conditions. Since the data obtained show no significant dependence on experimental geometry as well as no polarization preferences of different spectral features, conventional backscattering geometry without polarization analysis was employed.

3. Characterization of the Raman Spectra

The Raman spectra of amorphous hydrogen-free diamondlike films prepared by both deposition methods under consideration were, in general, quite similar. As an example, Fig. 1 demonstrates the Si-background subtracted Raman spectrum for a film with moderate diamondlike quality. There are two broad spectral bands around 1550 and 500 cm^{-1} and some additional scattering between them. The "main" asymmetrical peak at about 1550 cm^{-1} is conventionally attributed to well-localized sp^2-bonded clusters residing in the network. The broader band in the 400-800 cm^{-1} region seems to originate from a wider spatial area of the structure characterized by different sorts of mixed bonds [8], and this spectral feature is also connected with the existence of geometrical anomalies. We have also observed significant Raman intensity up to about 6000 cm^{-1}, which is usually referred to as second-order or combination scattering.

In order to describe numerically the shape of the Raman spectra, we use the following parameters (see Fig. 1):

(i) The intensity ratio of the low-frequency scattering to the main spectral feature, $R=I_{500}/I_{1550}$. We consider this parameter because the low-frequency vibrations are known to be very specific for amorphous hydrogen-free diamondlike films compared to hydrogenated structures. Besides, theory shows that signals at these vibrational frequencies correspond to different types of ta-C modes, unlocalized and localized [13,14], and from this viewpoint the parameter R should follow the modification of the sp^2/sp^3 ratio, namely, the maximum of the intensity ratio defined above tends to correspond to the maximum of the diamondlike properties.

(ii) The normalized average slope S of the curve in the 1100-1300 cm^{-1} interval, defined as $S=(I_{1300}-I_{1100})/I_{1100}$. First, this parameter is introduced to substitute the conventional extraction of the D line typically located at 1300-1400 cm^{-1} [6,7], the slope S increasing with the D/G intensity ratio. Indeed, for sp^3-rich hydrogen-free samples the D/G intensity ratios are known to be quite small, which seriously reduces the accuracy of D-line parameters obtained. Another reason to introduce this parameter is the fact that the known theoretical function for ta-C vibrational density of states shows a negative slope in the 1100 to 1300 cm^{-1} region, and one can expect the slope S to decrease with the increase of diamondlike properties.

4. Correlation of the Raman Spectra with Deposition Energy

Fig. 2 demonstrates the numerical parameters extracted from Raman spectra as a function of the deposition energy for ion-beam deposition, and as a function of discharge voltage for plasma arc discharge. Most of the data points correspond to averaged values for sets of samples prepared with the same deposition energy. The presented data obviously show the correlation between the spectral parameters introduced above and the deposition energy or the discharge voltage.

The main features observed are qualitatively similar for both deposition techniques. Clear extremal behavior is observed for both numerical parameters R and S. The minimum of S as well as maximum of R correlate with some

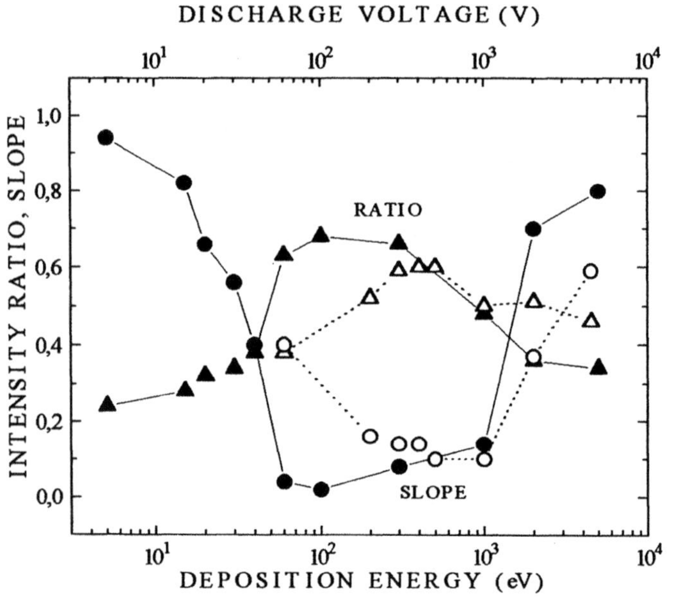

Figure 2. The intensity ratio R (triangles) and slope S (circles) vs the ion beam energy (solid symbols) and arc-discharge voltage (open symbols).

intermediate energy region, roughly, from 60 to 1000 eV for ion-beam deposition, and from 300 to 1000 V for arc discharge. The decrease of R and increase of S are observed either for lower or higher deposition energy. The comparison of these results with the known literature data shows that the extremal values of the numerical parameters correspond well to the optimal deposition conditions when considering such important film parameters as the sp^3 fraction, electrical resistivity, roughness, etc. [3,4,15].

It should be emphasized that the two deposition methods used are fundamentally different and a straightforward quantitative comparison of the deposition parameters is not valid. However, the shape of the curves for the numerical parameters R and S as a function of deposition energy or discharge voltage seems to be qualitatively the same. The curves are more smoothed in the case of arc discharge method mainly due to the broader energy distribution of the ions in this method compared to practically monoenergetic ions in the case of an isotopically separated ion beam.

5. Correlation of the Raman Spectra with the sp^3 Fraction

The sp^3 fraction of the network characterizes the approximation of the actual carbon material to the perfect diamondlike structure, and this is one of the most important film parameters to be measured. On the other hand, the problems facing Raman spectroscopy on the way of measurements of the sp^3 fraction are evident. Indeed, the Raman method probes essentially the sp^2-bonded clusters as well as the interaction between them because of much larger scattering cross-section of the graphitelike network. It means that the actual signal is essentially dependent on the structure and size of those geometrical anomalies, and one can, in general, obtain different Raman spectra for the same sp^3 fraction depending on a particular method of film preparation. Nevertheless, we expect the Raman spectra to correlate to some extend with the sp^3 fraction, and this desired correlation can be obtained empirically.

For films prepared with mass-separated ion beam deposition, the dependence of the sp^3 fraction η on the deposition energy E is available from ESCA [15] and EELS [4] measurements. Since the film growth is accurately controlled, the comparison of the results obtained in different laboratories is well-justified. Therefore, we used those known literature data to obtain the numerical parameters under consideration as a function of the sp^3 fraction via a known $\eta(E)$ function. As a result, Fig. 3 demonstrates the correlation between the sp^3 fraction and the spectral parameters. The presented curves allow diamondlike quality of the material to be numerically estimated. It should be emphasized that the dispersion of the data in Fig. 3 is mainly due to different conditions of deposition, and the experimental error is quite small. Also, the available literature data for a pulsed arc discharge method obtained by Stanishevsky et al. [11] are shown in Fig. 3 by triangles, and this demonstrates good agreement between the dependencies $R(\eta)$ and $S(\eta)$ for both deposition techniques.

The remarkable nonlinear character of the intensity ratio R versus the sp^3 fraction η should be mentioned. This specific behavior can be described assuming that there exist three different phases in the net characterized by different bond types, namely, bonds between sp^2 sites only, bonds between sp^3 sites only, and mixed bonds between different sites [8]. The relative number of such types of bonds can be roughly estimated as $(1-\eta)$, η, and $\eta(1-\eta)$, respectively. Then, we can obtain the signals at 500 cm^{-1} and 1550 cm^{-1} in relative units as

$$I_{500}=(1-\eta)a_1 + \eta(1-\eta)a_2 + \eta a_3 ,$$

$$I_{1550}=(1-\eta) ,$$

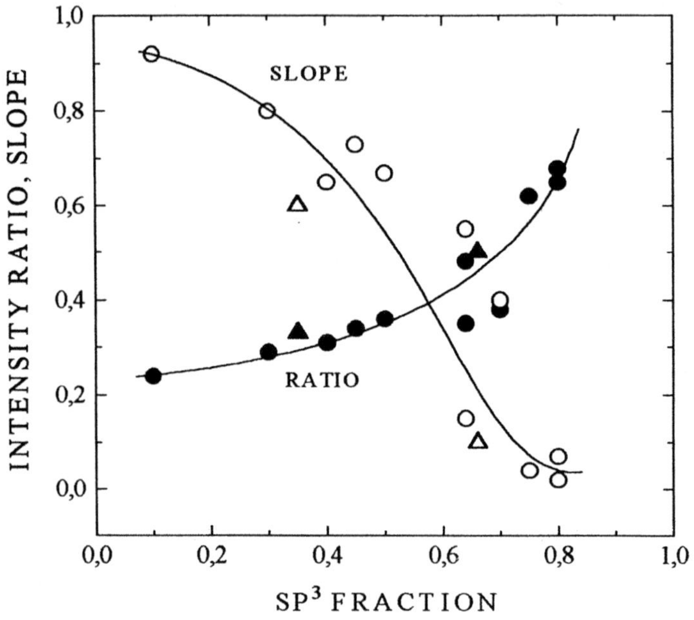

Figure 3. The intensity ratio R (solid symbols) and slope S (open symbols) vs the sp^3 fraction of the film material. The data of Stanishevsky et al. [11] are shown by triangles. The solid lines demonstrate the general behavior only.

assuming that the signal I_{1550} is mostly originates from well-localized sp^2-bonded clusters. The fit of the experimental data of Fig. 3 gives the values a_1=0.19, a_2=0.21, a_3=0.07. Although the employed model is extremely simplified, the obtained result is quite satisfactory. Indeed, in addition to nonlinearity of the dependence $R(\eta)$ defined by the relations above, the fit shows $a_3 << 1$, which means the much lower scattering cross-section for diamondlike network than that for graphitelike structure.

Analysis of the data obtained for ion-beam deposition demonstrates some distinctions between Raman spectra corresponding to close sp^3 fractions but very different deposition energies. In particular, for E=1000 eV, the Raman measurements show "better" structure (R=0.50, S=0.15) than for E=40 eV (R=0.35, S=0.55) although the sp^3 fraction is known to be very similar (η=0.65). This "disagreement" with the electron spectroscopy results might be explained by assuming that the mean size of sp^2-bonded clusters is different for these two energies, namely, it is smaller for the case of higher deposition energy. Indeed, the D/G intensity ratio is known to decrease for smaller clusters [7], and this

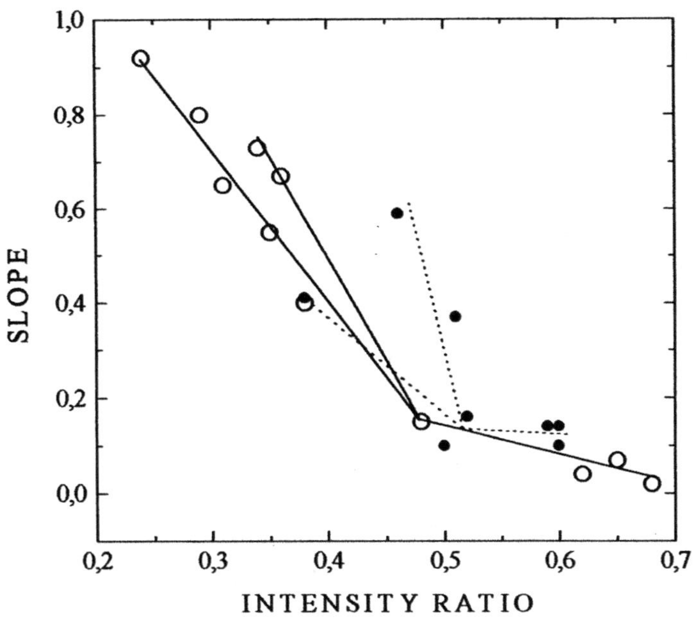

Figure 4. The slope S vs intensity ratio R for amorphous hydrogen-free diamondlike films prepared with ion beam deposition (open circles) and pulsed arc discharge (solid circles). The lines show the modification of the Raman spectra with the deposition energy, the upper branches corresponding to higher energy.

relates to the decrease of the introduced slope S. Then, for smaller mean cluster size, we clearly have more numerous mixed bonds responsible for the low-frequency vibrations [13,14], and the relative increase of the intensity ratio R seems to be also reasonable.

The use of the R-S plot appears to be useful to characterize the shape of a Raman spectrum, and hence the film quality. These plots for both deposition methods are shown in Fig. 4. The data presented demonstrate that the position of the films with highest diamondlike quality is well defined and occurs in the right-lower corner of the plot. The variations of the spectral parameters are more extensive in the case of ion beam deposition, which probably reflects the broader variations of diamondlike properties, and this is connected with details of the arc-discharge set-up [17]. The top-quality of films prepared by ion beam deposition seems to be slightly higher than that for the second deposition method, and this agrees with the previous remark about the energy-distribution width. The same reason can contribute to some distinction between the spectra related to the different deposition methods for high deposition energies.

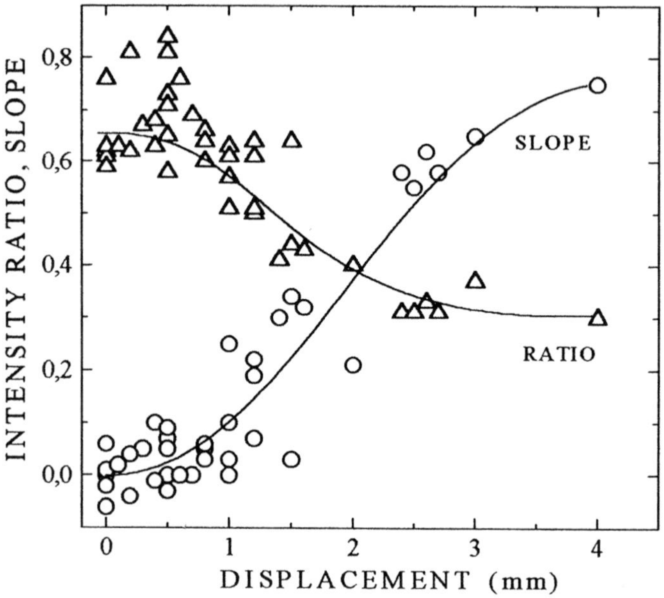

Figure 5. The intensity ratio R (triangles) and slope S (circles) vs the displacement of a probed area from the deposition center (ion-beam deposition, E=100 eV).

6. Spatial Properties and Influence of Impurities.

The spectral data shown in Figs. 2-4 as well as the ESCA data reported by Hakovirta et al. [15] were obtained for a central area of deposition. Now, we study modification of the Raman spectra with the position of a probed area. Systematic changes of both spectral parameters R and S were observed from point to point of a given sample. In particular, the function of a given spectral parameter versus the displacement of a probed area from the deposition center typically consists of two parts: smoothed changes probably following the sp^3-fraction of the material, and some additional modulation with a shorter spatial period. In the presented work we discuss the former smoothed spatial changes only. It should be mentioned that the latter spatial periodicity originates from interference effect, and the altitude of the variations does not exceed 10% for Si-substrate samples investigated in a backscattering geometry [12].

For the set of seven samples prepared by using a mass-separated ion beam with the optimum ion energy (E=100 eV), we measured the introduced spectral parameters R and S as functions of the displacement from the deposition center.

The data obtained are shown in Fig. 5. The decrease of the intensity ratio as well as the increase of the slope with the displacement from the film center are systematically observed. The reason for such behavior can be impurities implanted into the network during the deposition, the origin of the impurities being low-energy neutral carbon atoms and hydrocarbon molecules. It can be easily seen that the effect of these stray fluxes is more essential for lower deposition speed, i. e., for edges of deposition area. The similar effect is observed for samples deposited with other ion-beam energies as well as for a pulsed arc discharge method. However, this effect can be significantly reduced, for example, by using an extra-collimator or by improving vacuum conditions.

It should be emphasized that the observed spatial behavior is not the direct effect of film thickness. In particular, in the investigated set of samples prepared with a mass-separated ion beam at the optimum deposition energy, there were films of very different thickness of the central area. As a result, no essential dependence of the spectral features upon the film thickness was observed for the central area of the samples although, as was mentioned above, some minor influence of radiation interference on the shape of the spectra exists. The stability of the results with varying film thickness demonstrates that the numerical parameters under discussion reflect the properties of the bulk structure of the amorphous carbon films, and the influence of boundaries is small.

In order to demonstrate the effect of impurities on the shape of Raman spectra, we recorded Raman spectra for two special amorphous carbon films containing nitrogen and argon as an admixture. Indeed, the decrease of diamondlike properties when adding nitrogen into the film material was reported in literature [18]. To prepare these films, the pulsed plasma arc discharge method with gas ignition was used. We estimate the amount of admixture atoms for these two samples on the level of 9 at. % (nitrogen) and 1 at. % (argon).

The obtained data are collected in Table 1 in comparison with an ordinary sample prepared with the same deposition conditions. The presented data show the remarkable decrease of the intensity ratio as well as the increase of the slope for mixed films, which, in the framework of the previous discussions, can originate from the corresponding variations of sp^3-fraction. On the other hand, there can exist the additional effect of discrimination of low-frequency vibrations similar to that observed in hydrogenated samples.

Finally, in addition to 514-nm radiation, we recorded the Raman spectra with excitation by other Ar^+-laser lines. Quite clear resonance enhancement of the spectra with changing the laser radiation wavelength was observed. In particular, for the 488-nm excitation, the parameter R becomes considerably larger and exceeds 0.80 for samples prepared with optimum deposition energy of mass-separated ion beam. Nevertheless, the basic behavior of the spectral features

Table 1. Parameters of Raman spectra for samples with different composition

Sample	R	S
Carbon + nitrogen	0.38	0.48
Carbon + argon	0.40	0.40
Carbon	0.60	0.10

demonstrated in Figs. 2-5 remains the same for every laser wavelength employed and both methods of deposition. The resonance enhancement corresponds to the known experimental and theoretical results that the electronic structure of tetrahedral amorphous carbon possesses the optical gap of about 2.0-2.5 eV, i.e., exactly in the operational range of an argon ion laser.

7. Conclusions

The Raman spectra of amorphous hydrogen-free diamondlike films prepared with different deposition energies by using two methods, namely, a mass-separated carbon ion beam and pulsed arc-discharge deposition, were systematically investigated. For both deposition methods, the correlation of such parameters of the Raman spectra as the intensity ratio $R=I_{500}/I_{1550}$ and the average slope $S=(I_{1300}-I_{1100})/I_{1100}$ with the deposition energy, and hence the sp^3 fraction, electrical resistance, hardness, etc., was established. The data show a straightforward dependence between the parameters S and R, and the area corresponding to the best diamondlike films is well-defined in the S-R plot. The presented data demonstrate the possibility of effective quantitative Raman diagnostics of the amorphous hydrogen-free diamondlike materials.

8. Acknowledgments

This work was supported in part by the Nordic Council of Ministers and Academy of Finland. The Raman measurements were performed by using the equipment owned by the Chemical Instrument Center (University of Helsinki). The authors thank Prof. F. Stenman and Dr. A. Stanishevsky for useful discussions.

9. References

1. McKenzie, D.R., Muller, D., and Pailthorpe, B.A. (1991) Compressive-stress-induced formation of thin-film tetrahedral amorphous carbon, *Phys. Rev. Lett.* **67**, 773-776.
2. Anttila, A., Koskinen, J., Lappalainen, R., Hirvonen, J.-P., Stone, D., and Paszkiet, C. (1987) Comparison of diamondlike coatings deposited with C^+ and various hydrocarbon ion beams, *Appl. Phys. Lett.* **50**, 132-134.
3. Hirvonen, J.-P., Koskinen, J., Lappalainen, R., and Anttila, A. (1989) Preparation and properties of high density, hydrogen free hard carbon films with direct ion beam or arc discharge deposition, *Mater. Sci. Forum* **52-53**, 197-216.
4. Lifshitz, Y., Lempert, G.D., Gossman, E., Avigal, I., Uzan-Saguy, C., Kalish, R., Kulik, J., Marton, D., and Rabalais, J.W. (1995) Growth mechanisms of DLC from C^+ ions: experimental studies, *Diam. Rel. Mater.* **4**, 318-323.
5. Gilkes, R.K.W., Gaskell, P.H., and Robertson, J. (1995) Comparison of neutron-scattering data for tetrahedral amorphous carbon with structural models, *Phys. Rev. B* **51**, 12303-12312.
6. Yoshikawa, M. (1989) Raman spectra of diamondlike amorphous carbon films, *Mater. Sci. Forum* **52-53**, 365-386.
7. Tamor, M.A. and Vassell, V.C. (1994) Raman "fingerprints" of amorphous carbon films, *J. Appl. Phys.* **76**, 3823-3830.
8. Basca, W.S., Lannin, J.S., Pappas, D.L., and Cuomo, J.J. (1993) Raman scattering of laser-deposited amorphous carbon, *Phys. Rev. B* **47**, 10931-10934.
9. Fulin Xiong, Wang, Y.Y., Leppert, W, and Chang, R.P.H. (1993) Laser pulsed deposition of amorphous diamond-like carbon films with ArF (193 nm) eximer laser, *J. Mater. Res.* **8**, 2265-2272.
10. Khriachtchev, L.Yu., Hakovirta, M., Lappalainen, R., and Räsänen, M. (in press) Correlation of Raman spectra with carbon-ion energy for amorphous diamond-like films produced with a mass-separated ion beam, *Phys. Lett. A.*
11. Stanishevsky, A. and Khriachtchev, L.Yu. (in press) Correlation of Raman spectra with the diamond-like properties of amorphous hydrogen-free films prepared by a pulsed arc discharge method, *Diam. Rel. Mater.*
12. Khriachtchev, L.Yu., Räsänen, M., and Lappalainen, R. (in press) Influence of radiation interference on the shape of Raman spectra for amorphous hydrogen-free diamondlike carbon films, *J. Appl. Phys.*
13. Wang, C.Z. and Ho, K.M. (1993) Structure, dynamics, and electronic properties of diamondlike amorphous carbon, *Phys. Rev. Lett.* **71**, 1184-1187.
14. Drabold, D.A., Fedders, P.A., and Stumm, P. (1994) Theory of diamondlike amorphous carbon, *Phys. Rev. B* **49**, 16415-16422.
15. Hakovirta, M., Salo, J., Lappalainen, R., and Anttila, A. (1995) Correlation of carbon ion energy with sp^2/sp^3 ratio in amorphous diamond films produced with a mass-separated ion beam, *Phys. Lett. A* **205**, 287-289.
16. Anttila, A., Salo, J., and Lappalainen, R. (1995) High adhesion of diamondlike films achieved by the pulsed arc-discharge method, *Materials Letters* **24**, 153-156.
17. Lappalainen, R., Anttila, A., and Khriachtchev, L.Yu. (to be published).
18. Wang, X., Martin, P. J., and Kinder, T. J. (1995) Optical and mechanical properties of carbon nitride films prepared by ion-assisted arc deposition and magnetron sputtering, *Thin Solid Films* **256**, 148-154.

ESR Study of Neutron Irradiated Doped Diamond Films

V.S. Varichenko [1], A.A. Melnikov [1], N.M. Penina [1], M.A. Prelas [2], S. Khasawinah [2], T. Sung 2, and G. Popovici 2

1 Belarussian State University, Minsk 220080, Belarus
2 University of Missouri, Columbia, Missouri *65211,* USA

Introduction

The doping of diamond and diamond films with electrically active impurities can be accomplish during synthesis [1], or by ion implantation [2]. An alternate method of doping with impurities in diamond is transmutation doping. It is especially important for realization of a n-type in diamond, the creation of which is complicated. Recently an opportunity for Li doping of diamond films by means of the reaction $^{10}B\ (n,\alpha)^{7}Li$ [3,4] was shown. Transmutational doping is connected to formation of defects, rendering significant influence to the electrical and optical characteristics of the diamond films.
In this work paramagnetic defects in the doped ^{10}B initial diamond films, irradiated by neutrons and subjected to subsequent thermal annealing have been studied.

Experimental

Diamond films, grown in a hot filament reactor [5], contained *95 %* isotopically enriched ^{10}B with a total concentration of about 10^{20} cm^{-3}. Irradiation was carried out during 4 weeks by a flux of thermal neutrons $1.1\text{x}\ 10^{14}$ n/cm^{2}sec and fast neutrons (E> 0.1 MeV) $1.3\text{x}10^{14}$ n/cm^{2}sec. The concentration of the Li atoms after irradiation by means of the reaction $^{10}B\ (n,\alpha)^{7}Li$ was $6.3\text{x}10^{19}$ cm^{-3} and the total concentration of B was $4.2\text{x}10^{18}$ cm^{-3}. Annealing of the irradiated diamond films was carried out in argon at temperature l000°C for 8 hours and then in vacuum at temperature 1300°C - 1600°C for 20 mm.
The measurements of BSR spectra were carried out at room temperature on spectrometers "Radiopan" and "Varian" 3 cm range with rectangular TEl02 by the resonator and 100 kHz by modulation of a magnetic field. The signal of Cr^{3+} ions of a ruby standard sample permanently located in the cavity was used for the quality factor control and for calibration of amplitude of magnetic component HI of microwave field.

M. A. Prelas et al. (eds.), Diamond Based Composites, 323–328.

The measurements of absolute concentration spin were carried out with the use of the coal standard.

Results and discussion

The ESR spectrum from the as grown films is due to three centers. First, a single isotropic line with g = 2.0026 ± 0.0002 and width 3.42 G (Fig.la) which is characteristic of a defect structure with C-C dangling bonds [6]. Second a single line with g value close to that of the first center. The width of the second center is 11.4 G and it's intensity is significantly lower than of the first center. This center observed is particularly clear on the side wings of the total spectrum. A possible origin of the second center may be nitrogen combined with vacancy or, similar to the first center, dangling C-C bonds [6]. Areal concentration of the paramagnetic defects comprising the main contribution in the ESR spectrum has been measured as 1.6×10^{15} cm^{-2}. The third type of the observed paramagnetic defects is a characteristic of the centers with long relaxation times ($> 10^{-5}$ s), g-factor of which is equal to 2.0023± 0.0002.

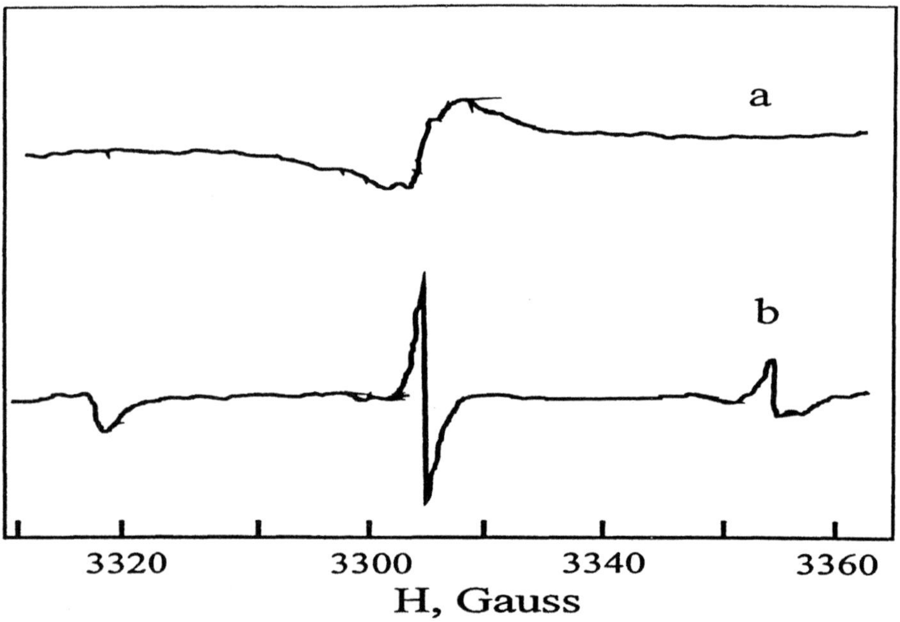

Fig.l. ESR spectra of unirradiated diamond film: a) signal is detected in-phase with 100 kHz modulation of magnetic field; b) signal is detected in 90° out of phase.

Spectrum ESR represents triplet of superfine structure with rather narrow symmetric central peak and a symmetric lateral components (Fig. lb). The asymmetry of the lateral component may indicate anisotropy of superfine structure. These centers can be detected only when the modulation phase differs from the signal phase by 90^0. The corresponding spectrum of the third center is the triplet which is often seen in polycrystalline diamond . A origin of this center is dispersed substitutional nitrogen atoms [7]. Areal concentration of these centers is very low, not exceeding a value of $5x10^{12} cm^{-2}$.

The dependence of the ESR signal on the intensity of microwaves was investigated for unirradiated diamond film (Fig. 2). At small power of microwave paramagnetic centers, saturated with growth of power, give input in EPR signal. With increased power nonsaturated centers give a substantial affect. It is clear, that the curve is influenced by centers with various relaxation times.

The irradiation by neutrons gives rise to a strong single line with g value 2.0015±0.0002 (Fig.3). The width of this line is 227±0.1 G. The spin concentration measured for this center has been measured as $4.4x10^{17} cm^{-2}$. Such paramagnetic center is a characteristic of mechanically damaged diamond powder [8]. The broadening of the spectrum may be caused by different processes. In samples containing nitrogen, the broadening may occur due to the interaction with ^{15}N *(I=1/2)* and the nearest ^{13}C nuclei [9]. Since the spin density in C-N complexes is higher for carbon atoms as compared with that of nitrogen's [10], the quadrupole, dipole [11] and exchange [12] interaction must be considered as origin of the broadening. Besides, the broadening may be additionally caused by local nonhomogenous magnetic fields due to ferromagnetic inclusions [13]. In case of the neutron irradiated diamond films the line broadening can be connected with relaxation processes which do not interfere with the fine and hyperfine structure of the point defects normally observed in neutron irradiated single crystals.

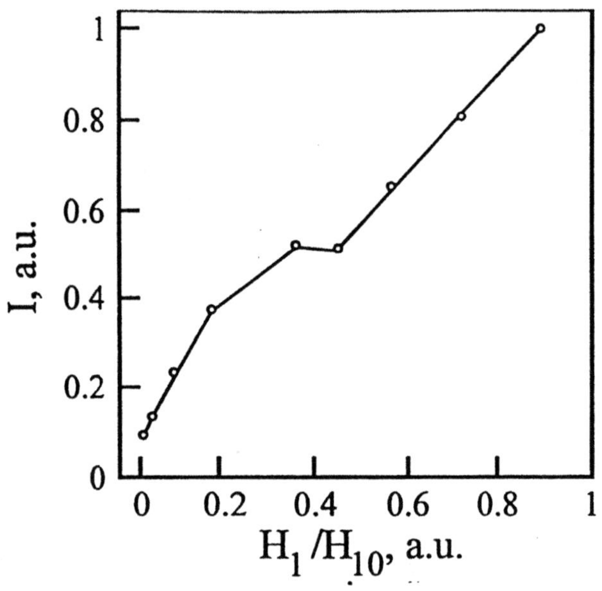

Fig.2. Change of ESR signal amplitude of unirradiated diamond film versus microwave power level.

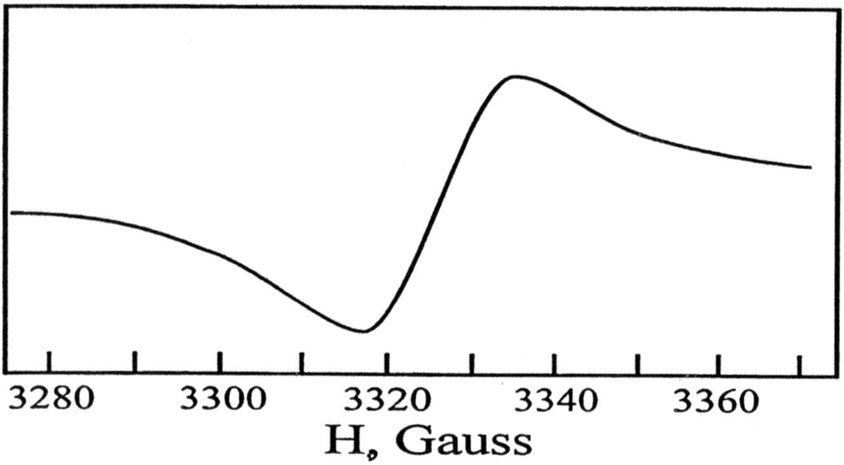

Fig.3. ESR spectrum of neutron irradiated diamond film.

After the subsequent annealing at 1000°C-1600°C ESR spectrum of neutron irradiated diamond film becomes similar to the spectrum of as grown sample (Fig.4). The line width of dominating isotropic center with g value 2.0023±0.0002 have become 2.1G and is not changing with annealing temperature. The decrease of the line width may be caused by an ordering of the internal structure of the irradiated area during annealing. The paramagnetic centers, which give input in a wider spectrum line, do not change their parameters during annealing. A real spin concentration of both types of paramagnetic centers remains equal to $1.6x10^{16}$ cm^{-2} up to the annealing temperature 1500°C. At the annealing temperature 1600°C their concentration decreases and is equal $6x10^{15}$ cm^{-2}. The concentration of the paramagnetic centers, due to nitrogen, with a large relaxation time decreases to the level lower than the registration ($2x10^{12}$ cm^{-2}).

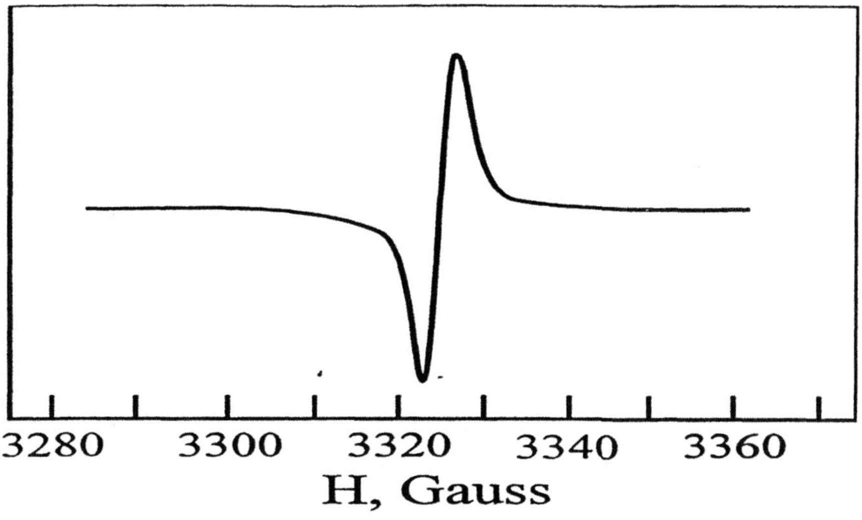

Fig.4. ESR spectrum of neutron irradiated diamond film after annealing at 1000°C.

Conclusion

Radiation effects of neutrons on ^{10}B doped diamond films grown by hot filament CVD have been investigated. Irradiation of the diamond films by neutrons causes changes to the ESR-line parameters (increase line width and g-factor change). After annealing at 1000°C-1600°C parameters of ESR-line approaches the parameters of ESR-line of the initial diamond film.

References

[1] Properties and growth of diamond, ed.: G.Davies, King's College, London, UK, 1994.

[2] J.F.Prins, Mat. Sci. Rep., 7 (7,8) (1992) p.271.

[3] B. Spytsyn, G. Popovici, M.A. Prelas, Int. Conf. on the Appl. of Diamond Films and Rel. Mat., ed.: T. Yoshikawa, M. Murakawa, Y. Tzeng, W.A. Yarbrough, Aug.25-27, Tokyo, Japan (1993) p.57-64.

[4] T. Sung, M.A. Prelas, A.V. Denisenko, N.M. Penina, V.A. Martinovich, E.N. Drozdova, A.M. Zaitsev, W.R. Fahrner, Conf. DF-95, September 10-15, Barselona, Spain (1995).

[5] S. Khasawinah, T. Sung, B. Spitsyn, W.H. Miller, G. Popovici, M.A. Prelas, E.J. Charlson, E.M. Charlson, J. Meese, T. Stacy, Diam. Mat., ed.: J.P. Dismukes, K.V. Ravi, Electrochem. Soc. Proc., 93-17 (1993) p.1032.

[6] P.R. Brosious, J.W. Corbett, J.C. Bourgoin, Phys.Stat.Solidi, A2 1(1974) p.677.

[7] D.P. Erchak, V.G. Efimov, I.I. Azarko, A.V. Denisenko, N.M. Penina, V.F. Stelmakh, V.S. Varichenko, A.M. Zaitsev, A.A. Melnikov, Diam. and Rel. Mat., 2(1993) p.1164.

[8] M.V. Vlasova, N.G. Kakazei, Electron spin resonance in mechanically damaged solid states, Naukova dumka, Kiev, 1979, 200p.

[9] J.H.N. Loubser, L. Du Preez, J. Appl. Phys., 16 (1965) p.457.

[10] M.J.A. Smith, B.R. Angel, J. Appl. Phys., 53(1982)

[11] N.D. Samsonenko, Soviet Phys. Solid State, 6 (1965) p.2460.

[12] J.H.N. Loubser, W.P. Van Runeveld, L. Du Preez, Solid State Commun., 3 (1965)307.

[13] H.B. Dyer, F.A. Raal, L.. Du Preez, J.H.N. Loubser, Phil. Mag. 11(1965) p.763.

Diamond Based p-i-n Transistor

A.A. Melnikov I, A.M. Zaitsev 1,2, A.S. Shulenkov 3, V.S. Varichenko I

I Belarussian State University, 220050 Minsk, Belarus
2 University of Hagen, 58084 Hagen, Germany
3 Institute RM, 220115 Minsk, Belarus

I. Introduction

Diamond based semiconductor structure with an active insulating zone is a very perspective component of diamond electronics. The work of such device is based on space charge limited current (SCLC) or electron-hole recombination and space charge limited double injection current in insulator. The main advantage of the structure is, first of all, a lower concentration of defects in the dielectric active zone than in doped one, that allows to receive high mobility of the carriers ~n diamond, and second, in elimination of restrictions, connected with deep acceptor level of B in diamond. P-i-M and p-i-n diodes, p-i-p structures, p-i-p field effect transistor (FET) with a Schottky barrier gate and p-i-p FET with a SiO_2 gate insulator have been already developed and described *[1-5]*. Usage i-diamond as working area leads to the improvement of working parameters of active diamond devices. In p-i-p FET the holes are injected from p-diamond and transported by SCLC mechanism in the i-region. However currents of double injection significantly exceed SCLC [6]. One of the variants of high power insulated gate p-i-n transistor is described [7].

In this report, diamond p-i-n FET with a Schottky barrier gate was fabricated for the first time.

II. Experimental

The cross section of the p-i-n FET with a Schottky barrier gate is shown in Fig. 1.

M. A. Prelas et al. (eds.), Diamond Based Composites, 329–334.

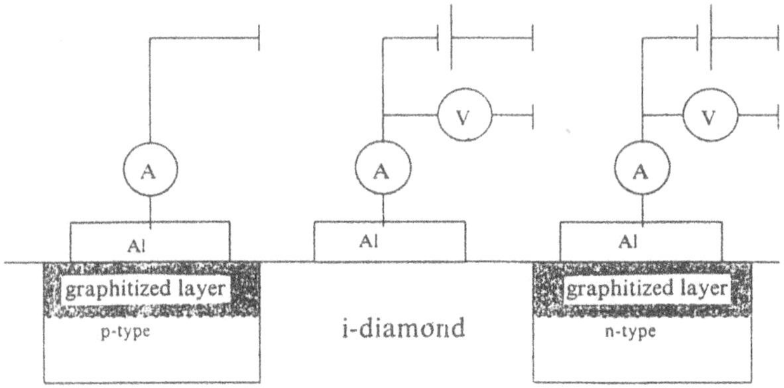

Fig. 1. Cross section of diamond based p-i-n transistor with a Schottky barrier gate.

The transistor is a planar p-i-n structure with i-zone covered by Al. The sizes p- and of n areas made 50 x 240 mkm and length of i layer was 20 mkm. The transistor was produced on a polished plate of a natural crystal of a type IIa using standard photolithography and ion implantation. p-type region was manufactured by boron implantation with distributed energy 25-150 KeV and total doze $2x10^{16}$ cm^{-2} and with subsequent thermal annealing at 1400 °C in vacuum. n-type region - by lithium implantation with energy 50 KeV and dose $4x10^{16} cm^{-2}$. Graphitized layer after annealing was removed by etching in a hot solution of H_2SO_4: CrO_3.
Selection of diamond substrates (IIa type natural diamond) has been performed using cathodoluminescence technique. Intensity of free exciton line was chosen as an indicator of crystalline quality of substrates. The choice of this parameter is caused by that the current of double injection grows with increase of life time of injected electrons and holes [6]. As the intensity of free exciton directly reflects life time of nonequilibrum charge carriers, in crystals of diamond with intensive radiation of free exciton it will be possible to observe high enough current of double injection. p-i-n transistor was made

on diamond, spectrum cathodoluminescence which is presented on Fig. 2. It is characterized by presence of intensive free exciton and A-band, which points at rather low concentration of levels recombination in a substrate.

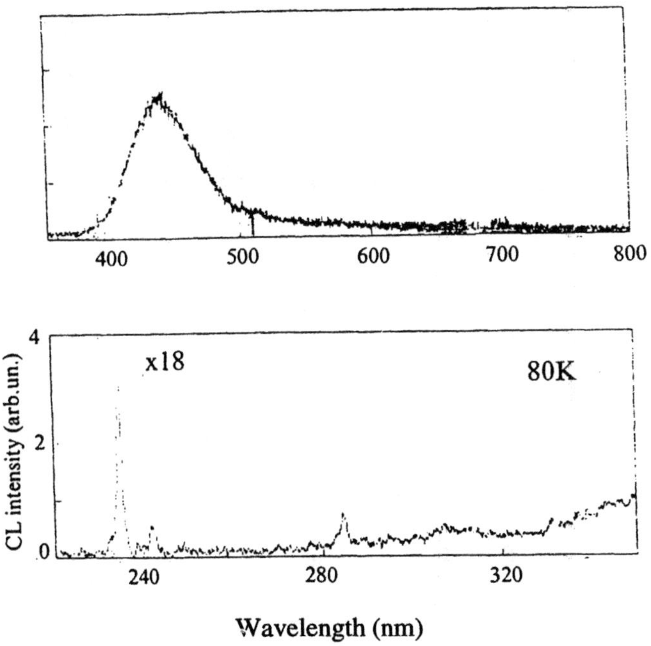

Fig.2. CL spectrum of natural diamond substrate on which has been fabricated planar p-i-n transistor.

III. Results and discussion

Earlier, as it was shown *[2,5]*, the operation of such p-i-n diodes on diamond is based on double carrier injection in insulating diamond with a limitation of the current by space charge and by electron-hole recombination. Drain current-drain voltage (I_d-V_d) and drain current-gate voltage (I_d-V_g) characteristics of the p-i-n transistor were plotted in Fig.3 and Fig.4, respectively.

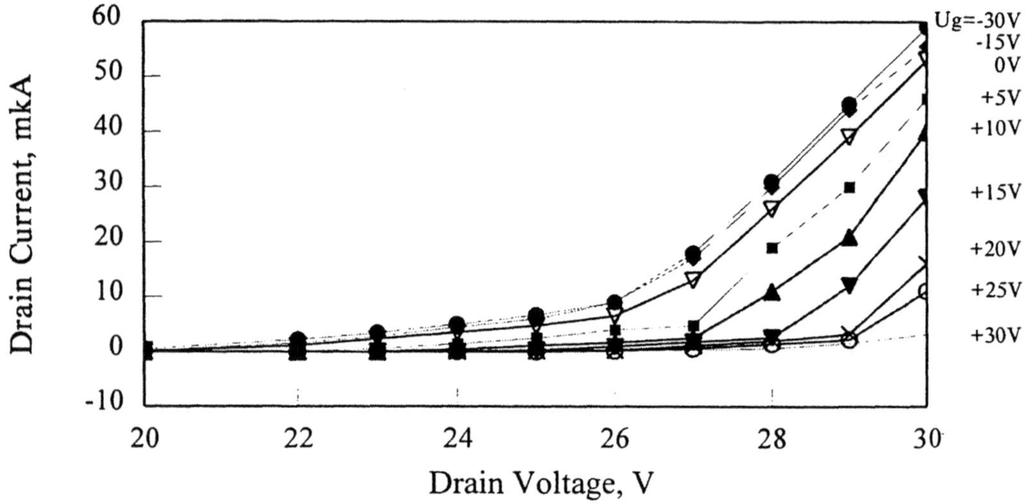

Fig.3. Drain current (I_d) - drain voltage (V_d) characteristics of the p-i-n transistor.

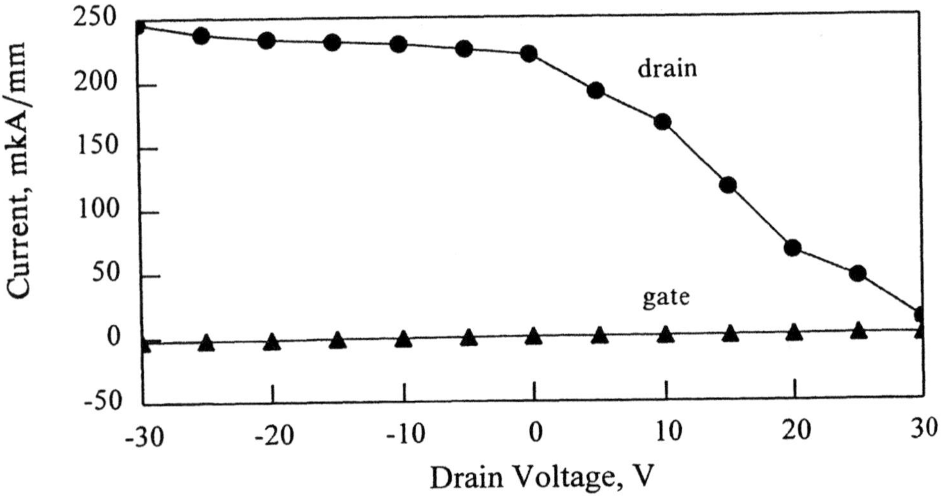

Fig.4. Drain current (I_d) - gate voltage (V_g) and gate current (I_g) - gate voltage (V_g) characteristics at of the p-i-n transistor at $V_d = 30$ V

It is possible to operate I_d by V_g in a depletion mode with the transconductance ~8 mkS/mm at positive V_g and with the transconductance 0.8 mkS/mm at negative V_g.

In made p-i-n transistor an opportunity of control of a current of double injection is connected with the different height of the potential barrier of the n-diamond - i-diamond for electrons and p-diamond - i-diamond for holes. The parameters of a potential barrier s-diamond - i-diamond depend mainly on concentration and distribution of traps in i-area and concentration of charge carrier [6,8,9]. Distinction of potential barriers p - i-diamond and n - i-diamond specifies a various kind I-V characteristics of p-i-p and n-i-n structures. In case p-i-p structure on the given crystal its opening occurs at voltage 10-30 V and output in a mode SCLC 20-40 V [1,9], in case n-i-n structure injection current does not occur before voltage 100 V. I-V characteristics p-i-M and n-i-M structures, shown on Fig.5, significantly differ. The direct currents in a p-i-M structure much exceed currents in n-i-M structure. It specifies the fact, that height of a barrier n - i-diamond for electrons is higher, than barrier p - i-diamond for holes.

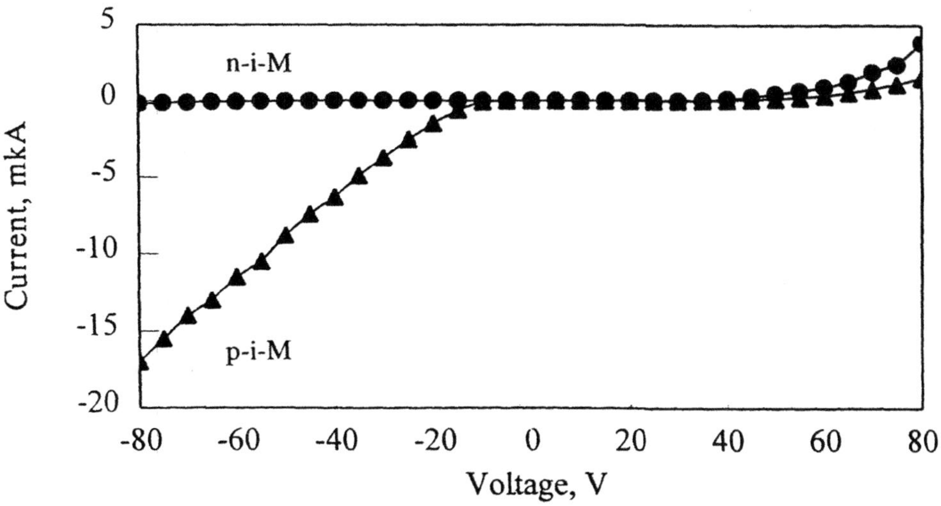

Fig.5. I-V characteristic of p-i-M and n-i-M structures.

While applying a positive voltage to the gate of p-i-n transistor, the height of p -diamond barrier increases, that results in the reduction of a holes flow through i area. The quantity of the holes, reaching of n-area, Significantly decreases, that causes reduction of injection is electrons and fall of double injection currents. While applying a negative voltage to the gate, the increase of a holes flow, reaching of n-area, influences a high barrier for electrons for n - i-diamond in a smaller degree. It results in the reduction of the transconductance of the transistor. As p-i-n diodes work up to high temperatures [2], the opportunity of control of currents of double injection in the diamond p-i-n transistor opens a new way for high power high temperature diamond transistors.

IV. Conclusion

Diamond p-i-n FET with a Schottky barrier gate was fabricated for the first time. The change of voltage applied to gate of the structure leads to the control of double injection current passing through i-region. Diamond p-i-n FET with a Schottky barrier gate operates in depletion modes.

Acknowledgment

The authors would like to thank V.Dravin for ion implantation. This work was partially supported by DAAD and INTAS project No.94-1982.

References

1. A.V. Denisenko, A.A. Melnikov, A.M. Zaitsev, V.I. Kurganskii, A.J. Shilov, J.P. Gorban, V.S. Varichenko. Mat. Sci. And Eng., Bl 1, (1992), 273.
2. A.A. Melnikov, A.V. Denisenko, A.M. Zaitsev, W.R. Fahrner, V.S. Varichenko. Proc. Applied Diamond Conference1995, Applications of Diamond Films and Related Materials: Third International Conference, Gaithersburg, MD, USA, August 21-24 1995, edit. A. Feldman, W.A. Yarbrough, M . Murakawa, Y. Tzeng, M .Y. Yoshikawa, NIST Spec. Publ. 885, 639
3. Diamond: electronic properties and applications. Edit.: L.S. Pan and D.R. Kania, Kluwer Academic Publishers, *(1995),* 472 p.
4. K. Miyata, K. Nishimura, K. Kobashi. IEEE Trans. Electron. Dev., v.42, No 11, (1995) 2010.
5. V.B. Guseva, E.A. Konorova, Y.A. Kuznetsov, V.F. Sergienko, Sov. Phys. Semicond., 12, (1978), 290.
6. M.A. Lampert, and D. Mark: Current Injection in Solids, Academic Press, New York, 1970.
7. A.Q. Huang. IEEE Electron Dev. Lett., v.16, No9, (1995), p.408-410
8. A.V.Denisenko, A.A.Melnikov, A.M.Zaitsev, B. Burchard, W.R. Fahrner. Diam. Films & Tech., 5 *(1995).*
9. M.S. Rusetski, A.V.Denisenko, A.A.Melnikov, A.M.Zaitsev. Proc. ISDF3, 16-19 May 1996, S. Peterburg, Russia. In print

NATURE OF THE ORIGIN OF NON-LINEAR CURRENT-VOLTAGE CHARACTERISTICS IN POLYCRYSTALLINE DIAMOND MATERIALS

N.D.SAMSONENKO, N.I.NOSANOV, S.N.SAMSONENKO
Department of Electrical Engineering, Donbass State Academy of Construction and Architecture, P.O. Box 6416, 340102, Donetsk-102, Ukraine

1. Introduction

At present, materials with nonlinear current-voltage characteristics (I-V characteristics) are widely employed in electronics and electric engineering. It has been shown that natural and HPHT synthetic diamonds have nonlinear I-V characteristics[1]. The nonlinearity of the I-V characteristics appears in the single crystals of the diamond at sufficiently high electric fields. But these electric fields are less than they have be for a diamond. The causes, that lead to the decrease of the break down field and to nonlinearity of the I-V characteristics still have not been investigated in detail.

2. Real structure natural, synthetic single crystals of diamond and polycrystalline diamond materials

Polycrystalline diamond materials consist of small single crystals. Typically, diamond single crystals contain a plenty enought zero-dimentional and extented structural and impurity defects. There are vacancies, interstital atoms of the carbon, admixture atoms, plaiteletes, dislocation, dislocation loops, dislocation intergraine borders there. Earlier we have shown, that dislocations and dislocation systems are main defects, that define the electronic properties of a diamond [2].

3. Energy spectrum of the electrons in real crystals of a diamond

We have calculate energy spectrum of electrons in a core of dislocations. Here we used the method molecular orbital (MO LKAO) to diamond claster with 60^0 - dislocation. These calculations point on separation of the levels from allow bands of a diamond and formation of the dislocation energy bands in bandgap of a diamond [2,3]. The dislocations create two dislocation filled bands (DFB), two dislocation empty bands (DEB) and alteration of the place of the ceiling valence band and bottom of the conductivity band. The brockene carbon bonds in the core of the dislocation form by Shockley [4] dislocation acceptor levels (DAL). They are plaice between I DFB and I DEB.

M. A. Prelas et al. (eds.), Diamond Based Composites, 335–339.

4. Polycrystalline diamond materials and their electric characteristics

They make polycrisralline diamond materials (PDM) by caking of the small-sized crystals of the natural or synthetic diamonds at big pressure (50 - 80êbar) and high temperature (1300 - 2000 Ñ).

For the study real structure of the PDM use is made of X-ray diffraction and electron spin resonance (ESR).

Figure 1 is presented typical I-V characteristics of the PDM. Here it is possible to see that all dependences of a current from voltage have nonliner character. We have established that main contribution to nonlinearity of the I-V characteristics gives a volume of the specimens.

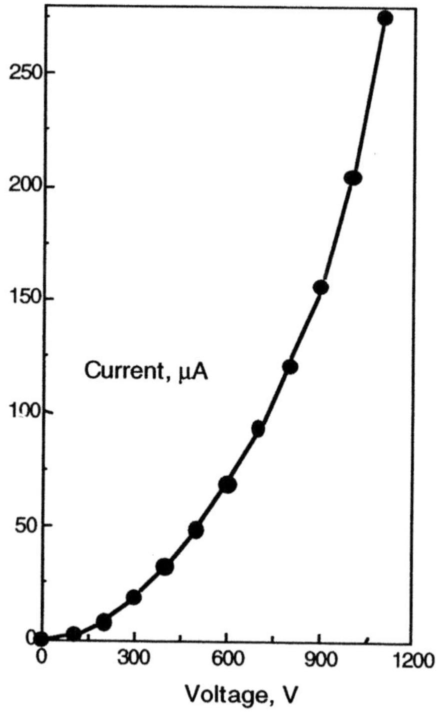

a

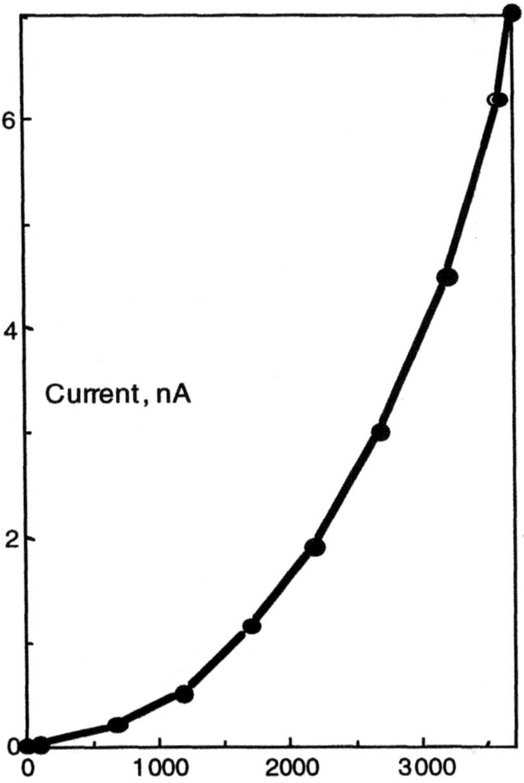

b

Figure 1. I-V characteristics for polycrystalline dimond materials: a - specimen with $\rho=5.66\cdot10^{8}$ Ohm·cm; b - specimen with $\rho=1.32\cdot10^{14}$Ohm·cm.

In paper [5] we have shown on direct connection of the density of the dislocations with nonlinear I-V characteristics of the polycrystalline diamond films. The higher the density of dislocation is the below voltage is at which appears the nonlinearity I-V characteristics. ESR-studies of the PDM have shown, that increase of the numbers paramagnetic centres from brocken carbon bonds decreases electric resistivity of the specimens ($1.27\cdot10^{18}$spin/g - $R \approx 10^{6}$ Ohm; $2.59\cdot10^{19}$spin/g - $R \approx 10^{3}$ Ohm). It is points on connection of the paramagnetic centres with electric active centres.

5. Discussion of the results

Experimental results point, that nonlinearity electric characteristics of diamond materials is connected with extent structural defects. We consider that such defects are dislocations and other dislocation structures.

For this problem we propose to use tunneling effect of electrons - effect by Zinner. A probability of the tunneling of the electrons may be written by following equation :

$$W \sim \exp\left(-c \cdot \varepsilon_G^{3/2} / E \right), \qquad (1)$$

where ε_G - energy bandgap, Å - of electrical field, c - constant, that is connected with effective mass of carriers. For a diamond with $\varepsilon_G = 5{,}5$ eV the effect of the tunneling and nonlinearity of the I-V characteristics should be observed at Å > 10^5 V/cm. Experiments give $E < 10^4$ V/cm.

The changes of energy spectrum of electrons in core of a dislocation have located character. The lenght of such energy spectrum is defined by length of the defects. The distances between defects in geometrical space of the crystal can be from 10 Angstrem unit up to 1000 Angstrem unit. It is shown on Figure 2 example of the two neighboring extent defects in electrical field. For simplicity here we consider only I DFB and I DEB. Experiments and calculations shows, that tunneling in the diamond take place in electrical fields for split in 2- 2,5eV. Electrons may pass from I DFB to I DEB another defect by tunneling. It lead to the increase of the hole current in valence band of a diamond.

6. Conclusion

The researches show, that electron properties of the monocrystals and PDM are stipulated with extent structural defects.

Nonlinearity of the dependence of a current from voltage in monocrystals and PDM can be justified by the effect by Zinner at participation located energy spectrum of the electrons, that lead to the local narrowing of the bandgap of a diamond 2 - 2.5eV. So, the nonlinearity comes at lower voltage, than for a ideal diamond.

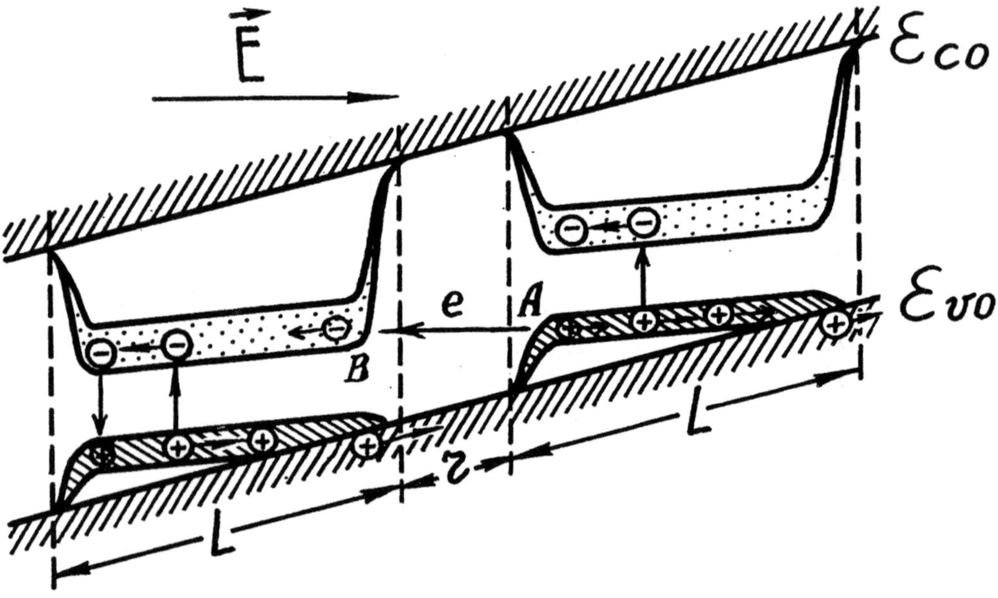

Figure 2. Scheme of the energy structure of the two neighboring extent defects in electrical field. AB is tunnel transitions of the electrons from I DFB of the one extent defect to I DEB of the other defect. L - is length of a extent defect, r - is distance between the defects.

7. References

1.Tkashenko V.I., Kvaskow V.B. (1990) Electronic devices on diamond base, in V.B.Kvaskow (ed.), Diamond in electronic technics. Energoatomizdat, Moscow, pp.22-33.

2. Samsonenko N.D. (1985) Electron properties of the deffects in diamond, Doctor of Science Dissertation, Moscow, 377p.

3. Samsonenko N.D., Timchenko V.I. (1986) The relations of the electron energy spectrum of dislocations with optical and electrical properties of the real crystal of diamonds, Optical spectroscopy and ESR of the admixtures and defects in diamond. ISM Academy Ukraine,Kiev, pp.8-13.

4. Shockly W. (1953) Imperfaction in nearly perfect crystals, Phys.Rev.,**91,** 228.

5. Samsonenko N.D., Nosanov N.I., Samsonenko S.N., Selskay I.V., Kashirin V.V. (1996) Reciprocal connection of some electrical parameters diamond films with their dislocation structure, Abstract Third international symposium on diamond films, St.Petersburg.

RADIATION INDUCED MODIFICATIONS OF DIAMOND

S.A. KHASAWINAH*, M.A. PRELAS, G. POPOVICI**,
University of Missouri, Columbia, Missouri 65211
* Present Address: Boston University, Boston, MA 02215
**Present Address: University of Illinois, Urbana, IL

ABSTRACT

A technique for radiation induced modifications is introduced. The process can be thought of as internal and uniform ion implantation. It has been shown that ion implantation can modify the surface properties of a material. However, its disadvantages are small and nonuniform interaction volumes. The proposed technique bypasses these limitations and could be suitable for large area modifications. This process could be used to form new composites and compounds. Additionally, the electronic and optical properties of materials could be specifically tailored. Promising experimental data using CVD diamond is presented.

KEY WORDS: Diamond, radiation effects, materials modification

1. Introduction

Since the 1960's ion beam bombardment was proposed as a technique to modify the surface properties of materials. Subsequently, the use of ion implantation as a research tool has encompassed many topics, ranging from deep levels in semiconductors to the phase study of phase equilibria in metallurgical systems. Commercial applications have primarily been in the semiconductor electronics field, where implantation is an industry wide process for the precise introduction of dopant species [1, 2, 3, 4]. Advantage of ion implantation is that it used to independently control amounts and depth of a particular impurity and can be carried out at shallow depths. The disadvantages are that the nonuniform bulk modification, small interaction volume and overall cost of process An alternative radiation induced modification technique is introduced in this paper. This consists of doping the material to be modified with an element that has a sufficiently large cross section for the absorption of a neutron and the subsequent release of energetic particles. The potential advantages of this technique are the suitability for bulk modification, large interaction volume, and suitability for scale up and mass production. and cost relative to existing technology. Promising results have been obtained using CVD (chemical vapor deposition) diamond substrates. The resistance of intrinsic CVD diamond increased after irradiation. Furthermore, a new carbon phase formed during the treatment. Preliminary results indicate that this phase may serve as a binder for diamond-based composites. As has been proposed with ion beam modification [1, 4], this

M. A. Prelas et al. (eds.), Diamond Based Composites, 341–347.

technique is proposed as a means of developing new composites and compounds. Additionally, modifications can be useful for mechanical, electronic and optical applications.

2. Background; Interaction of Radiation with matter

2.1. NEUTRONS

Neutrons from a reactor spectrum have a continuos energy distribution. In general, when neutron phenomena are studied, the neutrons are placed in groups according to their energies. The two group approximation, albeit an oversimplification, is useful for establishing the basic mechanisms of neutron-matter interaction. In this categorization scheme neutrons are classified as thermal neutrons and fast neutrons. The effects of each group on the target material will be discussed.

2.1.1. *Thermal Neutrons*

Thermal neutrons are defined as neutrons having an energy of .0253 eV. In order for neutrons to eject atoms from their lattice sites the incident neutron energy must exceed the displacement energy required to remove the atom from the lattice site. In the case of diamond this energy is approximately 37 eV [5]. Clearly, thermal neutrons cannot be the direct cause of primary displacements.

A mechanism in which thermal neutrons can have significance is that of transmutation. Transmutation occurs when an atom absorbs a neutron and results in the energetic release of an alpha particle and an ion whose mass number and atomic number differ by 4 and 2, respectively, from the original target atom. The energies of the resulting species are typically on the order of megaelectron volt. An example of this process follows:

$$^{10}B + .0253eV\ n \rightarrow 1\ MeV\ ^{7}Li + 1.7\ MeV\ ^{4}He \qquad (1)$$

In order for the process of transmutation to have significance the cross section for this reaction must be sufficiently large. The number of transmuted atoms can be calculated from the following formula:

$$N_t = N_h(1 - e{-}\Phi\ \sigma_{th}), \qquad (2)$$

where N_t =number transmutated host atoms, N_h =initial number of target atoms, σ_{th}=thermal neutron microscopic cross section and Φ = thermal neutron fluence

2.1.2. *Fast Neutrons*

Fast neutrons are defined as those neutrons with energies in excess of .1 MeV. As stated, an individual carbon atom in the diamond lattice requires 37 eV to be displaced. Clearly, fast neutrons can displace atoms from their lattice sites. If it is assumed that the displacement energy is isotropic, the number of primary knock-ons or displaced carbon atoms per unit volume in the bombardment can be calculated from the following relationship:

$$N_p = \Phi_f n_0 \sigma \ [5], \qquad (3)$$

where N_p = number of primary knock-ons, Φ_f = fast neutron fluence, n_0=target atoms/unit volume and σ = cross section for collisions.

In the case of diamond, the atomic number density is 1.76×10^{23} cm^{-3} and the cross section for collisions is approximately 4 for neutrons with energies between .1 and 1 MeV. The relationship indicates that one primary knock-on is produced for each incident neutron.

The energy transferred to the target atoms can be approximated from the following relationship:

$$\Delta E = 2ME/(M+1)^2 \ [6], \qquad (4)$$

where ΔE = average energy transferred to target atom, M = target atom atomic mass and E = incident neutron energy

If diamond atoms were the target and the fast neutron fluence was 2×10^{19} n/cm^2 then one can predict the creation of approximately 10^{21} vacancies based on this simplified model.

2.2. IONS

There are two mechanisms accounting for the energy loss of ions in matter; electronic stopping and nuclear stopping. The specific energy loss mechanisms will be discussed for each of these processes.

Electronic stopping is the prevalent mechanism for light ions processing energies of 100 keV or greater. The energy loss mechanisms can be categorized into four groups. First, a direct kinetic energy transfer to target electrons can occur. This is mainly due to electron-electron interactions. Next, the excitation of band-electrons and conduction electrons is possible. In other words, the weakly bound or unlocalized target electrons can be excited. Alternatively, the actual target atoms can be excited or ionized. Finally, the excitation, ionization or electron capture of projectile itself may occur [3].

The implications of this is that electronic energy efficiently coupled into the lattice and can displace individual atoms. In addition to the creation of new defects, ionization will liberate holes resulting in color centers at pre-existing defect sites, such as impurities. These color centers may serve as optical absorption bands and charge trapping centers [2].

Nuclear stopping is the dominant mechanism for lower energy ions. This occurs near the end of the ion range where nuclear collisions can displace lattice ions and potentially influence many tens of lattice sites. In many cases, the secondary collisions similarly destroy the original lattice and hence large cascades of damage are formed. For example of one assumes a typical displacement energy of 25 eV, then the deposition of 50 keV could imply 2000 disturbed ions. The large number of displacements and there subsequent diffusion or aggregation could lead to the creation of new phases and features that otherwise would not be thermodynamically possible at the temperature and pressure of the reaction [3].

3. Experimental

Polished silicon (100) substrates were seeded with 0.25 mm diamond powder. Diamond films were grown in hot filament reactors described elsewhere. A series of samples were doped in situ using a solid source of 95% isotopically enriched ^{10}B. Another series of films were prepared in another reactor and were not doped. The as-grown samples were annealed in oxygen at 500 ^oC and then chemically cleaned. These as grown samples were subsequently analyzed using Raman spectroscopy, Fourier Transform Infrared spectroscopy, Electrical Measurements, Scanning Electron Microscopy , Transmission Electron Microscopy, and electrical measurements. The samples were then subjected to moderately high neutron fluences (approximately 10^{19} n/cm^2/s thermal and 10^{19} n/cm^2/s fast neutrons). The samples were then characterized with the same aforementioned techniques.

4. Results and Discussion

Raman and electrical measurement results will be presented here. The Raman spectra of the unirradiated samples contained the characteristic diamond peak at 1333 cm^{-1}. The Raman spectra of the irradiated samples displayed a peak at a peak at 1333cm^{-1} and an additional 1629 cm^{-1}. This peak was observed for the doped and undoped samples alike. However, the intensity of the peak was seen to increase with increasing initial boron concentrations. This can be attributed to the increased number of transmuted atom and this process, as discussed above, will result in the release of two energetic ions within the material. This peak was nonexistent in the unirradiated samples. The feature at 1629 cm^{-1} was attributed by the authors to be regions of disorder [7]. They calculated that the size of the disordered clusters was on the order of 20 Angstroms in diameter. They surmised that such a peak should also be seen in materials that are formed highly non equilibrium conditions, such as nanocrystalline diamond manufactured in the Institute of superhard Materials, Kiev, Ukraine. They were able to show by taking the Raman spectra of both the irradiated diamond and the nanocrystalline diamond powder that these regions of disorder are in fact are contained in both nanocrystalline diamond and irradiated CVD diamond. They concluded that such regions were not graphite since graphite is more stable than diamond, and would not disappear upon annealing (Khasawinah 1995). Morelli et al. [5] also observed a peak at 1620 cm^{-1} upon irradiating natural diamond. They explained this as regions of disorder. They point out that this line is characteristic of "glassy" carbons and microcrystalline graphite of less than 25 Angstrom grain size as observed by Nemanich et al. [5, 8].

Structural changes, in which crystalline phases are converted to amorphous, amorphous phases are converted to crystalline, or ordered crystalline forms evolve to random crystalline forms have been known to geologists for over one hundred years in consideration of the existence of amorphous phases in radioactive minerals [9]. In the nuclear industry, where there have been attempts to coat nuclear fuel with oxides, amorphization of the oxides was observed [10]. A study of sedimentary materials under irradiation by natural uranium found that considerable changes take place in the physical and chemical properties of the organic matter as a result of irradiation from uranium. Graphite and diamond occurred at the contacts with inclusions of uranium minerals. These uranium bearing sedimentary rocks had never been exposed to high pressures or high temperatures. They concluded that the close association of these minerals with uranium indicated that they were formed as a result of irradiation of organic matter by emissions from the radioactive materials [11]. Kaminsky [12] proposed that carbonados were also formed by this mechanism. Daulton [13] conducted a high resolution transmission electron microscopy (HRTEM) study to determine whether diamond was present as a minor phase. Diamond grains approaching 40 nm were observed. They concluded that the observation of diamond in carburanium residue provided additional evidence that carbonaceous material disrupted by energetic particles can crystallize as diamond. Furthermore, this process was believed to have small formation efficiencies since the fraction of diamond in bulk carburanium was less than 0.003% by weight. The irradiation of coallike material and amorphous hydrogenated carbon results in structural rearrangement; C-H bonds are broken, resulting in the formation of both sp2 and sp3 C-C bonds [14].

The bulk resistances of the diamond samples was measured before and after irradiation. The resistance increased by two orders of magnitude after irradiation. This result was unexpected in light of what is known of the behavior other dielectrics after irradiation and even diamond after being subjected to high ion fluxes[15]. This trend was also reported by Mainwood et. al. [16]. A possible explanation of this phenomena is that the as-grown material has many defects of a certain type and the irradiation induced defect are of the other type thus providing compensating the intrinsic defects.

4.1. IMPLICATIONS FOR MATERIALS MODIFICATION

As was discussed above, there is evidence to support the formation of a radiation induced phase change. Additionally, this process was enhanced by the initial addition of isotopically enriched ^{10}B. This radiation-induced phase may be useful for the formation of diamond-based composites [17]. It is proposed that such radiation induced modifications be exploited for other materials. It has been proposed that ion implantation can be used for the modification of surface properties. So it seems quite feasible that this proposed process, which can actually be viewed as internal ion implantation, ca n be used not only for surface property modification but also bulk property modification.

Radiation induced modifications of materials proceed in two fashions; first, by utilizing the inherent particle material interactions to yield phases changes, defect and extended defect formation and porous material formation, second, using the energetic particles as driving mechanisms for the introduction of foreign atoms within a host material. The latter can be accomplished either radiation enhanced diffusion or recoil. Radiation enhanced diffusion of dopants within a host material is stimulated by the defects created in the bulk by irradiation. Next, additives can be incorporated into a material by recoil, i.e. the transfer of kinetic energy from the energetic ions to the additive atoms. This process is attractive since it may lead to the ability to form new compositions improbable by thermodynamics and inaccessible by conventional treatments. Ideally, a composite could be formed created by choosing the elements that compose it, without the conventional constraints of phase and atomic solubility rules [3].

Radiation induced modifications in materials can lead to enhanced friction, wear and hardness. The introduction of the additives and the associated stresses can inhibit the motion of the dislocations. Hardness and resistance to surface initiated fracture are both increased. Another possible benefit to the addition of the additives is enhanced corrosion resistance. This is possible since the number of surface sites which are suitable for chemical attack are reduced. Reduced corrosion resistance may result if the radiation produces a lower density amorphous material [3]. However, this may be avoided by annealing the damage effects in the lattice. An added feature of this process is that it could be used as an as-finished processing step. Since the treatment occurs at low temperatures there is low probability of major dimensional changes thus alleviating the need for grinding or straightening operations.

Additionally, radiation induced modifications can be used for the control of electrical resistance. As was found with diamond, the conductivity can be reduced by introduction of trapping levels. This will allow the production of higher resistance insulators. Additionally, the formation of the different levels can be exploited in optical applications such as the production of lasing regions or optically active materials.

5. Conclusions

Based on experiments conducted with CVD diamond and the known ion beam modification of materials, a method to modify materials has been introduced. This process would allow the uniform modification of bulk materials and the addition of multiple additives simultaneously. Materials could be potentially engineered for a variety of applications including; mechanical, electronic and applications. This process may allow the formation of new compounds and composites. This process will allow uniform modification of the bulk. Additionally, this process may be used for large area processing making it suitable for scale up.

6. References

1. Kelly, R. and Auciello, O (1984) *Beam Modifications of Materials*, Elsevier, Oxford.

2. Townsend, P. D., Chandler, P.J. and Zhang, L. (1994) Optical Effects of Ion Implantation, Cambridge University Press, Cambridge.

3. Picraux, S.T. (1984) Physics of Ion Implantation, in Kossowsky, R. and Singhal, S.C. (eds.), *Surface Engineering*, Martinus Nijhoff Publishers, Dordrecht, pp. 3-47.

4. Baumvol, I.J.R. (1984) Ion Implantation Mettalurgy, in Ziegler, J.F. (ed.), Ion *Implantation Science and Technology*, Academic Press, Orlando, pp. 261-310.

5. Morelli, D. T., Perry, T. A. and Farmer, J. W. (1993) *Phys. Rev. B* **47** 131-139.

6. Seitz, F (1949) *Discuss. Faraday Soc.* **5**, 271.

7. Khasawinah, S.A., Popovici, G., Farmer, J., Sung, T. , Prelas, M.A., Chamberlain, J. and White, H. (1995) *Journal of Materials Research* **10**(10), 2523-2528.

8. Nemanich, R. J. and Solin, S. A. (1979) *Phys. Rev. B* **20** 392.

9. Pabst, A. (1952) *Amer. Mineral* **37**, 137.

10. McCarthy, G.J. (1977) *Nuclear Technology* **32**, 92.

11. Dubinchuk, V.T., Kochenov, A.V, Pen'kov, V.F., Sidorenko, G.A. and Uspenskiy, V.A. (1976) *Doklady Akademii Nauk SSSR* **231,** 114-117.

12. Kaminsky, F.V. (1987) *Doklady Akademii Nauk SSSR* **294,** 439.

13. Daulton, T.L. and Ozima, M. (1996) *Science* **271,** 1260-1262.

14. Ergun, S., Donaldson, W.F., Breger, I.A. (1960) *Fuel* **39**, 71.

15. Prawer, S., Hoffman, A. and Kalish, R. (1990) *Applied Physics Letters* **57**(21), 2187-2189.

16. Mainwood, A., Allers, L, Collins, A. T., Hassard, J. F., Howard, A. S., Mahon, A. R., Parsons, H.L., Sumner, T., Collins, J. L., Scarsbrook, G. A., Sussman, R. S. and Whitehead, A. J. (1995) *J. Phys. D* **28** 1279-1283.

17. Popovici, G. (1996) Diamond-Based Composites, Presented at NATO Workshop, June 21, 1996 , St. Petersburg, Russia.

NANOPHASE DIAMOND CONTAINING MATERIAL BASED ON DETONATION SYNTHESIS POWDERS

V.B. SHIPILO, I.M. STARCHENKO
Institute of Solid State and Semiconduction Physics, Minsk

E.V. ZVONAREV, V.T. SENYUT
Powder Metallurgy Research Institute, Minsk

In the manufacture of diamond materials the challenge is to create highly dispersive diamond containing compacts with the nanophase structure useful as abrasive tools in optics and microelectronics for obtaining high purity of the machined surfaces. Ultradispersive diamonds (UD) of 4-6 nm in size synthesized by detonation can serve as starting material [1].

Graphitization, i.e. reverse transformation of diamond into graphite is highly essential while compacting the material based on UD powders. We have reported earlier [2,3] that the introduction into the UD powders of compounds, enhancing formation of glass with oxides of functional covering of UD impedes the graphitization process to temperatures 1700-1900 ^{0}C. This is because the glass melt formed by sintering isolates the UD powder particles, thus, preventing the oxygen penetration to the UD surface and moderating substantially the graphitization of diamond. Besides, the glass phase formed between the grains facilitates sintering of the samples.

While sintering satirically the compacts from diamonds the metallic binders are of frequent use [4]. In this work we investigated the effect of' metallic binders on sintering of compacts. The tin-lead eutectic, copper and aluminum was employed as additions. The binders concentration was varying from 5 to 50 mass %. The sintering of compacts was carried out at T = 200-1200 ^{0}C and t = 10-30 sec. The X-ray structural analysis implies that the graphitization starts already at temperatures of 300-400 ^{0}C for the samples sintered with eutectic addition at t = 10 sec, the increase in the exposure time also leads to the degradation of diamond.

The samples sintered with addition of 5 mass % of aluminum at temperatures T= 700-1000 ^{0}C showed the density of 2.8-3.0 g/cm-3 and tile hardness - 10 GPa. Further elevation of temperature was attended by graphitization of compacts and, as a result, the

M. A. Prelas et al. (eds.), Diamond Based Composites, 349–351.

reduction of density (Fig. 1).

The compacts with copper binder revealed sufficiently high mechanical strength. The X-ray structural analysis of the samples demonstrated neither graphitization no UD grain recrystallization.

It is known [4] that the finishing treatment of nonmetallic parts by diamond abrasive tools with metallic binder can lead to the damage of machined surface. Therefore, the use of organic binders in this case is favored.

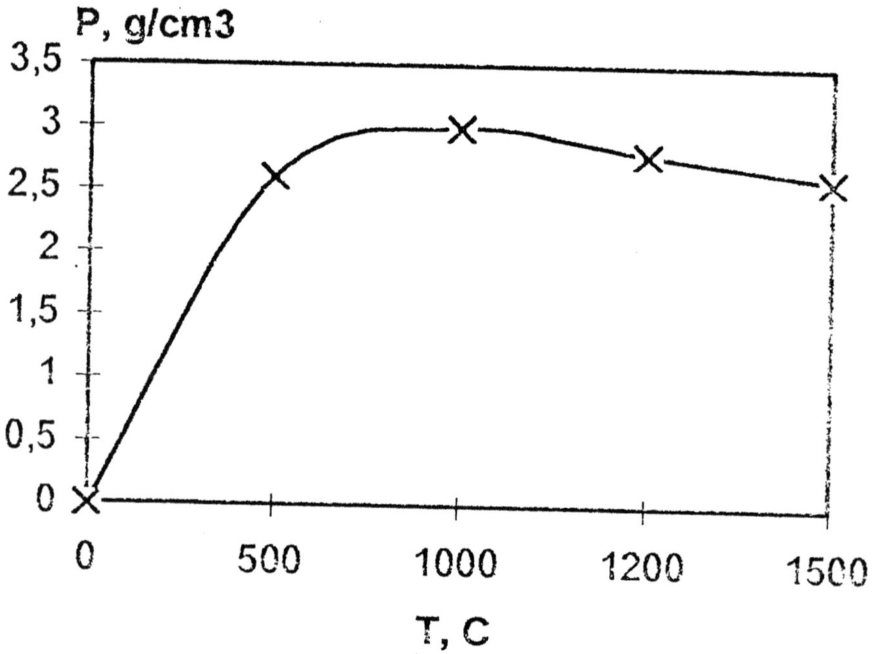

Figure 1. Density of compacts based on aluminum binder as a function of temperature.

The UD powders were sintered with organic additions based on formaldehyde resin with the additions concentrations ranging from 10 to 30 mass %, the temperature of compacting - from 150 to 400 ^{0}C, the exposure time - from 10 to 300 sec. The X-ray analysis data exhibited insignificant growth of diamond grain already at temperatures of 250-300 ^{0}C and the exposure time 60-180 sec and no graphitization present. The compacts density amounted to 2.2-2.4 g/cm-3 and microhardness 4-6 GPa. The increase in temperature and time of sintering resulted in the fall of density to 2.1-2.0 g/cm-3 (Fig. 2) due to the graphitization of the samples. Thus, it has been established that the application of organic additions reduces essentially the temperature of the compacts sintering with conservation of the starting highly dispersive structure of UD powders to permit their use on finishing treatment of the nonmetallic parts surfaces.

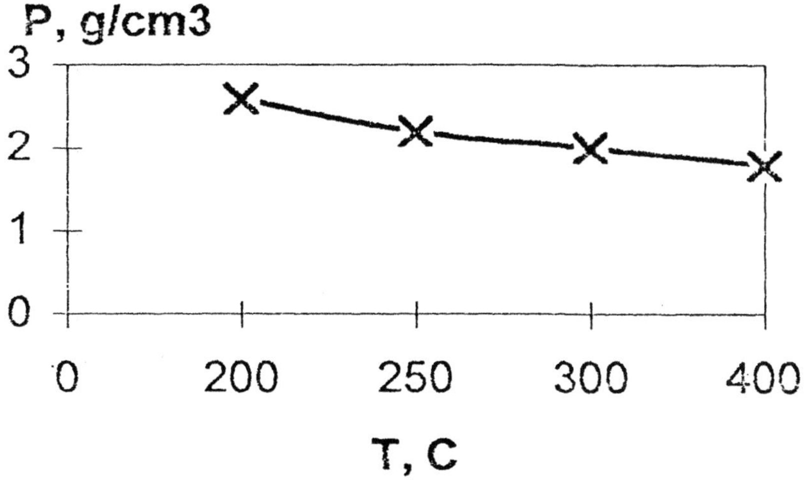

Figure 2. Density of compacts based on formaldehyde resin binder as a function of temperature.

REFERENCES

1. Lyamkin, A.M., Petrov, E.N., Eremin, A.P. et al. (1988) Fabrication of Diamonds from explosive material, *Dokl the USSR Acad. Sci.* 302/ 3, 61 1-613.
2. Shipilo, V.B., Starchenko, I.M., Gubarevich, T.M. et al. (1995) On sintering of UD powders under high pressures and temperatures, *Powder Metallurgy* 18, 126~130.
3. Shipilo, V.B., Starchenko, I.M., Zvonarev, E.V. et al (1995) Sintering of ceramics from UD diamonds under high pressures and temperatures, *Prospects for Diamonds application in Engineering And Electronics Industry* 5, 64-65.
4. Rogov, V. V. (1985) *Finishing diamond-abrasive machining of nonmetallic parts,* Navukova Dumka, Kiev.

Cluster polymers composites on basis of diamond containing nanocarbon of explosive synthesis.

A. P. Voznyakovskii, V. YU. Dolmatov.
Technology Institute, JSC "Diamond Center"
St. Petersburg, Russia.

The creation of cluster composites, that is composite polymer materials, with ultradispersed phase, is at the moment considered the most perspective way of creation of rubbers of a new generation [1]. One probable way of achievement of this purpose is synthesis of ultradispersed clusters [2]. In this connection researches in the development of methods of synthesis of such phases, and methods of their distribution in bulk of polymers are carried out very actively . In the last decade considerable study is being given to the process of explosive synthesis of diamond containing nanocarbon (DC). Particles of DC are defined as fractal clusters, and registered by X-ray analysis the sizes of particles of diamond allotropy range from 6 to 12 nm. Thus the input of a cluster in an elastomer matrix can be carried out by its synthesis in situ, or through introduction in rubber separately as a synthesized modifier. The comparative analysis of results of the elastomers using these two probable mechanisms has shown an advantage of in the separately synthesized modifier method.

We established, that rubbers from synthetic polyisoprene, modified by ultradispersed clusters, have superior properties to natural. We emphasize that the development of the techniques for synthesis of cluster fillers are multistep and, in our opinion, are not mature enough for industrial introduction in foreseeable future. Nevertheless, the results of both methods of synthesis of ultradispersed clusters, and ways of their distribution in the volume a polymer are important.

In the last decade considerable study has been given to the process of explosive synthesis of ultradispersed diamonds (UDD) [3] is investigated. UDD represents a product of explosive decomposition in an inert atmosphere blend of powerful condensed hydrocarbon explosive substances with a negative balance on oxygen. The process undergoes a condensation of gaseous products of detonation owing to the influence of high pressure in a shock wave front. In the initial stage of synthesis at a low density of products of detonation and pressure there is the condensation of free carbon in the

M. A. Prelas et al. (eds.), Diamond Based Composites, 353–359.

graphite allotropic state and formation of various hydrocarbon compounds, including incorporating heteroatoms of nitrogen, oxygen and hydrogen. Then, at a certain density of products in the detonation an increase of pressure begins and carbon is condensed not only in graphite state, but also in the form of diamond. The diamond formation processes results in the synthesis of a diamond - graphite composition that reaches 50 % diamond. With the purpose of unification of terminology used in the rubber industry, the diamond -graphite blend of explosive UDD shall be called diamond-carbon or DC. As consequence of nonequilibrium conditions in the synthesis of DC, the structure of condensed carbon is largely set by dissipative processes, which result in formation of advanced fractal structures of clusters [3].

The capacity of existing and projected units of DC can reach several tons per a year. This large amount of available material is the factor that stimulates a search of applications for DC, and UDD. This represents a study of the outlook of application of UDD for creation of cluster composites. This study does not take into account the serious problems of preservation of the UDD substance in the ultradispersed condition. So clusters of UDD with a range of dispersity between 30 and 60 nm exist only in suspension. In a dry condition, the dispersity is sharply reduced (up to 2000 - 3000 nm.). The translation of UDD material after drying again in ultradispersed condition is connected with significant experimental complexities. On the other hand it is rather reasonable to assume, that the clusters of UDD in DC are much higher in dispersity to dry UDD materials in the powder state. Based on this assumption, and also taking into account, that the traditional technology of preparation of rubber compositions in a greater measure is oriented for work with dry ingredients as the basic source of ultradispersed clusters it is preferable to choose just DC.

By use of DC for incorporation in rubbers, the greatest practical interest is in compositions based on polyisoprene rubbers. It is the result of this work that DC has been identified as a promising agent for improving the properties of synthetic rubber.

The majority of experiments deals with " rare - earth polyisoprene " (factory sample). However some data is available for SKI - 3.

Choice of a ratio of fillers: The presence of nondiamond components in DC basically admits its individual (without use of other types of technical carbon) use as a filler. However for composites, filled only by DC, physico - mechanical parameters were appreciably below the initial ones (samples 7 and 10 in tab. 1). The most probable reason of this decrease is non-uniformity in the distribution of DC in the bulk of the polymer.

As is known, uniformity of distribution of the filler in the volume of the polymer is increased by use of a filler with a wide range of particle sizes as compared to nonodispersed filler [4]. Taking into account, that during the application of a mechanical load, the dispersed phase is the concentrator of internal tensions, resulting in cracking of composites. It is possible to expect a reduction in the of probability of formation of a critical main crack with a uniform distribution of filler in the volume of a composite. This is due to a consequence of uniform redistribution of tension in a sample and low energy of fine cracks, formed around of particles of filler. Besides the use of a filler with sharply varying polydispersion parameters, the topological picture of distribution of the filler in the volume will be characterized by the presence of ensembles formed by large particles of one filler surrounded by fine particles of another filler. Thus power fields of filler of the various size and superficial activity owing to imposing and interference will cause a larger isotropy of properties, than in case of using only DC. Hence, the optimization of composition of filler on a low dispersed component can allow the greatest possible degree of the strengthening effect.

Choice of a technique for filling: Two techniques of filling rubber with DC;

"Dissolved" technique- The variant simulates an opportunity for input of DC directly in solution with the polymer in current methods adopted for commercial use. *In* this variant at the beginning a solution is prepared of polyisoprene in isopentane. Then the received solution is placed in the apparatus, developing high shift efforts at hashing. In parallel a suspension of DC in isopentane is prepared which is stabilized by powerful ultrasonic agitation. The suspension of DC is placed in the apparatus that contains the solution of polyisoprene and is mixed for 15 minutes with a three minute break every 5 minutes. Polymeasures are dried up, entered by necessary ingredient and transferred on vulcanization. Physico -mechanical characteristics were measured. (Table 1 samples 3 to 5 and 7). For comparison, data on the initial physico - mechanical parameters (rubber unfilled by DC) are shown for samples 1 and 2. An interval of filling of 0.5 to 10 weight % DC based on 100 weight % of rubber was investigated. In addition a variation of carbon black in an interval 0 to 50 weight % were investigated.

The analysis of data in table. 1 shows, that the " dissolved technology " introduction of DC in the volume of polyisoprene does not appreciably improve the properties of the resin. It is necessary to note that the most important parameter in this study is the resin's resistance to tear. This resistance to tear is probably connected to the complexity of achieving a uniform distribution of DC particles in the volume of a polymeasure, due to the impossibility of achievement at the given stage of work of steady stabilization of dispersion of DC in the hydrocarbon solvent.

Technique of "dry" input DC.; Essentially to overcome the problems of the "dissolved" technique the focus is shifted to the filling phase. In this connection we worked on DC filling using a method of rolling. The characteristic meanings of the physico - mechanical parameters of resin, obtained from the standard techniques, are shown in table I (samples 8 to 10). We have carried out optimization of DC filling and of filling with technical carbon.

The data of table 1 shows that the developed technique of filling of polyisoprene rubbers with DC has allowed appreciably improvement of complex parameters. Thus it is necessary to note, that the increase of strength parameters is accompanied by preservation of elastic - elastic properties. It is necessary to pay attention to the increase of resistance of tear, especially at increased temperatures.

Use of UDD in polymers; We worked on composites of polyisoprene with UDD, directly from use by us of DC (sample 6, table 1). However at the given stage of investigation it was not possible to receive a positive effect from the sample. First of all, it probably is connected to the complexity of stabilization of a suspension of clusters of UDD in the ultradispersed condition. Nevertheless, it is necessary to note the extremely high activity of UDD. An introduction even of about 1 % UDD in the volume of the polymeasure resulted in a significant change of its physico - mechanical characteristics.

Resistance to abrasive wear; Earlier we showed significant improvement of resistance to abrasive wear for industrial fluoroelastomers, modified by DC. [5] In this work, the influence of DC on resistance to abrasive wear was studied on examples of polyisoprene SKI - 3. The data are shown in table 2. As shown in table 2 the introduction of DC in SKI - 3 is results in some improvements. Thus, the connection of filling of rubber by DC with resistance to abrasive wear has merit. The optimum amount of DC in the case of polyisoprene lays in the range of 0 to 2.5 weight %, which correlates well with the physico - mechanical test data in table 1. An increase of filling of more than 2.5 % results in a sharp decrease of resistance to abrasion. Apparently it is connected to the presence in macromolecules of polyisoprene (as opposed to fluoroelastomers) double bonds. The last circumstance resulted in stronger adsorption interaction of macrochains of polyisoprene with clusters of DC and, hence, to a greater loss of number conformation in comparison with fluoroelastomers. Introduction of DC in quantity, exceeding the optimum, can result in formation areas of microstructured, and the occurrence of nonuniformity will result in increase of microtension (and, accordingly, to the reduction of local energy of mechanical destruction in the area of contact abrasive - resin).

Data in this work shows, that elastic-tensile properties most closely correlate with the serviceability of the resin (elongation 300 %, resistance to break and resistance wear, and also on resistance to abrasive wear) rubbers. Modification with DC surpass the initial properties, and the advantage of modified rubbers is displayed most appreciably at increased temperatures.

Table 1 Physico - mechanical properties of polyisoprene modified by diamond carbon

Parameters	Initial rubber		Introduction DC in polymerization product					Introduction DC with rolls		
	1	2	3	4	5	6	7	8	9	10
Quantity DC, %	0	0	0.5	5	5	1.2	10	2	5	10
Quantity carbon black, %	50	30	50	50	30	50	0	50	50	0
Modul at 300% elongation, Mpa	8.2	7.5	7.9	12.4	13	9.4	4.3	13.7	14.0	25
Resistance rupture Mpa										
23 ^{0}C	24.0	24.6	24.0	24.6	10.0	20.0	21.0	25.4	20.8	19.8
100 ^{0}C	14.3				3.6	14.0	13.2	17.2	15.7	15.1
Specific elongation, %										
23 ^{0}C	540	570	580	530	780	453	585	533	460	750
100 ^{0}C		500			760	685	710	580	573	880
Residual elongation, %										
23 ^{0}C	25		75			25	16	33	27	16
100 ^{0}C						35	-	-	-	-
Shore hardness	55	60	53	66	36	62	53	69	67	42
Elasticity recoiling, %										
23 ^{0}C	40	37	36	40	50	38	52	38	36	64
100 ^{0}C	52	49	53	48	47	55	68	50	48	70
Resistance tear, Mpa										
23 ^{0}C	95	75	80	19-40	10	45	16	108	73	14
100 ^{0}C	55	36	48		9	43	20	71	68	18
T.g,^{o}C	-72.5			-69.5						

Table 2. Resistance to abrasive wear of polyisoprene modified by diamond carbon

N/ n/n	DC mass%	Wear, conditional unit	Wear, relative unit
1.	0	0.01933	1.000
2.	1.5	0.01621	0.839
3.	2.5	0.02327	1.205
4.	5.0	0.02957	1.532
5.	10.0	0.03026	1.639

References

1. Grishin B.S., Pisarenko T.I., Evstratov V.F. (1991) Physical Modification of Elastomers, *Dokl. AN SSSR,* 321, No 2, 321-325.
2. Grishin B.S., Pisarenko T.I., Raevskii A.V., Erofeev G.I. , Synthesis and application of colloid clusters as physical modifyers of elastomers,*"Rubber-94":* (Moscow, 09/27/94 to 10/1/94), 2, 400-409.
3. Sakovich G.G., Brylyakov P.M., Gubarevich V.D., Yielding diamond clusters by explosion and their application, 1990 *Zhurn. VKhO im.. D.I. Mendeleev,* 35, 600 - 602
4. Lipatov S.Yu., Physical chemical principles of filling polymers, (1991) Moscow: Chemistry.
5. Voznyakovskii A.P., Dolmatov V. Yu., Levintova E.A., Gubarevich T.M., Composition materials on the basis of co-polymers and technical diamond carbon of explosive synthesis,*"Rubber-94":* (Moscow, 09/27/94 to 10/1/94), 2, 80 - 87.

SUPERHARD COMPOSITES OF CUBIC BORON NITRIDE

A.V. BOCHKO
Institute of Problems for Material Science
National Academy of Science, 252142, Kiev, Ukraine

1. INTRODUCTION

During past decades cubic boron nitride (cBN) and wurtzite boron nitride (wBN) have recieved considerable attention. These dense modifications of boron nitride possess hardness approaching to diamond and thermal stability better than that of diamond. They are chemically inert and do not react with iron. Superhard materials made on their base found wide application in many areas of metal cutting operations. The compacts sintered from cBN or wBN powder have a polycrystalline composite structure either of cBN or cBN plus wBN. At last case they have high fracture toughness. The cutting tools made of this material show better service life in interrupted cutting of cast irons and hardened steels. It is essential to have a proper understanding of this class of superhard material in order to realize its full potential not only in cutting applications, but due to their unique physical properties also in electronics.

2. BORON NITRIDE: CRYSTAL CHEMISTRY CHARACTERISTICS.

Boron nitride exists in hexagonal graphite-like form (gBN), hexagonal wurtzite-like form and cubic zinc blend form or cubic nitride form. Under ambient conditions gBN is a stable phase and wBN and cBN are metastable phases. The crystal chemical characteristics of various phases of boron nitride are given in Table 1.

3.METHODS OF SYNTHESIS OF cBN POLYCRYSTALS

Known processes can be generally classified in four categories:

a)catalytic conversion process, one-step process in which the catalyst, metal or alloy, aids in the transition of gBN to cBN simultaneously with the formation of the compact.

b)bonding medium process, two-step process in which the first step comprises the conversion of gBN to cBN and the second one comprises the formation of the compact from cleaned cBN crystals mixed with the metal alloy which aids in the bonding of the cBN to the compact.

M. A. Prelas et al. (eds.), Diamond Based Composites, 361–367.

c) direct sintering process, a two-step process which is the same as the process (b) except that compact is formed without addition of metal or alloy to aid in bonding cBN crystals.

d) direct conversion process, a one-step process in which substantially pure gBN is directly transformed to a cBN com- pact without the aid of the catalyst and/or bonding medium.

Table 1.Chemical composition for wBN powder

No	Denomination of a Index	wBN Standard	wBN Experimental
1.	Mass part of BN, % including:	98.0	99.25
1.1.	Mass part of gBN with disordered and crystallic structure, %, including:	2.0	0.2
1.1.1.	Mass part of gBN with crystallic structure, %	0.5	traces
1.2.	Mass part of wBN, %	95.0	99.0
2.	Mass part of mixtures,% including:	2.0	0.74
2.1.	Mass part of common carbon,%	0.30	0.10
2.2.	Mass part of boron anhydride,%	0.15	0.06
2.3	Mass part of iron, %	0.10	0.05
2.4.	Mass part of moisture, %	0.50	0.18
2.5.	The rest (not being determined) mixtures, %	0.95	0.35
3.	Size (middle), mkm	2.9	2.2

Hard phase BN compacts are of two general types: cluster compact and composite compact.

A cluster compact is defind as a cluster of abrasive crystals bonded together either in a self-bonded relationship (1) or by means of bonding medium disposed between the crystals (2) or by means of some combination of (1) and (2). For example, Borason [2], Elbor-R [3], Belbor [4], Hexanite-R [5], Amborit [6], Kiborit [7] and other are those cluster compacts.

A composite compact is defined as a cluster compact bonded to the substrate material, such as cemented tungsten carbide or cemented titanium carbide. The examples of such composite compacts are Compax BZN [8], DB 50 [9], Sumiboron [10], Wurzin [11], BPK [12], Composite 10D [14] and other.

The catalytic and bonding medium processes are generally disadvantageous because catalyst and bonding medium are lower in hardness than cBN and retaining in the resultant mass reduce the hardness and abrasive resistance of the masses.

The direct conversion process, which is theoretically and practicaly possible, has been found to have high losses in practice because it is difficult to achive consistently the sufficient number of crystal to crystal bonds distributed uniformly with in the compact. Without this, the strength and density of the compact are less than it is necessary. And the no less so significant progress in thisdirection was reached.

The considerable amount of high pressure sintering of cBN in presence of varions solvent catalyst has been done at the General Electric Company, and then at the De Beers Company. Many differut composites were created in the USSA. The best famous composites, which were created in "J.E." and De Beers are Borason and Amborite. Aluminium has prived to very effective solvent-catalyst for manufacturing of cBN composites. The residual binder phases in the Borason and Amborite range of products are aluminium nitride and aluminium diboride. The sintering of these composites is done over a pressure range 5-7 GPa, at temperature range 1500-2000°C.

Polycrystalline cBN, tool blanks consist of the layer 0.5-0.7mm of cBN crystals bonded to one another on a cemented carbide substrate. This composite structure is achived through high temperature - high pressure process, resulting in an extremely high wear and impact resistant product with very consistent physical properties from blank to blank. As the cobalt in the WC-Co cemented carbide substrate melts at high pressure - high temperature process, it begins to infiltrate the voids in the compact and acts as a solvent catalyst and converts any gBN presented to cBN, thus completing the cementation process within the existing cBN crystals.

Physical properties of hard and superhard materials are presented in Table 2.

4. DIRECT CONVERSIONS gBN--cBN AND gBN--wBN

The direct conversion under static pressure condition of gBN to more dense wBN and cBN phases at pressure of 10 Gpa and above described in details by Bandy and Wentorf [15]. More recently, numerous reports and patents have been concerning the direct conversion of gBN to cBN cluster compacts under pressure below 10.0 GPa. In the works of N.N. Sirota and A.M. Mazurenko [16,4] it was claimed that superhard material, which might be synthesized under pressure of 6.0 GPa and higher over the temperature range from 1800°C to 3000°C. Practically at the same time the works of the Japancese scientists were published [17]. In the publication [18] report direct conversion occurs under pressure more than 5.0 GPa (preferably 6.0 GPa and above) and sintering temperature more than 1100°C. The material, created by technology [4] received Belbor name and now is wide known in the industry.

Table 2.Physical properties of Hard and Superhard materials

Property	WC+Co ISO [9]	Syndite PCD25 [6]	Amborit PCBN [6]	Kiborit PSBN [12,13]	Composite05 PCBN [14]
Density, 10^3kg/m^3	14.7	3.86	3.42	3.4	4.3
Compressiv strength,GPa	4.5	7.61	2.73	2.9	2.3
Bending strength,GPa	2.7	1.19	0.57	-	0.47
Fracture toughness,MPa m	10.8	8.89	6.30	10.5	6.7
Knoop hardness, GPa	13	50	32	36	18.8
Young's modulus, GPa	620	810	680	880	620
Pousson's ratio, GPa	0.21	0.07	0.22	0.16	0.16
CTE, 10^{-6}K^{-1}	5.4	4.6	4.9		
Thermal conduc-tivity, W/m.k	100	560	100	100	-

Publication [19] the use of pyrolitic boron nitride (pBN) as the starting material for the synthesis of cBN cluster compacts in a direct conversion process practiced under pressure of 6.9 GPa and at temperature between 1800°C and 1900°C. The resulted product was characterized as the soft mass having large amount of unconverted gBN.

Publication [17,18] also reports the synthesis of wBN and cBN. There were no reported results of the successful formation of either wBN or cBN. And only the publication [20] of F.R.Corrigan reported the results of the successful formation of cBN cluster composite in a direct conversion pBN--cBN process under pressures and temperature between 2190°C and 2440°C without a trace of gBN. Thermal conductivity of this material was equal 470-700 W/m K (at 100°C). F.P.Corrigan proposed to use this cluster composite as a heat sink. This result was confirmed late in [21] where thermal conductivity was found equal 200-575 W/m K.

The direct conversion process of gBN to wBN by dynamical method was fixed for the first time in the end of the 60 th.[22]. Further on this process was examined rather good and at present time is considered to be finaly fixed and that it goes on by marthensite mechanism [23].

5. COMPOSITES ON THE BASE wBN

The first composite with the use of wBN was the composite (PHBN), made on the bases of the mixture of wBN+

Table 3. Phisical Properties of Superhard materials
(direct synthesis)

Property	Elbor-RM [3,14]	Belbor [4,14]	PHBN [24,14]	Hexanite-R [14]
Density,103kg/m^3	3.42	3.45	3.4-3.42	3.41-3.44
Compressiv strength,GPa	2.7	4.0-6.5	3.4-4.9	3.5-4.0
Bending strength,GPa	-	-	-	4.0-1.2
Fracture toughness,MPa	4.2	-	13-17	15-17
Knoop hardness, GPa	35	38	40.6-42.0	41.8
Young's modulus, GPa	840	800	800	750-820
Poisson's ratio, GPa	0.16	-	-	0.16
CTE,10-6K-1	-	-	-	1.8
Thermal conductivity,W/m.k	60-80	100	70	25-30

cBN powder [24]. The synthesis of the composite on the base of only wBN was the problem sometime, until it was solved by the synthesis of Hexanite-R [25]. Converted mechanism of wBN to cBN during synthesis of the composites on the base of wBN, as it is shown in [23] carries not marthensite, but diffusion character. Phisical properties of these composites are given in Table 3.

6.DISCUSSION

The comparison of the mechanisms of formation of the composites with the presence of the catalyst of or bonding binders and composites made by direct conversion of wBN to cBN and also the values of fracture toughness, presented in Table 2 and 3, allows to make analogy between hard alloys and superhard composite materials.

In this one and another case we have hard (WC) or superhard (cBN) particles, surrounded by less hard and more plastic matrix Co, AlN, AlB_2, when meeting with it the cracks, achieved critical size and capable to lead to the failure of the whole composite, brake.

In case of Hexanite-R this analogy is less obvious than for example, in the case of PHBN [24], where wBN transform into cBN completely, but in this case such approach is also correct. Increased resistance of Hexanite-R to impact loadings, so very high fracture toughness, can be also explained, by the fact that within stractures with wurtzite structure wBN differs by increased extention of coordination tetrahedron along the axis c (c/a = 1.6). The last circumstances allows to consider wBN not only as superhard material similar to cBN, but more plastic comparatively to cBN. Exept the granes of wBN more smole than the granes of recrystalized cBN, that usely given more high fracture toughness.

7. CONCLUSION

Presented overview of the composites on the base of cubical and wurtzite boron nitride touched only the main picture about this class of materials.

Performed analysis of their formation mechanism and physical properties allowed to identify those peruliarities, which make them irreplaceable in metal machining and also notice those, which may be used in electronics.

REFERENCES

1. Kurdyumov A. V., Pilyankevich A. N. (1979) Phase transitions of carbon and boron nitride, Naukova Dumka, Kiev.
2. Wentorf R.H., Delai A.J. (1966) US Patent 3 233 988.
3. Lysanov V. S. (ed.) (1978) Elboron, In machine-building, Mashinostroenie, Leningrad.
4. Mazurenko A. M. (1972) Investigation of kinetic of multimorphic transition of hexagonal BN to the cubic modification and elaboration of the super pressure apparatus. Abstract of Thesis for the Degree of Candidate of Physics-Mathematical Science.
5. Frantsevich I. N., et. al. (1980) Supperhard materials, Naukova Dumka, Kiev
6. Bosson P.K. (1990), Ind. Diamond Rev., 50, 228.
7. Novikov N. V. (ed.) (1986) Synthetic Supperhard Materials, Naukova Dumka, Kiev, p.277.
8. Wentorf R.H. and Rocco W.A. (1973) US Patent 3 745 623.
9. Hooper R.M. et al. (1989) Ind. Diamond Rev., 49.
10. Tabuchi, Hara A., Yazn S. et al. (1978) Performance of Sumiboron BN200, Sumitomo Electric Tech. Review, 8,57-65.
11. Araki M., Sawaoka A. (1984) Cutting tool inserts of hexagonal boron nitride, J. Ind. Explos. Soc. Jap., 44, 4, 230-235.
12. Shulzhenko, et. al. (1983) Synthesis, baking and properties of the cubic boron nitride, Naukova Dumka, Kiev.
13. Grishenko E. I., et. al., (1993) Grinding nickel alloys by the cubic boron nitride tool, Naukova Dumka, Kiev.
14. Bochko A. V., et. al. (1991) Technological features of Mechanical Processing by the tool of multicrystal supperhard materials, Naukova Dumka, Kiev.
15. Bundy F.P., Wentorf R.H.(1963) Direct transformation of hexagonal boron nitride to denser forms, J. Chem. Phys., 38, 1144-1149.
16. Sirota N.N.(1973)Process for Production of Cubic Boron Nitride, British Patent 1317 716.
17. Wakatsuki et al. (1972) Synthesis of polycrystalline cubic boron nitride, Mat. Res. Bull, 7, 999-1004.
18. Ishinose et al. (1974) Synthesis of polycrystalline cubic BN, Proc. of 4-th Int. Conf. on High Pressure, Kyoto, Japan, p.p. 436-440.

19. Wakatsuki et al. (1974) Synthesis of polycrystalline cubic boron nitride, Proc. of 4-th Int. Conf. on High Pressure, Kyoto, Japan, p.p. 441-445.
20. Corrigan FR..(1978) Cubic boron nitride (cBN) compact and direct conversion process for making same from pyrolitic boron nitride, Patent GB 2 002 333.
21. Ositinskaya T. D., et. al. (1989) Thermal conduction and some structural features of multi-crystals of the cubic boron nitride supperhard materials, 6, 23-26.
22. Dremin A. N., et. al. (1967) Formation of the wurtzite modification. of the boron nitride by shock wave compression, Dokl. AN SSSR, 172, 5, 1066
23. Kurdyumov A. V., et. al. (1994) Multimorphic modifications of carbon and boron nitride, A Guide Book, Metalurgiya, Moscow.
24. Vereshchagin L. F. (1982) Synthetic diamonds and Hydro-extrusion, Nauka, Moscow.
25. Alexceevsky V.P., Bochko A.V. et al. (1975) Method for producing of polycrystalline boron nitride, USA Patent 3876751.

ISSUES IN c-BN COMPOSITES

Peter J. Gielisse
Florida Agricultural and Mechanical University
Florida State University
College of Engineering
2525 Pottsdamer Street
Tallahassee, FL 32310, USA
gielisse@eng.fsu.edu

ABSTRACT

Cubic Boron Nitride is proposed as a highly promising second phase in the development and application of advanced composites. Examples are given of its application in conventional and nanocomposites towards mechanical properties and thermal conductivity enhancement.

1. Introduction

Cubic boron nitride, a high bandgap light element semiconductor of promise, has not yet been as successful as its monoatomic analogue diamond, except perhaps in ferrous abrasive applications. Its outstanding mechanical, physical and thermal properties, second in value to those of diamond in virtually all areas, combined with a high temperature (1400°) oxidation and chemical reaction inertness and relative ease in surface finishing, make its use as a reinforcement in composites highly promising. From a processing point of view, the higher ionic character in BN is favored over that of the prototype covalent diamond. It is the objective of this brief presentation to call attention to the unique possibilities for development of a range of future applications with c-BN as the enhancing component in multiphasic composites of various types.

2. Cubic boron nitride in conventional composites

Composites in general and ultrahigh strength reinforcement-based composites in particular, can be readily categorized into three distinct and completely different types. They are schematically illustrated in Fig. 1. The conventional composite structures make use of basically three types of reinforcements. In particle reinforced composites, size range (0.1 - 10)μm, the demand is for ultrahigh elastic modulus and strength reinforcements, that will form strong particle-matrix interfaces at as low a processing temperature as possible. Cubic boron nitride, due to its outstanding physical properties, lends itself particularly well to the formation of metal, ceramic or polymer matrix composites.

M. A. Prelas et al. (eds.), Diamond Based Composites, 369–375.

Boron nitride tends to display a high surface energy with many metals. Certain metals such as *Ti, Ta, Cr* and *V*, often in a noble metal base, will tend to reduce liquid contact angles, while at the same time gettering undesirable and interfering elements, such as oxygen, from the interface surface. The *Cu-Ti* system has shown excellent wetting characteristics and was at one time used for the formation of ohmic contacts in the manufacture of *c-BN* single crystal thermistors[1]. Wettability of *c-BN* to glasses is not very good, while the behavior with candidate ceramic materials is essentially unknown. The possible existence of an *AlN - BN* solid solution makes an *AlN* matrix attractive. The availability of ultrahard BN in both the wurtzite (hexagonal) type nanophase and the microcrystalline zincblende (cubic) structure broadens opportunities for novel materials design.

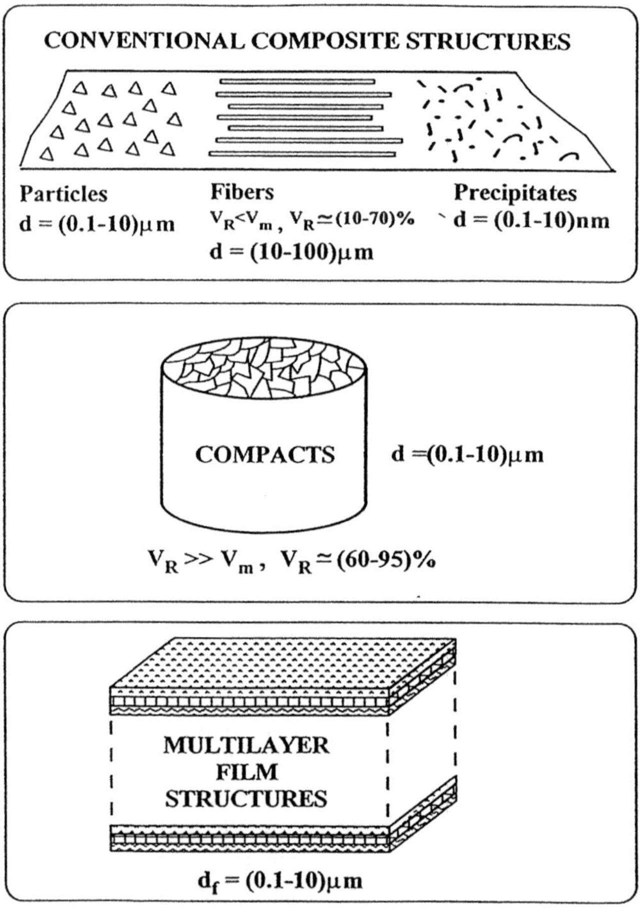

Figure 1. The three different types of *c-BN* composites presently under development (schematic).

3. Cubic boron nitride in nanocomposites

The use of microdispersants could be useful in developing structural materials with enhanced fracture toughness. Submicron or ultrafine dispersants can act as the source of crack deflection toughening, a mechanism in which the propagating crack is deflected by the dispersant particle or by the stress field that it creates in the surrounding volume.

Nanodispersants, on the other hand, are likely to act as impedance sources for dislocation movement thus effecting strengthening and hardening of the composite. This will be of particular value in the development of metal matrix *c-BN* composites.

The availability of *c-BN* in various size categories, as micro dispersants and as UFP's that simultaneously could realize several types of enhancement is another plus. It would provide opportunities for the tailoring of specific engineering characteristics, such as toughness and strength within any one composite system. The possibility of developing true *BN* nanocomposites has become much more realistic since the availability of the normally shock synthesized hexagonal wurtzite type high strength polymorph of *c-BN* , known as w-BN. Both structural types are four coordinated differing only in stacking sequence, ABAB... for wurtzite and ABCABC... for zincblende. Due to the small differences in atomic positions, internal energy differences between the polymorphs and thus their physical properties are expected to be small. An example with reference to hardness may be found in Fig. 2. Note the difference in hardness values between the two polymorphs, allowing for material design possibilities.

4. High thermal conductivity composites

The impact of *c-BN* particle size and morphology also needs to be analyzed with reference to other than mechanical properties of the composite, specifically thermal conductivity. A high strength - high conductivity (thermal and electrical) composite is much desired in many application areas. The availability of high strength (> 1 GPa) high electrical conductivity (70% of that of copper) and high thermal conductivity (300 - 400 W/m K) composites in wire or sheet form could revolutionize many aspects of electro-magnetic based industries. High thermal conductivity composites are, furthermore, much needed as rapid heat conductors and heat sinks in electronic modules to alleviate the problem of the ever increasing thermal loads.

Recent studies, see [3], have shown that small particles can introduce a thermal boundary resistance, R_B, which, under certain circumstances, will reduce the composite thermal conductivity, κ, to below that of the matrix, in spite of a very high intrinsic thermal conductivity of the reinforcement.

For a given heat flux, Q, the temperature gradient that develops across a material's interface is expressed as,

$$Q = \frac{\Delta T}{R_B} \tag{1}$$

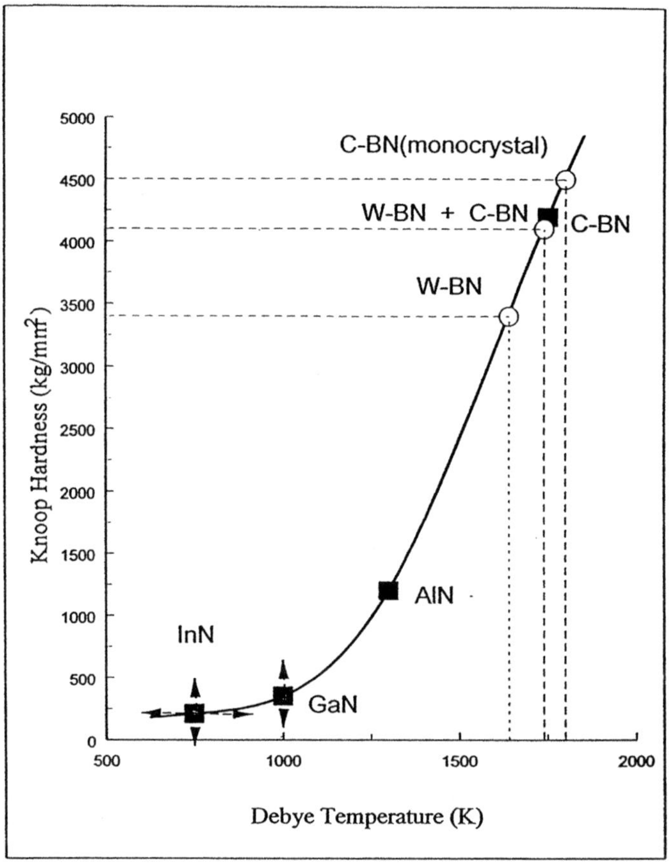

Figure 2. Hardness vs. Debye Temperature for light element nitrides. Values for *w-BN* from [2].

The composite's thermal conductivity relates, according to the effective medium theory, to, R_B as,

$$\frac{\kappa}{\kappa_m} = \frac{[\kappa_p(1+2\alpha)+2\kappa_m]+2\Phi[\kappa_p(1-\alpha)-\kappa_m]}{[\kappa_p(1+2\alpha)+2\kappa_m]-\Phi[\kappa_p(1-\alpha)-\kappa_m]} \tag{2}$$

in which Φ is the reinforcement concentration and, α, a dimensionless parameter defined by the ratio of the so-called Kapitza radius $a_k = R_B \cdot \kappa_m$ to the (spherical, monodispersed) particle radius, a, or

$$\alpha = \frac{R_B \cdot \kappa_m}{a} \tag{3}$$

The Kapitza radius [4] relates the ratio of the mean free path of the phonons, λ, to the probability, η, of the phonon transmission across the interface, or $a_k = \lambda/\eta$. The Kapitza radius can thus be conceptualized as the thickness of the thermal resistance layer around the particle, of value R_B. For large particles, $a >> a_k$, R_B will be small (α small), while for small particles the reverse will be true and a high thermal boundary resistance may dominate the phonon transfer process. One can define a critical particle radius, a_o, for the condition $\kappa = \kappa_m$ as,

$$a_o = \frac{R_B \cdot \kappa_m}{1 - \kappa_m/\kappa_p} \tag{4}$$

The R_B value has been determined for a few material systems only. The diamond-aluminum system has been quoted as 21.7 x 10^{-9} m^2K/W and the gold-diamond system as 25.0 x 10^{-9} m^2K/W [3]. As can be expected there are no data for *c-BN* - metal or *c-BN* - ceramic systems. Since phonon wavelength is directly related to the Debye frequency or temperature, a reasonable estimate of R_B would be made by using the ratio Debye temperature *c-BN* (1900 K) to diamond (2340 K) as a multiplier for the diamond - copper system, assuming the boundary resistance of diamond with copper to be the same as with gold. This yields a value of 0.87 for the ratio and an R_B value of 0.22 x 10^{-9} m^2K/W. The results of initial evaluations of this type are plotted in Fig. 3. The critical particle size value in the *c-BN* -copper system is about 30 μm. As size is reduced below the critical value, the particles will increasingly act as thermal resistances or thermal traps. At values $\alpha \geq 0.6$, the composite will display a thermal conductivity less than that of the matrix. Using a combination of UFP's (nm) for strengthening purposes, and larger (μm) sized particles for improved conductivity in composites, appears a possibility.

Kim et al. [5], have called attention to a second phase morphology effect. Apparently, composites with isolated, i.e. not interconnected, second phase particles have a higher thermal conductivity than those with interconnected morphology.

Kapitza conductances between diamond and several metals have been measured for a few metals [6]. Some metal-diamond interfaces showed measured Kapitza conductances of more than an order of magnitude larger than values obtained from lattice dynamics theory. Heat transfer across particle interfaces in composites is clearly in need of further development.

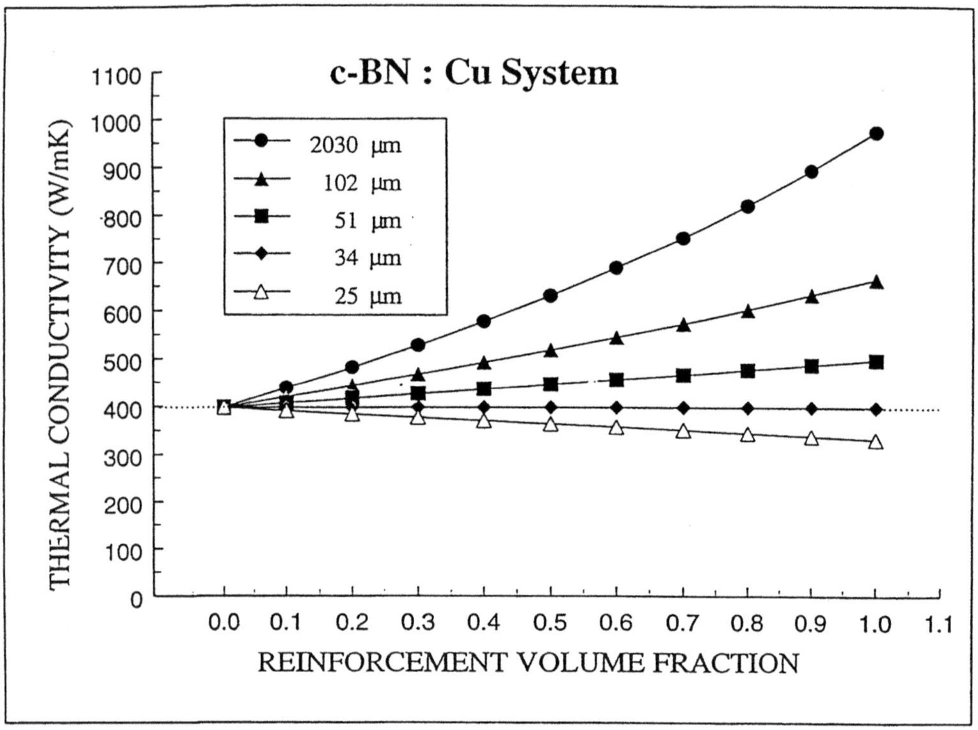

Figure 3. Change in thermal conductivity with second phase volume fraction in the *c-BN* : Cu system.

5. Conclusion

It could be shown that the very high hardness, high thermal conductivity and high bandgap sphalerite (*c-BN*) and wurtzite (*w-BN*) polymorphs show excellent possibilities for incorporation as reinforcements and property modifiers in future high performance composites. Initial results indicate that present strengthening and (thermal) conductivity models cannot generally explain the experimentally derived values. It could be shown that incorporation of $< 30\ \mu m$ *c-BN* particles in a metal matrix can lead to a reduction of the thermal conductivity to below that of the matrix material.

6. References

1. Gielisse, P. J. and Doser, M. (1969) *Thermistor Device and Method of Producing Said Device*, U.S. Patent 3,435,398.

2. Bochko, A. V. (1996) Private Communication at Third International Symposium on Diamond Films, ISDF-3, Saint Petersburg, Russia.

3. Davis, L. C. and Artz, B. E. (1995) *Thermal Conductivity of Metal-Matrix Composites*, J. Appl. Phys. **77**, 4954-4960.

4. Kapitza, P. L. (1941) J. Phys. USSR **4**, 181.

5. Kim, W. J., Kim, D. K. and Kim, C. H. (1996) *Morphological Effect of Second Phase on the Thermal Conductivity of AlN Ceramics*, J. Am. Ceram. Soc. **79**, 1066-1072.

6. Stoner, R. J., Maris, H. J., Anthony, T. R. and Banholzer, W. F. (1992) *Measurements of the Kaptiza Conductance between Diamond and Several Metals*, Phys. Rev. Letters **68**, 1563-1566.

Workshop Co-Directors

Mark A. Prelas, Professor, Nuclear Engineering Dept., University of Missouri-, Columbia, MO 65211 USA

Vladamir I. Ivanov-Omskii, Professor and Head of Laboratory,A.F.Ioffe Phys.-Technical Institute - A. F. Ioffe Phsico Technikal Institute, Politeknicheskay 26, St. Petersburg, 194021 Russia

Workshop Organizers

Mark A. Prelas, Professor, Nuclear Engineering Dept., University of Missouri-, Columbia, MO 65211 USA

Vladamir I. Ivanov-Omskii, Professor and Head of Laboratory,A.F.Ioffe Phys.-Technical Institute - A. F. Ioffe Phsico Technikal Institute, Politeknicheskay 26, St. Petersburg, 194021 Russia

Galena Popovici, University of Illinois, Coordinated Science Lab, 4401 West Springfield Avenue, Urbana, Illinois 61801

Sussmann Ricardo, DEBEERS Ind. Diamond Division (UK) LTD, Charters Sunninghill - Ascot, Berkshire SL5 9PX, United Kingdom

Participants

Bentz, Julie A., University of Missouri,EBE W1066 Nuclear Engineering,Columbia, Missouri 65211 USA

Blaut-Blatcher, A., Institute of Physical Chemistry, 31, Leninsky Pzospekt, Moscow 117915 Russia

Bochko, Anatoly Vasivievich, Institute of Problems for Material Science, National Academy of Sciences of Ukraine, 3, Krzhizhanovsky Str., Kiev, 252142, Ukraine

Borisovich, Lodygin Artem, A.F. Ioffe Physical-Technical Institute, St. Petersburg, Politechnicheskaya St., 26, 194021 Russia

Bronislavovick, Shipilo Viktor, Institute of Solid State Semiconductors, Physics of Academy of Science of Belarus, 17 P. Brovka, Minsk, Belarus

Cederstrom, Ian, Superfarad Nenika AB, GranBackens, 17335 Saltsibaden, Sweden

Clarke, Roy, University of Michigan, 500 East University Street, Ann Arbor, Michigan 48109-1120

Constantiu, Morosanu Octavian, Institute of Physics and Technology of Material, MAGURELE PO Box Mg7, Bucharest, Romania

Dann, Dr. Allister W.E., BNFL, R101 Rutherford House, Risley, Warrington, Cheshire WA36AS, United Kingdom

Davidson, Jimmy Lee, Vanderbilt University, P.O. Box 99, Station B, Nashville, Tennessee 37235 USA

Dolmatov, Valezy, TSC Diamond Center, 33 a/1 p2. Sovietsky, St. Petersburg 193177, Russia

Ekstrom, Thommy, Department of Inorganic Chemistry, Stockholm University, 5-10691 Stockholm, Sweden
Galouchko, Ivan, Institute of Physical Chemistry RAS, 31 Leninsky Pl. 117915, Moscow, Russia
Gielisse, Peter J., Florida State University/Florida A and M, College of Engineering, 2525 Pottsdamer Street, Tallahassee, Florida
Georgievich, Babev Vladimir, Moscow State University, Department of Physics, Moscow, Russia
Grieger, Ronald D., Norton, 1 New Bond Street, Worcester, Massachusetts 01615 USA,
Gunes, Mehmet, Izmir Institute of Technology, Gaziosmanpasa Bulv. No:16, Cankaya, Izmir 35210, Turkey
Gurievich, Yatrebov Sergey, 194021, St. Petersburg, 194021, Russia,
Herndon, H. Brooks, Corda Technologies Inc., 1600 Baltimore Avenue, Kansas City, Missouri 64108 USA,
Ivanovich, Tkachev Viktor, Moscow Institute of Radio Engineering, Automation and Electronics, 117454 Vernadsky Av. 78, Moscow, Russia
Izumi, Tomio, Department of Electronics Tokai University, 1117 Kitakaname, Hiratsuka-shi, Kanagawa, Japan 259-12
Jennings, Howard, British Nuclear Fuels PLL, Hinton House, Warrington, Cheshire, United Kingdom Kamo, Mutsukazu, NIRIM, 1-1 Namiki: Tsukuba, Ibarak, 305 Japan,
Khasawinah, Dr. Salim, Boston University, Manufacturing Engineering, 15 St. Mary's Street, Boston, Massachusetts 02215
Khriachtchev, Leonid, Laboratory of Physical Chemistry, University of Helsinki, Helsinki, Finland FIN-00014,
Levashov, Evgeny, Moscow Steel and Alloys Institute, Leninsay Prospect, 4, Moscow 117936, Russia
Manfredotti, Claudio, Experimental Physics Department, Via Giuria 1, Torino, Italy
Mazur, Jacek, Institute of Physics, Silesian Technical University, Krzywoustego 2, Gliwice 44-100 Poland,
Melnikov, A.A., Belarussian State University, F. Skorina Str. 4, Minsk, Belarus,
Mihaylovich, Starchenko, Igor, Institute of Solid State and Semiconductor Physics, Academy of Science, 17 Brovka Street, Minsk, Belarus 220072
Mikchailovna, Shishonoic, Elena, Institute of Solid State and Semiconductor Physics, Academy of Sciences of Belarus, 17 P. Brovka , Minsk, Belarus,
Mountford, Andrew, BNFL, Consultancy Services, Risley, Cheshire, WA3 6AS, United Kingdom,
Perov, Polievkt I., Institute of Engineering and Electronics, 11 Mohovaya Street, Moscow 103907 Russia,
Piuir, Niedzielski, Technical University of Lodz, Wstafanowskiego 7175, Institute of Material Science, Lodz 90-924 Poland
Popovici, Galena, University of Illinois, Coordinated Science Lab, 4401 West Springfield Avenue, Urbana, Illinois 61801
Ralchenko, Victor, General Physics Institute, 38 Vavilov Street, Moscow 117942,

Russia
Rutledge, Karen McNamara, Boston University, 15 St. Mary's Street, Manufacturing Engineering, Boston, Massachusetts 02215 USA
Samsoneuko, Nikolai D., Doubass State Academy of Construction and Architecture, 2, Derzhovina Street, 23 Makeyevka, 339023, Donetsk Region, Ukraine
Siklitsky, Vadim, Ioffe Physical-Technical Institute, Politechnicheskaya St. 26, St. Petersburg 194021 Russia
Sildos, I., Institute of Physics, Estonian Acad. Sci.., Rua Street 142, Tartu, EE2400 Estonia
Smorgonskaya, Dr. Emelia A., A. F. Ioffee Phyical-Technical Institute, Polytechnicheskaya 26, St. Petersburg 194021 Russia
Sokotowska, Aleksandra, Warsaw Technical University, Sr. Narbutta 85, Warsaw, Poland
Spitsyn, B. V., Institute of Physical Chemistry, 31 Leninsky Pr., Moscow 117915 Russia
Stanislaw, Mitura, Technical University of Lodz, Institute of Materials Science, ul. Stefanowskiego 1, Lodz 90-924 Poland
Stoyan, Vladimir, Institute of Physical Chemistry, RAS, 31 Leninsky Prospect, Moscow 117915, Russia
Sung, Talun, University of Missouri - Columbia, E2433 Engineering Building East, Columbia, Missouri 65211 USA
Sussmann, Dr. Ricardo, DEBEERS Ind. Diamond Division (UK) LTD, Charters Sunninghill - Ascot, Berkshire SL5 9PX, United Kingdom
Szmidt, Jan, Institute Microelectronics and Optoelectronics, Warsaw University of Technology, Koszykowa 75, Warsaw 00-662 Poland
Tolmatchev, Alexi V., St. Petersburg State Technical University, Politechnicheskaya St., 29, St. Petersburg 194021 Russia
Uzbanovich, Vladimir S., Institute of Solid State and Semiconductor Physics, of Academy of Sciences of Belarus, 17, P. Brovka Street, Minsk 220072, Belarus
Varichenko, V.S., Belarussian State University, F. Skorina Pr. 4, Minsk, Belarus
White, Dongming Yuan, FAMU/FSU College of Engineering, Department of Mechanical Engineering, 2525 Pottsdamer Street, Tallahassee, Florida 32310 USA
Yelisseyev, Dr. Alexander, Institute of Mineralogy and Petrography, 3 University Avenue, Novosibirsk 630090 Russia
Zbrzezny, Adam, University of Western Ontario, Engineering Science Building, Room 3042, Materials Engineering, London, Ontario, Canada

Workshop Program

Date	Time	Presentation	Speaker
June 19			
	13:00- 19:00	Registration	
June 20			
		Morning Session	Chair: M. Prelas and V. Ivanov-Omskii
	9:00	Introduction	M. Prelas and V Ivanov-Omskii
	9:30	Discussion of Russian & FSU Program on DBC	S. K. Gordeev; Central Material Research Institute Russia
	10:10	Phase Transition in C:N Films Under Shock Wave Compression	M. B. Guseva;
	10:35	Chemical Vapor Deposition of Diamond Films on Diamond Compacts	V. G. Ralchenko
	10:55	Break	
	11:20	The Diamond and Hard Alloy Based Composite Material	V. S. Urbanovich
key	**11:40**	**Superhard Composites**	**Anatoly Vasivievich Bochko**
	12:20	Cluster polymers composites on basis of diamond containing nanocarbon of explosive synthesis	V. Yu Dolmatov
	12:40	Lunch	
		Afternoon Session	Chair: Pol Perov
Key	**14:20**	**Advanced Composites on Diamond**	**S. K. Gordeev**

Key	**15:00**	**Diamond CVD Materials**	**R. Sussmann**
	16:10	Composite materials based on cubic boron nitride	V. B. Shipilo
	16:40	Future developments in DBC synthesis	Shulhzenko A.A.
Key	**17:00**	**Nanodiamond composites**	**G. Popovici**
	17:20	Basis for Superhard Composites	Samsoneuko, Nikolai D.
	17:40	Poster Papers	
	18:00	Supper	
	19:00	Round Table Discussion: Synthesis and Processes	Leader: S. K. Gordeev, P. Geilisse
June 21			
		Morning Session	Chair: V Ivanov-Omskii
	9:00	Discussion of Western Program on DBC	P. Geilisse
Key	**10:00**	**Impurity Additions to Diamond**	**T. Sung and M. Prelas**
Key	**10:30**	**Radiation Effects on Diamond**	**S. Khasawinah and M. Prelas**
key	**10:50**	**Allotropic forms of carbon nitride**	**Sokotowska, Aleksandra**
	11:20	Hydrogen in Diamond	K. McNamara
	11:40	Break	
	12:00	**Issues in cBN Composites**	**P. Geilisse**
	13:00	Lunch	
		Afternoon Session	Chair: R. Sussmann
Key	**14:20**	**cBN/Si(100) hetero Structures and Composites**	**Roy Clarke**

Key	**15:00**	**Diamond Composites for Grinding Applications**	**R. Grieger**
	15:40	AlN Composite Films	B. V. Spitsyn
	16:00	Break	
Key	**16:40**	**Copper nanoclusters in DLC**	**V. I. Ivanov-Omskii**
	17:00	On the Elastic Constant, Tensile Strength and Fibrous Possibilities of CVD Diamond	J. Davidson
	17:20	Round Table: Status of Diamond (and related materials) Based Composites	Leader: P. Gielisse
	18:20	Adjourn	

Index of Authors

Index

GPSR Compliance

The European Union's (EU) General Product Safety Regulation (GPSR) is a set of rules that requires consumer products to be safe and our obligations to ensure this.

If you have any concerns about our products, you can contact us on ProductSafety@springernature.com

In case Publisher is established outside the EU, the EU authorized representative is:

Springer Nature Customer Service Center GmbH
Europaplatz 3
69115 Heidelberg, Germany

Batch number: 10755284

Printed by Printforce, the Netherlands